Immunotherapy – A Novel Facet of Modern Therapeutics

Sujata P. Sawarkar • Vandana S. Nikam •
Shariq Syed
Editors

Immunotherapy – A Novel Facet of Modern Therapeutics

Springer

Editors
Sujata P. Sawarkar
Department of Pharmaceutics
SVKM's Dr. Bhanuben
Nanavati College of Pharmacy
Mumbai, Maharashtra, India

Vandana S. Nikam
Department of Pharmacology, STES's Smt.
Kashibai Navale College of Pharmacy
S. P. Pune University
Pune, Maharashtra, India

Shariq Syed
School of Pharmacy
Anjuman-I-Islam's, Kalsekar
Technical Campus
Mumbai, Maharashtra, India

ISBN 978-981-15-9037-5 ISBN 978-981-15-9038-2 (eBook)
https://doi.org/10.1007/978-981-15-9038-2

This Springer imprint is published by the registered company Springer Nature Singapore Pte Ltd.
The registered company address is: 152 Beach Road, #21-01/04 Gateway East, Singapore 189721, Singapore

Contents

About the Editors

Sujata P. Sawarkar is Professor and Head of Department, Pharmaceutics, SVKM's Dr. Bhanuben Nanavati College of Pharmacy, India. Before joining academics Dr. Sujata Sawarkar was associated with major pharmaceutical companies in India at managerial position, in the field of Research and Development, formulation development of conventional and novel dosage forms for regulated and domestic market. Several oral solid dosage forms and sterile products developed by her team have received approval in US and EU and have been commercialized. She has about total 22 years of research and teaching experience. Her research interests include Formulation Development of novel drug delivery and targeted systems based on nanotechnology for oral, ophthalmic, colonic and vaginal delivery, development of evaluation techniques for novel drug delivery systems, Translational research. Dr. Sujata Sawarkar has been Executive Member of Controlled Release Society (India Chapter) since 2014 and Secretary of Controlled Release Society (India Chapter) CRS IC for 2017-2019. She has been invited reviewer for Indian Drugs, AAPS PharmSciTech, International Journal of Nanomedicine Dove Press, Drug Design, Development and Therapy, Journal of Bioequivalence & Bioavailability, European Journal of Pharmaceutics and Biopharmaceutics and Journal of Cosmetic Dermatology. Dr. Sujata Sawarkar has received research project grants worth about 85 lakhs rupees (8.5 million INR i.e. about 1.19 million USD) from Industry and Government funding agency. She has presented about 50 research papers in national and international conferences like CRS Inc. and AAPS. She has several papers published in peer reviewed journals like AAPS PharmSciTech, International Journal of Pharmaceutics, Critical Reviews in Therapeutic Drug Carrier systems, Drug Development and Translational Research, Frontiers in Pharmacology, Expert Opinion on Drug Delivery. She has coauthored 5 book chapters and two books. Dr. Sujata Sawarkar has been awarded IDMA award for reviewer, Research and Industry Outcome by SVKM's Dr. Bhanuben Nanavati College of Pharmacy in 2018 and 2020, Received Travel Grant from Controlled Release Society and All India Council of Technical Education for attending and presenting paper at 42nd and 46th Annual Meeting & Exposition of the Controlled Release Society in July 2015 and July 2019 respectively. She has several papers published in peer-reviewed journals like AAPS PharmSciTech, International Journal of Pharmaceutics. She has co-authored 2 book chapters and one book.

Vandana S. Nikam is presently Associate Professor and Head of Department, Pharmacology, at STES's Smt. Kashibai Navale College of Pharmacy, S. P. Pune University, India. Before her present association, she was a Post-Doctoral Fellow at Max Planck Institute for Heart and Lung Research, Bad Nauheim, Germany. Dr. Nikam has pursued her Master's degree in Molecular Bioengineering from the Technical University of Graz, Austria and a Doctoral degree from School of Medicine, University of Giessen, Giessen, Germany. Dr. Nikam's research interests and expertise include stem cells, tissue remodeling and regeneration, toxicology, understanding molecular and cellular biology of underlying diseases. She has received an honorarium from United Therapeutic, NC, USA, Fellowship award for Master of Science in Molecular Bioeng. (M.Sc.) from Technical University of Graz, Austria and received the scholarship from Afro- Asiatische Institute, Graz, Austria. She has several papers published in peer-reviewed journals like British Journal of Pharmacology, Eur Respir J, AAPS PharmSciTech, J Proteome Res, Am J Respir Cell Mol Biol and Pneumologie. She attended and presented her research work at several international conferences like American Thoracic Society Meeting (ATS) 2006, 2009, Symposium of the Austrian Pharmacological Society (APHAR) 2004, European Respiratory Society (ERS) 2006, Austrian Academy of Sciences Proceedings 2008, American Association for Cancer Research (AACR) Annual Meeting, 2013.

Shariq Syed is presently Interim Dean and Associate Professor at AIKTC School of Pharmacy and Advisor to Board, Medley Pharmaceuticals. Earlier, he has served as Senior Research Investigator, Bristol Myers-Squibb, USA (2007-2012). At Bristol Myer Squibb, he was part of Ipilimumab drug development team, and has collaborated with Dr. James Allison, immunologist and Nobel Laureate of the year 2018. At Bristol Myer Squibb, he was responsible for all key aspects of Clinical Pharmacology in Oncology therapeutic area for large and small molecule drugs/ compounds, Co-lead Clinical Pharmacology development of Brivanib®, a novel Kinase inhibitor for Liver cancer and contributed to NDA submission package of Yervoy®, a breakthrough therapy in immune-oncology approved for skin cancer. Dr. Shariq's research interest and expertise include Clinical Pharmacology, Pharmacokinetics, understanding drug exposure-response relationship (PK-PD) in early Proof of concept (POC) trials. He has received Pharma Ratan Young Achiever Award for his contribution tothe field of Immuno-oncology and award from Rho-Chi Honor society (Honor Society in Pharmaceutical Sciences). He is an invited reviewer of the European Journal of Drug Metabolism and Pharmacokinetics, Current Clinical Pharmacology. He is a member of American Association of Pharmaceutical Scientists, elected member of American College of Clinical Pharmacology, elected Member of American Society of Clinical Pharmacology and Therapeutics, a member of the Indian Society of Clinical Research. He has published several research articles in peer-reviewed international journals.

Immunotherapy: A Concept

1

Vritika Kulwal and Sujata Sawarkar

Abstract

Immunotherapy has emerged as a promising therapeutic strategy to combat cancer, autoimmune diseases and infectious diseases in the last few two decades. Immunotherapy is a type of treatment which involves induction, enhancement, or suppression of the immune system. This modern approach has changed the paradigm and revolutionized the concept of modern therapeutics. The approach lies on understanding the molecular basis of the disease and making medicine personalized, accurate, and precise. The therapy relies on the host's immune system and can be broadly categorized as activation or suppression immunotherapies. Health authorities worldwide have accepted and acknowledged the significance of immunotherapy. As a result of this numerous therapeutic moieties based on immunotherapy have received regulatory approval and are being used in clinical practice. This chapter discusses a broad overview of the concept and principle of immunotherapy and its utility in a wide plethora of diseases.

Keywords

Immunotherapy · Adaptive · Activation · Cytokine · Lymphocytes

Immunotherapy is a type of treatment which involves induction, enhancement, or suppression of the immune system. Immunotherapy is a therapeutic strategy that targets or manoeuvres subjects/patient's immune systems (Naran et al. 2018). Immunotherapy aims to utilize the host's immune system to eradicate diseased cells. In recent times immunotherapy has been envisaged for many diseases that

V. Kulwal · S. Sawarkar (✉)
SVKM's Dr. Bhanuben Nanavati College of Pharmacy, University of Mumbai, Mumbai, India
e-mail: sujata.sawarkar@bncp.ac.in

S. P. Sawarkar et al. (eds.), *Immunotherapy – A Novel Facet of Modern Therapeutics*,
https://doi.org/10.1007/978-981-15-9038-2_1

may or may not be of genetic or of infectious origin. The therapy is broadly categorized into two classes. Immunotherapies that evoke immune response are termed as activation immunotherapies while those which suppress immune response are known as suppressive immunotherapies (Wraith 2017). Activation therapies are commonly used for diseases like cancer while suppressive immunotherapy is used for autoimmune diseases. Additionally, immunotherapy is also classified based on passive (adaptive, antibody-based) and active (vaccine, specific allergen based) approaches. Passive immunotherapy includes supplementing immune components like antibodies or immune cells to patients. In this approach, the host's immune response is not invoked whereas in Active immunotherapy, involves triggering immune response wherein antibodies are produced and T-cells are activated. Immunotherapy is now being seen as a promising approach for certain ailments wherein other therapies are failing, however the origin of immune-based therapy is way back in the nineteenth century.

1.1 History and Origin of Immunotherapy

Manipulation of the immune system to treat diseases first started when there was development of vaccination strategy for the protection of humans against smallpox infection. The idea was to bring subjects in contact with the cowpox virus so that immunity is boosted, as it was observed by the British Lady Montague in Turkey. Dr. Charles Maitland conducted the first trial in immunotherapy in the eighteenth century. This concept was further undertaken by Edward Jenner, who developed the true cowpox vaccine which became the first-ever clinically effective vaccination. Bavaria was the first country in the world where smallpox vaccination was made mandatory by law in the year 1807. Studies in the field of immunology and infectious disease were conducted and observed a growth in the subsequent years but exploring the field of immunotherapy in cancer began in the late nineteenth century. It was more than 135 years ago that two German physicians Buschand Fehleisen independently observed that after infection with erysipelas, there was tumour regression in cancer patients. In 1868, Busch observed tumour shrinkage after intentionally infecting a cancer patient with erysipelas. Further Fehleisen, in 1882, repeated this exercise of infecting with erysipelas and also identified *Streptococcus pyogene* sas the agent causing erysipelas. In the year 1891, William Coley an American surgeon at Memorial Hospital in New York observed Regression of unresectable sarcoma due to erysipelas infection. More than 1000 patients were observed with regression and cure. Therefore this method gained wide acceptance. But later on, the use of toxins was discontinued because of the failure to follow good scientific protocols and inconsistency in results. Also the development of radiation therapy and chemotherapy for cancer, further led to loss of interest in this therapy. Later in 1976 Morales et al. proved the efficacy of the bacterium Bacillus Calmette-Guérin (BCG) in treating superficial bladder cancer. The clinical trial was conducted on the basis of study conducted by Old et al. in 1959 in which anti-tumour effects were exhibited by BCG in a mouse model. Apart from the work on BCG, Old also discovered tumour necrosis factor in 1975. Coley and Old were referred to as the 'Fathers of Immunotherapy', due to their discoveries and work in this therapeutic. Interferon-alpha 2 (IFN-a2), an antitumor cytokine, was the first approved immunotherapy-based drug in the year 1986 (Ring and Gutermuth 2011; Oiseth and Aziz 2017).

1.2 Theory of Immunotherapy

Immune system has a significant role and impact on well-being of humans wherein immune-mediated diseases pose serious medical problems. Therefore to treat immune system-related diseases immunotherapy treatment is required which involves either the enhancement or augmentation of the immune system. Immunotherapy treatment involves clinical strategies that modify the activities of immune system components causing activation and improvement of immune function responsible for the prevention or treatment of diseases. For understanding how immunotherapeutics works we need to first understand the immune system and then how these Immunotherapeutics modify the immune system for the treatment of these diseases.

The immune system is divided into two parts, i.e. innate immunity and adaptive immunity wherein the two differ in their speed of action, their effectors, their specificity for antigens and fast-acting capabilities. The innate immune system lacks specificity for antigen and is fast acting as well as does not generate memory. Granulocytes, macrophages, and natural killer cells are the major effectors of innate immunity. On the other hand, adaptive immunity is very specific to particular antigen. Therefore adaptive therapy takes a definite lag time to elicit its response. Another important feature of adaptive therapy is that it can recollect and remember the previous immune response. Hence when there is recurrent exposure to that particular antigen, the host's immune system responds quickly. The main effectors for adaptive immunity include T and B lymphocytes. The details of innate and adaptive immunotherapy have been described in subsequent chapter on Immunotherapy intended for autoimmune diseases.

Another important aspect that is pertinent and readers must understand at this juncture is immune tolerance. Immune tolerance is the process where the immune system fails to exhibit an immune response to antigens. Immune tolerance can be of two types, i.e. natural tolerance or self-tolerance where they fail to attack the body's own proteins and antigens while induced tolerance is where the immune system shows tolerance to external antigens which can be in the case when the immune system is deliberately manipulated to protect the body from allergic reactions as well as enable the acceptance of transplanted organs and also to protect the body from the harmless bacteria in the intestine.

T-cell tolerances are of two types, i.e. Central tolerance and Peripheral tolerance. Central Tolerance refers to the deletion of autoreactive T-cells. AIRE (autoimmune regulator) is a protein that exposes the T-cells to all healthy proteins of the body and their receptors bind these epitopes so tightly that the T-cells attacking those self-antigens undergo apoptosis. T-cells surviving this selection and apoptosis, travel all over the immune system.

Peripheral tolerance: In this type, the cells that leave the thymus can respond to self-antigens. T-cell tolerance is maintained via various mechanisms such as

1. AIRE present in some antigen-presenting cells (APC) in organs of the peripheral immune system eliminate any remnant T-cells which fail to get eliminated in the thymus.
2. Lack of co-stimulation: The activation of the T-cell to effector cells to show immune response must require not only binding of the T-cell receptor to the epitope (MHC peptide) to produce signal one but also a second signal is required. This signal is produced on the binding of B7 molecule on APC with CD28 molecule of T-cells. This second signal is called co-stimulation. So even if they encounter self-antigens, the T-cells only respond when they receive the second signal. Also binding of the T-cell with the self-antigen in the absence of the second signal leads to apoptosis of T-cells. This is because the self-antigens are unable to provide the second signal or provides an unidentified second signal that turns T-cells to regulatory T-cells which suppress the immune response. This leads to self-tolerance.
3. Some organs or tissues are hidden behind anatomical barriers, for example, interior of the eye, testes, brain, which protect the T-cells from reaching them.
4. T-cells activity is suppressed by Regulatory T-cells.
5. Some of the body cells express the Fas ligand, (FasL). Activated T-cells always express FasL (type II transmembrane protein). On encountering these cells, there is the binding of Fas to FasL which triggers their death by apoptosis.

Inadequate response as a result of immune tolerance leads to various disorders. The failure of immune tolerance for self-antigens leads to autoimmune diseases whereas in the case of cancer the immune tolerance prevents the cancer cells from being killed. Thus primary aim of immunotherapy is either to hamper and counteract the effect of immune tolerance to treat diseases like cancer or increase the effect of immune tolerance to treat autoimmune diseases (Jewett and Tseng 2017; Immunologic Tolerance n.d.).

The therapeutic moieties in immunotherapeutics can be broadly classified as immunostimulants and immunosuppressants.

1. *Immunostimulants*: These drugs increase the immunity of the host. They have been put in clinical practice in the treatment of infections, cancer, and immunodeficiency conditions.
   a. Immunostimulant drugs of synthetic origin (Brunton et al. 1992; Patil et al. 2012):
      - *Levamisole*: Levamisole is a synthetic drug mainly inducing B and T-lymphocytes, monocytes, and macrophages. It is used in adjuvant supportive therapy along with 5-Fluorouracil after surgical resection in patients with Duke's stage C colon cancer. Its disadvantages are allergy, nausea, flu, and muscle pain.

- *Thalidomide*: Thalidomide or Immunoprin which was once a prohibited drug has now been explored for its immunomodulatory properties. Thalidomide was found to reduce the number of circulating TNF-α in patients suffering from erythema nodosum leprosum. In contrast, it increased TNF-α in HIV seropositive patients. Furthermore, its therapeutic effects were observed in patients with severe rheumatoid arthritis and angiogenesis.
- *Immunocyanin*: Immunocyanin is a stable form of haemocynin, a copper-containing protein, found in molluses and arthropods used in the treatment of urinary bladder cancer.
- *Bestatin*: Bestatin, a dipeptide [(2S, 3R)-3-amino-2hydroxy-4-phenylbutanoyl]-L-leucine, is an immunostimulant, which stimulates or increases the activity of T and B lymphocytes. It was observed in an experimental setup that, Bestatin efficiently prevented the metastasis of P388 leukemia when the antibiotic was constantly injected. This was due to the binding of Bestatin to lymphocytes and macrophages, and thus activating them, which further led to an increase in T-lymphocytes. The dipeptide was shown significant immune-restoration activity in cancer and HIV affected patients (Tsuruo et al. 1981; Mathe 1991).

b. *Bacterial product:*
Bacillus Calmette-Guérin (BCG) is an attenuated live culture bacillus of Calmette and Guérin strain of *Mycobacterium bovis*. BCG was found effective in treating as well as for prophylaxis of urinary bladder, papillary tumours, and chronic infections related to immunodeficiency. BCG was found to enhance specific immunity responsible for the mitigation of tumour in the experimental animal model, and counter the effect of immunosuppressive drugs. Since IL-1 stimulates the maturation of CD4+ T-cells, therefore it activated macrophages producing IL-1 (Jewett and Tseng 2017; Brunton et al. 1992).

c. Passive immunotherapy (Naran et al. 2018; Brunton et al. 1992; Makkouk and Weiner 2015; Cytotoxic n.d.): Passive-mediated immunotherapy involves direct administration of immune components such as antibodies or immune cells, to the patients. The immune components are produced ex vivo. This therapy does not depend on the immune system of host. Passive immunotherapy is categorized into two types, Antibody-based therapy and Adaptive T-cell therapy.
   - Antibody based therapy

   Monoclonal antibodies (mAbs) are used as immunotherapeutic agents. mAbs either bind to mediators like growth factors and cytokines or to their receptors. By this action, they prevent these mediators from reaching their targets. Fc (fragment crystallisable) receptors contribute significantly to the protective effects of the immune system. mAbs are targeted to those cells comprising of Fc receptors on their surface or to the effector moieties. Fc receptors interact with antibody-coated pathogen or tumor. Further signals from antibodies are generated when receptors pertaining to programmed cell death, get crosslinked. mAbs produce antitumor activity via antibody-dependent cell-mediated cytotoxicity, phagocytosis, and complement-dependent cytotoxicity. Studies have also shown that in some cases, mAb-induced lysis of tumour cell can generate an adaptive immune response by enhancing the tumor antigens uptake and cross-presentation by dendritic cells (DCs).

- Adoptive T-cell therapy

Enhancing T-cell activation: For successful activation of T-cells, it requires two signals, one is the binding of the T-cell receptor (TCR) with major histocompatibility complex (MHC peptide complex) and another is the binding of T-cell co-receptors with counter receptors on Antigen-presenting cells (APC). T-cells, quite often lose their functionality due to frequent antigen exposure and inflammation. T-cell exhaustion leads to many chronic infections and cancer. Therefore it is essential to prevent this exhaustion and restore immune responses for effective mitigation of infection or malignancies.

Immune checkpoint inhibitors: Immune checkpoints include T-cell surface molecules, i.e. cytotoxic T-lymphocyte antigen 4 (CTLA-4), programmed cell death protein-1 (PD-1), T-cell immunoglobulin and mucin domain-containing protein 3 (Tim-3), and lymphocyte activation gene3 (LAG-3). These molecules play a critical role in preventing autoimmunity by regulating the T-cell response. Intratumoral lymphocytes expressing these markers in the tumor microenvironment results in immune exhaustion. Also these molecules are highly expressed on Regulatory T-cells (Tregs) and play a key role in suppressing the effector T-cells. These molecules can therefore be considered as targets for reversing immune tolerance. Various immune checkpoint inhibitors currently used in clinical practice for the treatment of cancer include anti-CTLA-4 mAb Ipilimumab and anti-PD-1 mAbs, i.e. MDX-1106 and CT-01 act by reversing immune tolerance.

Activation of Cytotoxic T-cells: Cytotoxic T-cells are cells that attack and kill target cells such as cancer cells and virus-infected cells. Antigen along with class 1 MHC molecules when presented on the infected cell, are detected by these cytotoxic CD8+ cells destroying the cells. Therefore activation of these cytokines would be a good therapeutic strategy. T-cells can be activated to attach to the antigen present on infected cells thereby killing the cells. Bispecific antibodies (bsAbs) have been designed wherein they have two binding sites, i.e. the antigen-binding site and T-cell activating site. Bispecific T-cell engagers (BiTEs) is a type of bsAbs, usually used for Cancer as well as studies for inflammatory conditions. It has both, an antigen-specific region and a T-cell binding region. Therefore it simultaneously binds to the cytotoxic T-cells and the antigens on tumour cells. This creates a link between the T-cell and the target cells, thus leading to activation and proliferation of the T-cells resulting in cell death Catumaxomab was the first bispecific antibody used for treating malignant ascites. Dual affinity re-targeting antibodies (DART's) is another form of bsAbs, investigated for the treatment of HIV, activating the T-cells thereby killing HIV-infected cells. Also BiTEs targeting the HIV-1 envelope protein gp120 and CD4 have shown promising results in vitro and ex vivo. Drawbacks associated with these agents is that they are rapidly cleared due to their small size and therefore have short half-life and hence needs to be continuously administered.

Chimeric Antigen Receptor T-cell Therapy (CAR T-cell Therapy): CAR (chimeric antigen) T-cell are peculiar immune therapeutic moieties comprising of recombinant receptor along with single-chain variable fragment (scFv) where the CARs lead to generation and proliferation of tumour targeting T-cells in the tumour microenvironment. Introduction of DNA fragment into T-cells stimulates the generation of Chimeric Antigen receptors on their surface. Such T-cells are now termed as chimeric antigen receptor T-cell which is reinfused into patients. Some of the FDA approved treatments include Tisagenlecleucel (Kymriah™) for the treatment of patients up to 25 years of age with B-cell precursor acute lymphoblastic leukaemia, Axicabtagene ciloleucel (Yescarta™) for adult patients with relapsed or refractory large B-cell lymphoma.

- *Passive Immunization*: Passive immunization is provided to the patient deficient with antibodies, due to conditions such as congenital or acquired immunodeficiency, or where treatment using passive antibodies is necessary (Jewett and Tseng 2017).

Human Immunoglobulin: Human immunoglobulins are injected in cases of decreased humoral immunity. They are commonly given in measles and infectious hepatitis. Human Immunoglobulins are sourced and isolated from patients who have been previously contracted with these diseases and have produced protective antibodies.

Antivenins or Antitoxins: These immunoglobulins derived from horses are immunized with toxins or venoms. They induce immunity in recipients by neutralizing toxins such as snake venom or diphtheria toxins.

Rho (D) Immunoglobulin: Rh disease occurs when an Rh-negative woman (one whose red blood cells is deficient with Rho(D) antigen) gets sensitized to the antigen on exposure to Rh-positive blood of her foetus. As a result of this, on subsequent pregnancies, the baby is born with the haemolytic disorder as the anti-Rh antibody from the mother passes through the placenta to the foetus causing massive destruction of foetal erythrocytes. In such cases, early diagnosis is done and Rh-negative mothers who will give prospective birth to Rh-positive infants are administered anti-Rh antibodies. Injection of these antibodies results in the generation of anti-idiotypic antibodies that obstruct the interaction of B-cells with Rh antigen. Also these antibodies rapid clearance of foetal red blood cells from mother systemic circulation by liver macrophages. As a result, it prevents inflammatory reactions which are necessary for antibody responses.

d. *Active Immunotherapy*: Active immunotherapy involves the induction of the patient's immune response resulting in the development of antibodies and T-cells which are the immune effectors' cells Active immunotherapy includes the administration of vaccines and allergen-specific T-cells.

- *Vaccines*: These are part of an active immunization programme wherein non-pathogenic, non-infectious, attenuated version of the disease-causing microbes (pathogens) are administered to the recipients. This provides a stimulus for the T-cells activation and for developing immunologic memory. Most of the vaccines are targeted for infectious disease. In addition to this new generation of vaccines are designed for Cancer or Autoimmune diseases to provide limited or complete protection. The fundamental difference between viral targeted vaccines and cancer vaccines is that these vaccines majority of times, rather than prevention, they stimulate the immune system to counter-attack already existing disease in patients. In the case of cancer vaccination therapy, peptides, and co-stimulatory ligands present on antigen-presenting cells (APCs) stimulate T-cells to attack tumour cells. Hence the patients are immunized with the aid of APCs which express tumour-specific antigen. In the first generation of these vaccines whole cancer cells or tumor-cell lysates were used as antigen along with various adjuvants, where they were dependent on the host APCs for processing and presenting tumour-specific antigens. In the second generation, specific APCs were incubated with antigen ex vivo and APC's presented the antigens and they were further reinfused into patients.

Also, isolated immune cells from the patients are exposed to cancer antigens, and after activation are introduced back into the patient's systemic circulation for suppression of the cancer cells. Some of the approved therapeutic cancer vaccines include Sipuleucel-T is the first FDA approved vaccine for Cancer (Naran et al. 2018; Brunton et al. 1992; Schuster et al. 2006).

2. *Immunosuppressants* (Choudhary n.d.): Immunosuppressive agents are drugs that attenuate the immune response. This type of therapy is required for diseases where the immune system attacks self-antigens like autoimmune disorders and some allergic reactions. It is also used to suppress the normal immune system, which is required in conditions like organ transplantation where normal immune reactions are not required.
   a. *Corticosteroids:* These were the earliest immunosuppressive agents used in clinical practice. Corticosteroids exert their actions through either genomic or non-genomic mechanisms. Glucocorticoids on internalization in the cytoplasm interact and form complexes with glucocorticoid receptors (GCR) and this complexes directly or indirectly with transcription factors such as activator protein 1 (AP1), Nuclear factor kappa B (κB) and interferon regulator factor 3 (IRF 3) which are involved in the regulation of pro-inflammatory genes. This negative regulation by GC/GCR complex causes reduced transcriptional activities of GCR activating genes resulting in anti-inflammatory and immunosuppressive actions. Non-genomic mechanisms also lead to reduced T-cell activation (Hardinger et al. 2016; Stahn et al. 2007).
   b. *Inhibitors of Cytokine Production and Functions (T-cell Inhibitors)* (Hardinger et al. 2016; Kapturczak and Kaplan 2004):
      - Calcineurin Inhibitor: Calcineurin is involved in the activation of T-lymphocytes by catalysing some of the intracellular processes related to T-lymphocyte activation. Therefore binding of the calcineurin inhibitors to

the intracellular proteins called immunophilins blocks the effect of calcineurin leading to reduced interleukin-2 production and T-cells proliferation and therefore showing immunosuppressive action. There are two calcineurin inhibitors, i.e. Cyclosporine and Tacrolimus. Cyclosporine inhibits calcineurin by binding with cyclophilin. Tacrolimus inhibits calcineurin by binding with immunophilin and is more potent than cyclosporine as immunosuppressant.

- mTOR Inhibitor: Sirolimus and Everolimus are the types of mTOR inhibitors. Similar to Tacrolimus, they bind to FK proteins but they do not inhibit calcineurin. Instead, they inhibit the mammalian target of Rapamycin (mTOR), which is a protein kinase responsible for cell cycle progression. Inhibiting mTOR leads to suppression of cytokine driven T-Lymphocyte proliferation, by halting the progression of the cell cycle from G1 to S-phase, and activation ultimately leading to immunosuppression.
- Co-stimulation Blockers: Immune Response induced by T-lymphocytes requires two signals, i.e. Signal 1 which occurs by reaction of MHC complex, present on APC, with T-cell receptors and Signal 2 which is a co-stimulatory signal which occurs via interaction of CD28 molecules on the T-lymphocyte with CD80 and CD86 molecules on the antigen-presenting cell (APC). Further T-cell activation expresses additional co-stimulatory molecules such as CD152 and CD154. CD154 (CTLA4) interacts with CD80 and CD86 on APC to diminish immune response. Antibodies for CD80, CD86, and CD152 can serve as potential therapeutic agents. CTLA4-Ig is a chimeric protein that is a constant region of IgG containing the binding region of CTLA4 and therefore inhibits CD28 and further inhibiting T-cell activation and hence can act as a therapeutic agent. Belatacept, a human CTLA4-Ig fusion protein, is a second-generation Co-stimulation blocker approved for renal transplantation. Other molecules which can act as potential therapeutic agents include h1F1 and h3D1 humanized mAb antibodies against CD80 and CD86 molecules, LEA29Y which is a second-generation CTLA4Ig, anti-CD154 humanized monoclonal antibodies, using an OX40-Ig fusion protein for blocking the OX40-OX40L pathway, which is involved in the expansion of Effector T-cells as well as promoting the generation of Memory T-cells, blocking of CD27: CD70 pathway (which is involved in T-cell development and activation), 4-1BB (a co-stimulatory molecule involved in T-cell stimulation), TIM family proteins (involved in enhancing T-cell activation) and LFA-1:ICAM and VLA-4:VCAM pathway (Brunton et al. 1992; Hardinger et al. 2016; Kinnear et al. 2013).

c. *Anti-proliferative/Anti-metabolic Agents/Cytotoxic Agents* (Nelson et al. 2003; Goumas et al. 2010):

- Azathioprine: Azathioprine a purine antimetabolite and an imidazolyl derivative of 6-mercaptopurine. This on oral administration metabolizes to 6-mercaptopurine which on incorporating into the DNA causes the death of rapidly dividing cells of the bone marrow and intestine. Therefore this drug is used for organs transplant rejection and for autoimmune diseases such as rheumatoid arthritis, Crohn's disease, ulcerative colitis, and chronic graft-versus-host disease (GVHD). However, it has been observed that long term treatment may lead to squamous carcinoma of the skin and lymphoma or bacterial infections.

- Mycophenolate Mofetil: It is an inhibitor of inositol monophosphate dehydrogenase, which is an enzyme required for de novo guanosine nucleotide biosynthesis. As a result of impaired enzyme activity, DNA and RNA synthesis are hampered. Since T-cells and B-cells are dependent on the de novo pathway for DNA synthesis whereas other cells can recycle the nucleotides by salvage pathway and therefore Mycophenolate Mofetil causes T-cell inhibition. It is used for preventing organ transplantation rejection and is usually given along with GCS and cyclosporine. It is also being studied for the treatment of autoimmune diseases.
- Cyclophosphamide: Cyclophosphamide is an alkylating agent which on forming covalent bonds with DNA causes mutation and fragmentation of DNA resulting in cell death. This leads to cellular immunity suppression and inhibition of antibody and autoantibody production. They are used as a treatment for Systemic Lupus Erythromatous (SLE) and vasculitis. Adverse effects associated with cyclophosphamide are leukopenia, sterility, haemorrhagic cystitis, and malignancy, including leukaemia and transitional cell carcinoma.
- Methotrexate: Methotrexate functions by inhibiting dihydrofolate reductase, resulting in inactive oxidized folates accumulation and therefore inhibiting nucleotide synthesis thereby killing cells in S-phase (DNA synthesis). Methotrexate is used for treating rheumatoid arthritis, psoriasis, Systemic lupus erythematous and in anticancer therapy. Methotrexate should be used with caution as long term treatment is associated with hepatic toxicity leading to liver cirrhosis. Being teratogenic, it should be avoided in pregnant women.

d. *Antibodies/Biologics* (Jewett and Tseng 2017; Brunton et al. 1992; Hardinger et al. 2016; Drosos 2002; Visser et al. 2017): They are of two types monoclonal and polyclonal antibodies. They are mostly used for preventing organ transplant rejection by targeting the T-cell signalling to control the T-cell mediated allo-immune responses.

- *Polyclonal Antibody*: These are obtained by repeatedly injecting the human thymocyte or the human lymphocyte in animals such as horses, rabbits, sheep, or goats after which the serum immunoglobulin fraction is purified. Though being highly efficacious they have batch to batch variation in terms of toxicity and efficacy.

Anti-thymocyte Immunoglobulin: It is a purified gamma globulin obtained from the serum of rabbits after immunizing them with human thymocytes. Ant thymocyte globulin containing cytotoxic antibodies bind to CD2, CD3, CD4, CD8, CD11a, CD18, CD25, CD44, CD45, and HLA class I and II, which are antigens molecules expressed on the surface of T-lymphocyte. This binding of Antibodies causes cytotoxicity and therefore lysis or agglutination of T-lymphocyte thereby reducing the number of circulating lymphocytes and these antibodies also bind to molecules on the cell surface involved in regulating cell function, thereby blocking lymphocyte function.

It is used as induction immunosuppression, mostly used for transplant rejection.

- *Monoclonal Antibodies*: Monoclonal antibodies are single antibodies having defined specificity. They bypass and overcome the problem of polyclonal antibody, i.e. variability in toxicity and efficacy.

Muromonab (Anti CD-3 Antibodies): CD3 is a component of the T-cell receptor complex (TCR), besides the T-cell receptor, having the function of antigen recognition, cell signalling and proliferation. Muromonab-CD3 causes rapid internalization of the T-cell receptor by binding to the binds to the ε chain of CD3 molecule, thereby further blocking antigen recognition. Therefore administering this antibody depleted most of T-cells from the bloodstream and lymph nodes and spleen. There is a further reduction in the T-cell functions due to the deficiency of IL-2 production and other multiple cytokines, the exception being IL-4 and IL-10.

Chimeric antibodies (Basiliximab): Major limiting factor for MAbs is that it is of animal origin. Therefore to circumvent this problem, the antigen-specific portion of the animal origin monoclonal antibody is combined with a constant region of IgG1 human monoclonal antibody to form a chimeric antibody. Chimeric monoclonal antibody with IgG1 constant region is used for antibody-dependent cell-mediated cytotoxicity and those with IgA constant region exhibited anti-inflammatory effects. Basiliximab an IL-2 receptor antagonist is the only Food and Drug Administration (FDA) approved agent for renal transplantation as IL-2 is a major cytokine involved in organ rejection. The adverse events associated with it is allergic reactions.

Rituximab is a chimeric monoclonal antibody having a function of B lymphocyte depletion. This is used for B-cell lymphoma, rheumatoid arthritis and granulomatosis with polyangiitis.

Humanized Antibodies (Anti CD-25 antibody): These are the complementary defining region (CDR) grafted monoclonal antibody. CDR is the hypervariable peptide region of the antibody that binds with the antigen. These hypervariable regions are joined to the other sequences of the antibody.. Therefore, in these humanized antibodies the hypervariable region is of the rodents while the framework sequences and the constant regions are of human antibody. This is particularly used for organ rejection as well as other diseases such as rheumatoid arthritis, Crohn disease, systemic vasculitis, septic shock, various neoplasms, and viral infections. Daclizumab, an anti-IL-2 receptor monoclonal antibody, is the humanized antibody version of Basiliximab, containing 90% human antibody part and 10% murine antibody part, therefore reducing the animal content of the antibody to 10%. It prevents the adverse effect of Basiliximab, i.e. the cytokine release syndrome.

Campath 1-H (Alemtuzumab): It is a humanized mAb approved for the treatment of chronic lymphocytic leukemia. This antibody targets CD52 glycoprotein expressed on lymphocytes, monocytes, macrophages, and natural killer cells leading to apoptosis of targeted tumour cells causing extensive lympholysis. It has also proved to be effective in renal transplantation.

Efalizumab, an IgG1 humanized monoclonal antibody, is an anti-CD11a molecule binding to the CD11a chain of lymphocyte function-associated antigen (LFA-1). This binding prevents the interaction with LFA-1–ICAM (intercellular adhesion molecule) thus blocking adhesion and activation of T-cell. This is used for organ transplantation rejection as well as it has been found to be effective in treating psoriasis.

3. *Cytokine Therapy* (Brunton et al. 1992; Hardinger et al. 2016; Drosos 2002; Ohno et al. 2012; Biotechnolgy Forum n.d.; Santamaria 2013; Delves n.d.): Cytokines are soluble factors that act as a tool for communication between the immune system. They are responsible for immune homeostasis in the body. Modifications in their expression or dysregulation of the intracellular pathways leads to altered homeostasis which ultimately leads to various disorders such as autoimmune disorders, chronic inflammation and malignancies. Cytokine therapy functions by manipulating the activity of specific cytokines either by restoring or blocking their activity to treat the disease. Cytokines are classified into five types, i.e. Type 1 and Type 2 cytokines, Tumour necrosis factor (TNF) family cytokines, Interleukin 1 (IL-1) and Interleukin 17 (IL-17) family of receptors. Type 1 includes IL-2, colony stimulating factors and hematopoietic factors and they are involved as both positive and negative regulators of immune response. Type 2 contains interferons and cytokines of IL-10 family. IL-10 family contains IL-10, IL-19, IL-24 and IL-26 and IL-10 has an essential role in immunosuppressive as well as anti-inflammatory functions. TNF family includes molecules which modulate the development of the immune cells and also pro-inflammatory responses. IL-1 family is involved in pro-inflammatory and anti-inflammatory responses where IL-1, IL-18, IL-36 induces other pro-inflammatory cytokines and IL-37 and IL-38 are negative regulators. IL-17 contains IL-17 A and IL-17 E (IL-25) and their function is the induction of other cytokines such as TNF and IL-1. Therapeutically targeting these molecules or their receptors would lead to the treatment of inflammatory conditions, immune-related disorders and malignancies.

1.3 Application of Immunotherapy in Various Diseases

1.3.1 Immunotherapy for Cancer

Cancer is one of the most life-threatening disease. Despite the standard treatment available for cancer treatment, i.e. surgical resection, chemotherapy, and radiation therapy, cancer accounts for 25% of mortalities. The primary tumours are treated via combination of the standard treatments, but these combinations of standard treatments are ineffective in the case of metastatic cancer which has spread through disseminated cells. Therefore, there is a need to eradicate these disseminated cells which are present in the blood and has also spread to various organs. Therefore, immunotherapy acts as a promising alternative treatment that helps in stopping or slowing the growth of cancer cells as well as works in the prevention of the metastatic spread of the disease. Immunotherapy aims at restoring the immune system to fight the tumour/cancerous cells. The various types of immunotherapy used for cancer include passive immunotherapy or adaptive transfer of immune effectors which are generated ex vivo and active immunotherapy in which tumour antigens are introduced which results in immune effector proliferation in vivo. Passive immunotherapy includes tumour-specific monoclonal antibodies where Rituximab was the first FDA approved mAb for low-grade lymphomas, Adaptive T-cell transfer which involves removing lymphocytes from the patients bearing the tumour and then growing the cells ex vivo and the reinfusing these cells back into the patient. To enhance the specificity of the treatment to the tumour, tumour-infiltrating lymphocytes are used as there are reported assumptions that they are rich in antitumor T-cell populations. Other therapy is the CAR (chimeric antigen T-cell therapy), wherein the CAR genes are genetically transferred to T-cells that get activated and proliferated in vivo when they come in contact with tumour-specific antigen. Active immunotherapy includes vaccination, i.e. providing the immune system with tumour-specific antigens, peptides, or whole-tumour cells which leads to activation of the immune response towards cancer cells. Cervarix, which was granted FDA approval in 2009, is a prophylactic vaccine cervical carcinoma that is associated with HPV. Immunomodulatory therapy which includes T- cell activation and immune checkpoint inhibitors are also involved in cancer treatment. The immune checkpoint inhibitor anti–CTLA-4 mAb Ipilimumab Bristol-Myers Squibb), which is used for the treatment of metastatic melanoma, received FDA approval in 2011.Also the tumour cells exert immune resistance mechanism through PD-1/PD-L1 pathway to respond to endogenous immune antitumor activity. Therefore PD-1 blockade that targets either PD-1 or its ligand PD-L1 is showing promising results for Cancer. Nivolumab and Pembrolizumab are anti-PD-1 and PD-L1 therapies that are routinely used for previously treated metastatic melanoma and squamous non-small cell lung cancer (Makkouk and Weiner 2015; Schuster et al. 2006; Abcam n.d.).

1.3.2 Immunotherapy for Autoimmune Diseases

Autoimmune diseases like Rheumatoid Arthritis, Type 1 Diabetes, Multiple Sclerosis and Systemic lupus Erythromatous, psoriasis, Crohn's disease and Ulcerative Colitis represent a failure or loss of tolerance of the immune system for self-antigens. Currently, control of autoimmune diseases depends on the use of non-specific immunosuppressive drugs or the use of biologics which aim at blocking cytokine, T and B lymphocyte depletion or increasing the regulatory components. Detailed information is included in the subsequent chapter.

1. *Rheumatoid Arthritis*: Rheumatoid arthritis (RA) is a chronic, autoimmune inflammatory disease. Suppression of the immune system or restoration of immune tolerance by enhancing regulatory T-cell numbers or functions would lead to the treatment of Rheumatoid Arthritis. Regulatory T-cells (Treg) are key mediators involved in peripheral immune tolerance, which is required for maintaining immune homeostasis. Cytokines such as TNF-α and IL-17 present during inflammation are known to inhibit the suppressive action of natural Treg. Therefore either introducing new functional Tregs or restoring the function of Treg already present using biologic therapy is an attractive strategy for treating disease. The immunotherapy for rheumatoid arthritis includes cytokine targeting therapies like TNF inhibitors, IL-1 and IL-6 antagonist, IL-17 inhibitor, Anti-GM-CSF (granulocyte-macrophage colony stimulating factor) and anti-chemokine therapy. Other therapies include cellular therapies where regulatory T-cells are developed and introduced in the patients, B- and T-lymphocyte-targeting molecules and Kinase inhibitors. CD3 mAb has the potential to re-establish the control of regulatory T-cells and thus maintain immunologic tolerance and therefore treating Rheumatoid arthritis (Reynolds et al. 2014; Meier et al. 2013; Van Amelsfort et al. 2004; Mcgovern et al. 2012).
2. *Multiple Sclerosis*: Multiple sclerosis (MS) is an autoimmune disorder causing inflammation in the central nervous system. MS occurs due to a poorly regulated immune system. Effector T-cells, Natural Killer cells, B-cells are some of the components contributing to the development of MS. Some of the treatment strategies include decreasing Th1/Th17 cells, inducing Regulatory T-cells and targeting B-cells. Another factor responsible for MS includes Toll like receptors (TLR) While activation of, TLR4, TLR7, TLR9 are responsible for the development of MS, activating TLR3 causes the treatment of MS by increasing the anti-inflammatory agents. Therefore TLR-agonists and antagonists molecules are used as therapeutics for the treatment of multiple sclerosis. Some of the FDA approved drugs include alemtuzumab, natalizumab, and ocrelizumab (Baecher-Allan et al. 2018; Marta 2009).
3. *Systemic Lupus Erythematosus*: Systemic lupus erythematosus (SLE) is an autoimmune syndrome that causes dysregulation in the immune responses. Corticosteroids and Immunosuppressive drugs are the main treatment approach, but patients not responding to these treatments and also various side effects associated with these treatments have led to the development of new immunotherapies. These are termed as biologics that include biological agents that target and inhibit cytokines, co-stimulatory molecules, B-cells. Other therapies include cell-based and peptide immunotherapy (Visser et al. 2017; Mok and Shoenfeld 2016).

1.3.3 Immunotherapy for Allergy

Allergy is a disease of the immune system with the most common type of allergies being allergic rhinitis, conjunctivitis, allergic asthma, atopic dermatitis, eczema, urticaria, angioedema, food allergies, drug allergies and insect sting allergies.

Advancement in immunology research has led to the understanding of cellular and molecular mechanisms related to allergy. Allergic response is usually characterized by continuously elevated levels of IgE antibodies against the allergens to which the patient is exposed by either ingestion, inhalation, or exposure to skin. Allergic sensitization involves antigen processing by antigen-presenting cell (APC) and is presented to the T-cell receptor in association with MHC protein. Further Activation of T-cells requires an additional co-stimulatory signal which occurs through binding of CD28 or CTLA4 on a T-cell to B-7 (CD80/86) on an APC. Therefore the helper T (TH) cells when stimulated by antigens further produce specific cytokines, i.e. TH 1 cytokines (interleukin [IL]-2, interferon [IFN]-gamma, and tumor necrosis factor [TNF]-beta) or TH 2 cytokines (IL-4, IL-5, IL-9, IL-10, and IL-13). Therefore a TH 2-like cytokine function is to induce IgE antibody (Ab) production and thus specific IgE levels determine the manifestation of an allergic reaction.

Allergy immunotherapy works by introducing the increasing doses of the allergens thereby inducing immunologic tolerance to that allergen and therefore would reduce the symptoms of allergy when exposed to the allergens in the future.

Two mechanisms that may contribute to this immunologic tolerance, i.e. by immune deviation and induction of regulatory T-cells. In immune deviation, the T-Helper Type 1 (TH-1) cells are stimulated at the expense of TH-2 cells. These TH-1 cells further produce interferon gamma (IFN-γ) which in turn stimulates B-cells thus producing IgG instead of IgE which does not generate any allergic reaction. Induction of regulatory T-cells produces interleukin (IL)-10 and transforming growth factor (TGF)-β which further redirects the antibody class switching in favor of IgG4 and IgA. These IgG4 antibodies work by blocking allergen-induced activation of mast cells and basophils by interrupting the presentation of antigen to Th2 cells in turn weakening the allergic reaction.

These allergens are introduced by two routes, i.e. by parenteral/subcutaneous route (SCIT) or sublingual route (SLIT).

In SCIT, increasing quantities of allergen are administered to slowly induce tolerance and evaluate the sensitivity of the patient. Further, the maximum dose tolerated or the maximum dose suggested is further given as the maintenance dose. While SLIT usually consists of drops or tablets. SLIT tablets for treating allergies to ragweed, dust mites and pollen grains have received FDA approval (Larsen et al. 2016; Wu 1954; Mohapatra et al. 2010).

1.3.4 Immunotherapy for HIV Infection

The Human Immunodeficiency Virus (HIV) infects the immune cells causing their depletion as well as the affects the uninfected immune system effector cells, leading to immunodeficiency. Studies have reported immunotherapy as a treatment approach for HIV infections. Natural Killer cells are the cytotoxic lymphoid cells which are major effector cells responding to the early HIV infection and eliminating HIV viruses. HIV induces functional defects and adversely affects these cells and therefore strategies designed in improving their function in the HIV-infected patients will be valuable. The cytotoxic action of the natural killer cells is regulated via the inhibitory and activating receptors on their surface. The inhibitory receptors include IL T2, killer Immunoglobulin like receptors (KIR) and CD94/NKG2A receptor and the activating rectors include NKp30, NKp44, NKp46, and NKG2D. MHC 1 class molecules on the surface of the target cells, i.e. the normal cells, triggers the NK cells inhibitory receptors signalling pathway leading to blockade of the cell lysis of the target cells. Whereas the virus-infected cells have low expression of MHC 1 class molecules which reduces the signalling via inhibitory receptors and in turn there is activation via activating receptors leading to lysis of virus-infected cells. Therefore NK cells are known to combat HIV viruses. There are also NKT cells present, which expresses T-cell receptors on their surface. Most of these NKT cells express invariant T-cell receptors and therefore known as invariant NKT cells (iNKT). These iNKT cells have reduced number and function in HIV patients, therefore immunotherapeutics such as vaccines containing adjuvants to increase NKT cells function and number can be used for HIV. Also Cytokine therapy of IL-2, IL-7, IL-15, IL-21, which stimulate and maintain T- cells, NK and NKT cells, have shown to be potential immunotherapeutics (Funke et al. 2011; Sirskyj et al. 2008).

1.3.5 Immunotherapy for Hepatitis

Infection due to both Hepatitis C (HCV) and Hepatitis B viruses (HBV) have increased in the past two decades. These viruses cause infections that eventually lead to varying degree of hepatic inflammation ultimately leading to liver cirrhosis and hepatocellular carcinoma. HCV treatment with newer therapies has led to 90% of the patients achieving successful clearance of the virus. Comparatively treatment with curative intent for HBV are still required and thus new therapies that either target the steps of HBV replication or the host immune system, are required. So immunotherapy for HBV virus aims at targeting the innate immunity as well as the adaptive immunity. Cytokines such as TNFα, IFNα, IFNγ and IL-1β have antiviral effects against HBV. For this purpose, Toll Like Receptors (TLR) and retinoic acid-inducible gene-I (RIG-1) agonist have been developed where RIG-1 triggers the activation of cytokines from the hepatocytes directly while TLR-7 and TLR-8 trigger the activation of cytokines from the neighbouring immune cells. T-cell plays role in controlling the HBV replication. For targeting adaptive immunity there are two strategies, i.e. either increasing the T-cells which are specific to defective HBV that are present in some chronic HBV patients or by producing new HBV specific T-cells which are then transferred into the patients. Therapy to boost HBV specific T-cells in chronic HBV patients includes checkpoint inhibitors (PD-1 and CTLA-4) and vaccines. Other strategies that are applied, wherein the cytokines are directed to the HBV-infected hepatocytes, for example, a T-cell receptor (TCR)-like antibody when conjugated with the cytokine IFNα is shown to target and deliver the antiviral cytokine specifically to HBV-infected hepatocyte. But the disadvantage of using this therapy is that there are rare chances of existence of HBV specific immune cells in HBV patients or they may be with exhausted phenotype or metabolic alterations and therefore the next therapy where there is adoptive transfer of engineered HBV specific T-cells (Maini and Gehring 2016; Bertoletti and Le Bert 2018).

1.3.6 Immunotherapy for Neurodegenerative Diseases

Immunotherapy is currently being extensively researched for treatment for various neurodegenerative diseases like Alzheimer's disease (AD), Parkinson's disease (PD), Huntington's disease, Dementia with Lewy bodies, Frontotemporal dementia and others. Neurodegenerative diseases usually cause accumulation of insoluble aggregates and/or their soluble oligomeric precursors which are either extracellular (amyloid-beta (Aβ) protein) or intracellular α-synuclein (α-syn), tau and other proteins in neurodegenerative diseases) Therefore immunotherapy for Neurodegenerative disorder aims at clearing these aggregates or oligomers as they are believed to be the main components of neurodegeneration. Immunotherapeutics, targeting Aβ, α-syn and tau, such as Active immunization, passive immunization, and T-cell-mediated approaches are developed for Neurodegenerative disorders.. Clinical trials of immunotherapeutics have suggested that this could prove to be promising therapeutics for neurodegenerative disorders which usually progress through accumulation and propagation of protein aggregates (Montoliu-gaya and Villegas 2018).

1.4 Conclusion

Immunotherapy-based drugs are now being approved for the treatment of a wide plethora of diseases. The therapy has shown promising results however Pharmacoeconomics of the same should be taken into consideration. Therapeutic concepts and practices are evolving for better mitigation of disease. Immunotherapy emerges as a viable alternative to existing measures. However extensive research is required to be done so that it gains acceptance worldwide.

References

Abcam (n.d.) Cancer immunotherapy and the PD1/PDL1 checkpoint pathway. https://www.abcam.com/cancer/cancer-immunotherapy-and-the-pd1pdl1-pathway. Accessed 8 July 2020

Baecher-Allan C, Kaskow BJ, Weiner HL (2018) Multiple sclerosis: mechanisms and immunotherapy. Neuron 97(4):742–768. https://doi.org/10.1016/j.neuron.2018.01.021

Bertoletti A, Le Bert N (2018) Immunotherapy for chronic hepatitis B virus infection. Gut Liver 12(5):497–507

Biotechnolgy Forum (n.d.) Recombinant cytokine therapy for immune and inflammatory disorders. https://www.biotechnologyforums.com/thread-1875.html. Accessed 16 Oct 2019

Brunton LL, Lazo JS, Parker KL (1992) Goodman and Gilman's the pharmacological basis of therapeutics, 11th edn. McGraw-Hill, New York

Choudhary D (n.d.) Immunosuppressants: drugs for autoimmune diseases; organ transplantation. http://www.doctoralerts.com/immunosuppressants/. Accessed 16 Oct 2019

Cytotoxic T lymphocytes (CTL) (n.d.). https://www.biology-pages.info/C/CTL.html. Accessed 7 July 2020

Delves PJ (n.d.) Immunotherapeutics—immunology; allergic disorders. MSD manual professional edition. https://www.msdmanuals.com/professional/immunology-allergic-disorders/biology-of-the-immune-system/immunotherapeutics. Accessed 16 Oct 2019

Drosos AA (2002) Newer immunosuppressive drugs their potential role in rheumatoid arthritis therapy. Drugs 62(6):891–907

Funke J, Dürr R, Dietrich U, Koch J (2011) Natural killer cells in HIV-1 infection: a double-edged sword. AIDS Rev 13:67–76

Goumas FA, Braun F, Broering DC (2010) Drugs that act on the immune system: immunosuppressive and immunostimulatory drugs, vol 32. Elsevier, Amsterdam. https://doi.org/10.1016/S0378-6080(10)32038-1

Hardinger KL, Agha IA, Brennan DC (2016) Immunosuppressive agents. Springer, Cham. https://doi.org/10.1007/978-3-319-28797-3

Immunologic Tolerance (n.d.). https://www.biology-pages.info/T/Tolerance.html. https://www.google.com/search?rlz=1C1CHBF_enIN793IN793&ei=iSCnXZjeNtq9rQGysY_oBw&q=%09https%3A%2F%2Fwww.biology-pages.info%2FT%2FTolerance.html.&oq=%09https%3A%2F%2Fwww.biology-pages.info%2FT%2FTolerance.html.&gs_l=psy-ab.3...3680.7477..7703...0.0..0.319. Accessed 16 Oct 2019

Jewett A, Tseng H (2017) Immunotherapy. In: Pharmacology and therapeutics for dentistry. Elsevier, St. Louis, pp 504–529. https://doi.org/10.1016/B978-0-323-39307-2.00035-7

Kapturczak MH, Kaplan B (2004) Pharmacology of Calcineurin antagonists. Transplant Proc 36:25S–32S. https://doi.org/10.1016/j.transproceed.2004.01.018

Kinnear G, Jones ND, Wood KJ (2013) Costimulation blockade: current perspectives and implications for therapy. Transplantation 95(4):527–535. https://doi.org/10.1097/TP.0b013e31826d4672

Larsen JN, Broge L, Jacobi H (2016) Allergy immunotherapy: the future of allergy treatment. Drug Discov Today 21(1):26–37. https://doi.org/10.1016/j.drudis.2015.07.010

Maini MK, Gehring AJ (2016) The role of innate immunity in the immunopathology and treatment of HBV infection. J Hepatol 64(1):S60–S70. https://doi.org/10.1016/j.jhep.2016.01.028

Makkouk A, Weiner GJ (2015) Cancer immunotherapy and breaking immune tolerance: new approaches to an old challenge. Cancer Res 75(1):5–11. https://doi.org/10.1158/0008-5472.CAN-14-2538

Marta M (2009) Toll-like receptors in multiple sclerosis. Ann N Y Acad Sci 462:458–462. https://doi.org/10.1111/j.1749-6632.2009.04849.x

Mathe G (1991) Bestatin, an aminopeptidase with a multi-pharmacological inhibitor function. Biomed Pharmacother 45:49–54

Mcgovern JL, Nguyen DX, Notley CA, Mauri C, Isenberg DA, Ehrenstein MR (2012) Th17 cells are restrained by Treg cells via the inhibition of Interleukin-6 in patients with rheumatoid arthritis responding to anti-tumor necrosis factor antibody therapy. Arthritis Rheum 64 (10):3129–3138. https://doi.org/10.1002/art.34565

Meier MF, Frerix M, Hermann W, Ladner UM (2013) Current immunotherapy in rheumatoid arthritis. Immunother - Futur Med 5(9):955–974

Mohapatra SS, Qazi M, Hellermann G (2010) Immunotherapy for allergies and asthma: present and future. Curr Opin Pharmacol 10(3):276–288. https://doi.org/10.1016/j.coph.2010.05.012. IMMUNOTHERAPY

Mok MY, Shoenfeld Y (2016) Recent advances and current state of immunotherapy in systemic lupus erythematosus. Expert Opin Biol Ther 16(7):927–939. https://doi.org/10.1517/14712598.2016.1171840

Montoliu-gaya L, Villegas S (2018) Immunotherapy for neurodegenerative diseases: the Alzheimer's disease paradigm. Curr Opin Chem Eng 19:59–67. https://doi.org/10.1016/j.coche.2017.12.006

Naran K, Nundalall T, Chetty S, Barth S, Marcello A (2018) Principles of immunotherapy: implications for treatment strategies in cancer and infectious diseases. Front Microbiol 9:23. https://doi.org/10.3389/fmicb.2018.03158

Nelson RP, MD J, Ballow M (2003) Immunomodulation and immunotherapy: drugs, cytokines, cytokine receptors, and antibodies. J Allergy Clin Immunol 111(2):720–732. https://doi.org/10.1067/mai.2003.146

Ohno M, Natsume A, Wakabayashi T (2012) Cytokine therapy. In: Glioma immunothery. Springer, New York, pp 86–94

Oiseth SJ, Aziz MS (2017) Cancer immunotherapy: a brief review of the history, possibilities, and challenges ahead. J Cancer Metast Treatment 3:250. https://doi.org/10.20517/2394-4722.2017.41

Patil US, Jaydeokar AV, Bandawane DD (2012) Immunomodulators: a pharmacological review. Int J Pharm Pharm Sci 4:30–36

Reynolds G, Cooles FAH, Isaacs JD, Hilkens CMU (2014) Emerging immunotherapies for rheumatoid arthritis. Hum Vaccin Immunother 10(4):822–837. https://doi.org/10.4161/hv.27910

Ring J, Gutermuth J (2011) 100 years of hyposensitization: history of allergen-specific immunotherapy (ASIT). Allergy 66(8):713–724. https://doi.org/10.1111/j.1398-9995.2010.02541.x

Santamaria P (2013) Cytokines and chemokines in autoimmune disease: an overview. Landes Bioscience, Austin, pp 1–7

Schuster M, Nechansky A, Loibner H, Kircheis R (2006) Cancer immunotherapy. Biotechnol J 1:138–147. https://doi.org/10.1002/biot.200500044

Sirskyj D, Thèze J, Kumar A, Kryworuchko M (2008) Cytokine disruption of the cc cytokine network in T cells during HIV infection. Cytokine 43:1–14. https://doi.org/10.1016/j.cyto.2008.03.001

Stahn C, Mark L, Hommes DW, Buttgereit F (2007) Molecular mechanisms of glucocorticoid action and selective glucocorticoid receptor agonists. Mol Cell Endocrinol 275:71–78. https://doi.org/10.1016/j.mce.2007.05.019

Tsuruo T, Naganuma K, Iida H, Yamori T, Sakurai Y, Tsukagoshi S (1981) Inhibition of lymph Node metastasis of P388 leukemia By bestatin in Mice. J Antibiot (Tokyo) 34(9):1206–1209

Van Amelsfort JMR, Jacobs KMG, Bijlsma JWJ, Lafeber FPJG, Taams LS (2004) CD4 ϩ CD25 ϩ regulatory T cells in rheumatoid arthritis. Arthritis Rheum 50(9):2775–2785. https://doi.org/10.1002/art.20499

Visser K, Houssiau FA, Antonio J, Silva P (2017) Systemic lupus erythematosus: treatment. https://www.eular.org/sysModules/sysFiles/ckeditor_4/plugins/doksoft_uploader/userfiles/18_main_CH21.docx_1.pdf

Wraith DC (2017) The future of immunotherapy: a 20-year perspective. Front Immunol 8:1–6. https://doi.org/10.3389/fimmu.2017.01668

Wu AY (1954) Immunotherapy—vaccines for allergic diseases. J Thorac Dis 4:2. https://doi.org/10.3978/j.issn.2072-1439.2011.07.03

Immunotherapy in Cancer: Immune Checkpoint Inhibitors; Changing Oncology Treatment Paradigm

2

Shariq Syed

Abstract

Immune checkpoint inhibitors that target T-cells to activate immune response against tumors have shown remarkable clinical responses emerging as new potent weapon against cancer. This therapy has led to durable responses in hard to treat tumor types with long-term remissions. The field of immune-oncology has greatly evolved in recent times primarily by our enhanced understanding of T-cell stimulation and checkpoint blockade, primarily of CTLA-4 and PD-1. Clinical responses although remarkable are, however, limited to limited pool of patients and indications. This calls for further understanding of underlying biological mechanism and function of an optimal immune response. As the immune response evolves, it is unlikely to have a single actionable biomarker to predict clinical response but rather we would need a panel of markers to guide in development of therapy. Clinically validated biomarkers would therefore be needed ultimately for optimal patient and regimen selection. Clearly, the way forward is deeper understanding of our immune system and its dynamic interaction with tumor environment. The magnitude of immune response and its regulation will have to be targeted through a combination approach to provide benefit to wide range of patients and tumor types. If done properly, there is a strong chance of turning this hope into reality.

Keywords

Checkpoint inhibitors · Antigen presenting cells · Immune Activation · Modulation · Immune response

S. Syed (✉)
School of Pharmacy, Anjuman-I-Islam's, Kalsekar Technical Campus, Mumbai, Maharashtra, India
e-mail: dean.sop@aiktc.ac.in

S. P. Sawarkar et al. (eds.), *Immunotherapy – A Novel Facet of Modern Therapeutics*,
https://doi.org/10.1007/978-981-15-9038-2_2

2.1 Introduction

Immune response as a treatment modality for response to cancerous growth has been investigated for several decades. In one of the earliest efforts, William Coley, a surgeon observed that cancer patients with post-operative infections tend to have a better clinical outcome, thus correlating immune response with cancer growth inhibition. In decades to follow, several immunotherapeutics were approved for cancer that included Bacillus Calmette-Guerin, interferon-α, and interleukin-2 (IL-2). Effects of Interleukin-2 in clinic were significant with regard to mechanism of action as it demonstrated durable effect in several tumors. These clinical studies acted as Proof of Concept (POC) that a cytokine capable of expanding T-cells demonstrated durable response in difficult to treat tumors such as advanced melanoma. In essence it also emphasized the important role of adaptive immunity in treating tumorous growth. In the ensuing decades, intense research activity was focused on developing mechanistic understanding in fields of immunology, virology, molecular biology, and cell biology. The decades of investment in basic sciences finally yielded two major breakthroughs in cancer immunotherapy: Chimeric Antigen Modified Receptor (CAR) modified T-cells and Immune activation using antibodies that act as immune checkpoint inhibitors.

2.2 Preclinical Evidence of Immune Modulation

Tumor microenvironment is very heterogeneous consisting of cancerous cells with multiple genetic alterations, fibroblasts, and host of other immune cells. They include cells with lytic capacity such as NK cells, macrophages, and most importantly T-cells. As a result of this lysis, tumor specific antigens expressed on tumor cells get bound to Major Histocompatibility Complex (MHC) on cells. T-cells get activated on recognition of tumor antigens leading to proliferation, differentiation and leading to destruction of cells expressing antigens. T-cell response, however, is a complex process consisting of stimulatory and inhibitory pathways which ultimately determines its response to cancer and eradication of tumors.

Recognition of antigen by T-cell receptor is just not sufficient to activate naïve T-cell which requires additional costimulatory signals. This costimulation is in essence a checkpoint to ensure whether T-cell activation is truly required. These costimulatory signals are provided by engagement of CD28 on T-cell surface with B7 molecules (CD80 and CD86) on APC. B7 molecules are typically expressed on subsets of hematopoietic cells such as dendritic cells that possess special ability for effective antigen presentation. Cancer cells typically do not possess B7 molecules and are thus as a result largely invisible to immune system. In an inflammatory response which leads to killing of tumor cells, APC presenting cells such as dendritic cells take up antigens and present it along with B7 molecules needed for effective activation of T-cells. Once the tumor specific T-cells are activated, they move to tumor site to mount an attack. Effective tumor site infiltration is a critical hurdle that must be overcome for effective tumor response. Tumor microenvironment can be a significant barrier put up by a host of cancer cells, stroma, regulatory T-cells, suppressor cells, and cytokines effectively blunting immune response. A major breakthrough in understanding of immune modulation occurred when a protein known as cytotoxic T Lymphocyte-associated protein 4 (CTLA-4) displayed potent inhibitory role in managing T-cell response. CTLA-4 in resting T-cells is located within the cell. Studies done by two groups, one led by James Allison and other by Jeffrey Bluestone showed that during the T-cell receptor (TCR) engagement, there is a costimulatory signal activated via CD28 binding with costimulatory molecules (CD80, CD86) (Chambers et al. 2001; Walunas et al. 1994). Following the activation, CTLA-4 moves to cell surface and outcompetes CD28 for binding with costimulatory molecules (CD80, CD86) due to much higher binding affinity. Because both CD80 and CD86 provide positive costimulatory signals through CD28, inhibition of both molecules by CTLA-4 is necessary for effectively attenuating T-cell activation. Both CD28 and CTLA-4 exhibit rapid binding kinetics with B7–1 which coupled with different binding strengths allows for swift competitive inhibition by CTLA-4. In addition to enhanced expressions, CTLA-4 contained in intracellular vesicles is rapidly transported to immunologic synapse. This movement to synapse is directly correlated with TCR signal strength. At synapse, CTLA-4 is stabilized by B7–1 binding allowing for accumulation and outcompeting with CD28. This ultimately leads to robust regulation of TCR signal amplitude and activity. Thus CTLA-4 primarily functions to regulate T-cell activity at sites of T-cell priming and in peripheral tissues. The binding of CTLA-4 with costimulatory molecules leads to inhibition of T-cell activation resulting in loss of proliferation and activation. Allison proposed that if this inhibition of T-cell activation could be blocked temporarily preferably using antibodies, this would allow the T-cell to activate and proliferate above the normal physiological levels. They further proposed to combine CTLA-4 inhibition with agents that directly kill tumor cells to release tumor specific antigens for presentation by APC which would improve antitumor response. To validate this hypothesis, CTLA-4 knockout animals were generated to understand the effect of CTLA-4 deficiency in a whole animal model. CTLA-4 knockout animals died due to hyperimmune activation leading to lymphocyte infiltration in several organs confirming the role of CTLA-4 as a non-redundant co-inhibitory protein.

In early POC preclinical studies to evaluate blockade of CTLA-4, effect of Anti-CTLA-4 Antibody were evaluated in xenograft tumor models. The tumors evaluated were wild type, unmanipulated established tumors. These studies conclusively showed that checkpoint inhibition using Anti-CTLA-4 Antibody led to durable regression of established tumors insyngeneic animal models (Leach et al. 1996). Thus the Anti-CTLA-4 Antibody had a dramatic curative effect on tumors. In addition, these studies suggested that CTLA-4 blockade provided enhanced immunity to secondary challenge. Animals that were cured of tumors were reinjected with tumor cells approximately two months after earlier tumors had regressed. Significant number of animals remained tumor free suggesting that tumor rejection mediated by CTLA-4 blockade results in immunologic memory. These definitive preclinical POC studies demonstrated that removing inhibitory signaling in costimulatory pathway enhances anti-tumor immunity. This is achieved by affecting T-cell response in two nonexclusive ways; non-reactive T-cells convert to active cells by lowering the threshold of activation or removing inhibitory signaling leads to sustained T-cell proliferation.

2.3 Early Clinical Evidence of Immune Modulation

After it was conclusively shown in animal models that immune checkpoint inhibition leads to regression of tumors, the next step was clinical validation of these preclinical results. Early Phase-I trials with Anti-CTLA-4 antibody, Ipilumumab (MDX-010) with multiple tumor types including melanoma showed remarkable results (Ribas et al. 2005; Hodi et al. 2003). In hard to treat tumors such as melanoma, a remarkable $>10\%$ objective response (OR) was observed. This regression of tumor was quite durable (>10 years) even after stopping therapy. In contrast to a typical chemotherapy response where the responses are observed early, there is a delayed response in case of immune checkpoint inhibition where there is initial progression or new tumorous growth appearing followed by regression. This atypical response, however, led to challenges in the manner efficacy could be evaluated for such treatment where the typical regulatory assessment involved calculating an objective response (OR) or progression free survival (PFS). In case of immune checkpoint inhibition, a more long-term efficacy assessment such as Overall Survival (OS) as a primary end point would be more appropriate. Following an impressive early phase trials that clearly demonstrated efficacy of Ipilumumab in hard to treat tumors such as melanoma, large scale phase three trials were conducted to confirm these early phase findings. Ipilumumab clearly extended PFS compared to peptide vaccine or standard dacarbazine chemotherapy in late stage melanoma patients becoming the first immunotherapeutic agent to be approved by FDA in 2011.

2.4 PD-1 Another Important Immune Checkpoint Player

Another important immune checkpoint that has a dominant impact on downregulating T-cell activation is programmed cell death (PD-1) receptor. Initially discovered role of PD-1 was cell death inducer as name suggest. However, further work shed light on another important role an as immune checkpoint. PD-1 receptor has two ligands, PD-L1 which is broadly expressed by somatic cells in response to proinflammatory cytokines, and PD-L2 which is more restricted to antigen presenting cells (APC) (Ishida et al. 1992).

CTLA-4 as discussed above is a downregulator of T-cell activation upon engagement with tumor specific antigen on presentation by APC which primarily happens at lymph nodes. Once the T-cell is activated by blocking CTLA-4, they circulate inside body to locate cognate antigen presented by cancer cells. Upon engagement with antigen, TCR activation also leads to expression of PD-1 which acts as negative regulatory receptor (Keir et al. 2006). There are two ways by which tumor cells can express PD-L1 providing them immunity from immune response; "innate immune resistance" and "adaptive immune resistance." Innate immune resistance refers to constitutive expression of PD-L1 by tumor cells due to gene amplification or aberrant oncogenic pathway activation. In contrast to innate immune resistance, adaptive immune resistance refers to PD-L1 expression in response to proinflammatory cytokines released by tumor/immune cells in response to immune response. While INF-g is primary cytokine thought to be responsible for PD-L1 expression, other cytokines which are resident of tumor microenvironment such as IL-1, IL-10, IL-27, and IL-32 can upregulate PD-L1 in tumor/immune cells. This deactivation of T-cells can be overcome by blocking PD-L1/PD-1 pathway.

2.5 Impressive Anti-Tumor Effect on PD-1 Blockade

Blocking of PD-1 or PD-L1 interaction becomes relevant to cancer cells as it leads to deactivation of immune response via preferential blockade of activated T-cells. The clinical effect of blockade of activated T-cells was tested in an early Phase-I trial using a fully human monoclonal Antibody, Nivolumab as PD-1 inhibitor. Nivolumab is a fully human IgG4 antibody which binds to PD-1 receptor with nanomolar affinity along with high specificity for PD-1. Impressive Objective responses (OR) in the range of 40% were observed in a wide range of tumors that included melanoma, renal cell carcinoma, and non-small cell lung cancer with low incidences of toxicity (Joseph et al. 2013). These impressive clinical responses with PD-1 blockade in heterogeneous tumor population led to several clinical trials with PD-1 and PD-L1 Antibodies. First FDA approvals for PD-1 inhibitors were granted in 2014 in refractory melanoma and non-small cell lung cancer in 2015 under accelerated and breakthrough pathways. Subsequently, in 2016 first PD-L1 antibody approved was atezolizumab for urothelial cancers followed by avelumab for Merkel cell carcinoma in 2017. The next immediate question is to identify the type of tumors that would respond to PD-1/PD-L1 blockade. As the body of evidence grows, responding tumors appear to be either carcinogen induced or viral infections driven. Common variants of Melanoma which are carcinogen induced have shown high response rates in range of 35–40% with PD-1 blockade. Another series of cancers associated with carcinogenic effect of cigarette smoking such as NSCLC have also shown impressive response rates of 20% with PD-1 blockade.

2.6 Predictive Biomarkers of Immune Response

There is a compelling need to identify biomarkers that would aid in patient and regimen selection to ultimately predict clinical outcome to immunotherapy. Although there are several strong candidates, a definitive single predictive biomarker is still lacking. Several candidate biomarkers are largely based on mechanistic understanding of anti-tumor immune response. Simplistically, T-cell should be able to infiltrate tumor microenvironment, get activated by immune modulating agents and recognize tumor derived antigens. Intuitively, extent of T-cell infiltration along with PD-1/PDL-1 expression in a tumor microenvironment would be a good predictor of immune response. Indeed, patients responding to immunotherapy had a higher degree of existing activated T-cells (CD8+) at tumor margins. Another range of biomarkers would be immune checkpoint expression along with their respective ligands: PD-1, PD-L1 or CTLA-4. Higher expression of inhibitory proteins such as PD-1 or CTLA-4 on circulating T-cells was associated with better clinical outcomes (PFS) after treatment with Anti-CTLA-4 therapy. Melanoma patients who had higher expression of CTLA-4 and PD-2 responded favorably to Anti-CTLA-4 antibody Ipilumumab compared to those with lower expression. A similar result was observed in lung cancer patients where patients with >50% PDL-1 positive showed an improved response (Daud et al. 2016). Although these results have tempted oncologists to use PD-L1 as a marker for selection of patients, the data is yet not conclusive. Interestingly, there are significant numbers of patients that have benefitted with Anti-PD-1 therapy in spite of lower expressions of PD-L1. This could be due to limitations of single biopsy sample not able to capture the dynamic expression of PD-L1 or heterogeneity of expression. Ability of cells to express PD-L1 is directly correlated with expression of Interferon Regulatory Factor (IRF-1). In clinical trials of melanoma treatment by different immune modulators (Anti-PD-1, Anti-CTLA-4), IRF-1 expression was found to be correlated with good response rate.

Cancerous cells are often product genetic mutations that result in novel proteins which can act as antigens. The uniqueness of antigen presentation can lead to effective immune response upon activation of T-cells. As the number of mutations in a cancerous cell increases, so is the possibility of having a unique antigen capable of invoking immune response. In melanoma which is known to harbor high rate of mutations, patients treated with Anti-CTLA-4 was significantly associated with clinical outcome. However, high mutational burden is not an exclusive predictor of clinical outcome. In the final step of immune response, Cytotoxic killer T-cells mount cell killing response involving perforin and granzyme. It was observed retrospectively that Ipilumumab responders had enriched perforin and granzyme transcripts compared to non-responders.

Several peripheral blood markers have shown promise as a prognostic marker for response. As such these markers are appealing because of ease of assessment and offering longitudinal evaluation. One such marker is serum lactate dehydrogenase (LDH) where elevated levels have indicated worst prognosis in case of melanoma. In one of the trials, melanoma patients with lower levels of baseline LDH had a better overall survival. Several other markers that have shown promise in clinical setting are absolute monocyte count, CD14+ monocytes, and absolute eosinophil count.

Clinical features can also be useful indicator of immune response resulting from immunotherapy. Analysis of patient subset that have responded to immunotherapy also tend to have immune related adverse event (irAEs). The nature of adverse event suggests immune activation by immunotherapy. In melanoma trials with PD-1 inhibitor Nivolumab, those patients that had experienced any irAEs had a significantly longer overall survival compared to those who did not (Freeman-Keller et al. 2016). Among the notable, cutaneous irAEs such as rash and vitiligo were strongly correlated to longer overall survival. irAEs can thus provide valuable prognostic information in a minimally non-invasive manner.

2.7 Resistance to Immune Checkpoint Inhibition

Although the overall response to immunotherapy in broad range of tumors has been quite remarkable, there are still certain type of tumors that do not respond (Primary Resistance) and in some cases those that respond initially develop an acquired resistance to intervention. As a result these non-responder patients endure toxicities and treatment cost with no clinical benefit. Interestingly, different responses have been observed for different tumors even within the same patient. Functioning of Immune system is dependent on host of different factors such as environment, genetic factors as well as interventions such as chemotherapy and radiation. Accordingly, the anti-tumor response within a patient is also dynamic and is affected by host of different factors leading to either primary or acquired resistance. Both of these types of resistance could be attributed directly to Tumor cell or Non-Tumor cell related factors. Multiple Tumor cell related factors have been identified such as (1) Activated MAP pathways, (2) expression of WNT/b-catenin pathway, (3) loss of IFN-y pathway, and (4) lack of T-cell response due to poor antigen presentation. Activation of oncogenic MAP pathway is known to inhibit T-cell recruitment and function. The interferon-gamma pathway appears to have both positive and a negative effect on anti-tumor immune response. Interferon-gamma expression upon T-cell activation has shown to recruit other immune cells, improved antigen presentation and have a direct pro-apoptotic effect on tumor cells. On the other hand, continuous Interferon-gamma signaling can lead to immunoediting of cancer cells.

Outside of tumor, several mechanisms can contribute to immunotherapy resistance. Tumor microenvironment contains a host of immune related components such as regulatory T-cells (T-regs), myeloid derived suppressor cells (MDSC) all of them can contribute to anti-tumor response. It is well known that T-regs can down regulate effector T-cells response by secreting inhibitory cytokines. Depletion of T-regs in tumor microenvironment animal model has shown to enhance or restore anti-tumor immunity. MDSC have lately shown to play a major role in regulating immune response. MDSC have been implicated in tumor metastasis and invasion. The mere presence of MDSC of tumor microenvironment is correlated with decreased efficacy of immunotherapy agents. Macrophages especially those associated with tumor (TAM) are known to provide anti-tumor immunity along with pro-tumorigenic effect.

In spite of remarkable anti-tumor response especially in hard to treat tumor such as melanoma, close to one third patients who had previously shown improvement with anti-CTLA-4 or Anti-PD-1 tend to progress. This acquired resistance to immunotherapy is in spite of continued drug therapy. Several mechanisms can contribute to this resistance that includes loss of T-cell function (Exhaustion), lack of antigen recognition, and development of mutant forms of tumor. Activated T-cells can undergo phenotypic changes leading to loss in their cytotoxic activity as a result patient develops tumor relapse. T-cell activation occurs after tumor antigen presentation and recognition. Any changes in these neo-antigens can lead to non-recognition by T-cells. Any genetic deletions, mutations or epigenetic changes can lead to loss of expression of neo-antigens making the tumor non-responsive to immunotherapy. Continued research in this area is shedding light on several other inhibitory checkpoints in tumor microenvironment such as LAG-3, TIGIT, and VISTA. Classical approach of combining therapies with different mechanism has shown promise with immunotherapy agents as well. Combination of anti-CTLA-4 plus anti-PD-1 has shown improved overall survival compared to monotherapy alone for patients with metastatic melanoma.

2.8 Clinical Strategies to Overcome Resistance

As we develop a better mechanistic understanding of drug resistance, combination strategies using multiple targeted approach is evolving and is being currently tested in clinic. The hope behind this approach is by of targeting different immune escape pathways is that it leads to improved patient outcomes.

One of the initial strategies was to combine anti-CTLA-4 along with anti-PD-1 which is now already approved in multiple tumor types. Long-term survival data with combination therapy is descriptively superior with dual blockade. Complimentary mechanism of dual blockade is a likely reason for beneficial effect. In addition, anti-CTLA-4 leads to depletion of T-regs along with broader antigen recognition. Combination with chemotherapy leads to tumor cell death leading to increased release of antigens. Similar is the effect of combination with radiation therapy leading to inflamed tumor microenvironment.

Overall immune response after T-cell activation largely depends upon extent of costimulation and coinhibition. T-cells can be activated using several costimulatory agonists including OX40, CD40, GITR, and ICOS. There appears a strong rationale for combining these with immune checkpoint inhibitors. Targeting suppressive signaling in tumor microenvironment (TME) along with depleting T-regs may improve response to immunotherapy. Colony stimulating factor-1 receptor (CSF1R) inhibition along with simultaneous anti-CTLA-4 or anti-PD-1 has shown promise in tumor inhibition in a preclinical pancreatic tumor model. CSF1R inhibition also leads to reprogramming of macrophages for better antigen presentation. Chemokines/cytokines can also modify TME via recruitment of inhibitory cells leading to drug resistance. They bind to their respective receptors on immune-suppressive cells including CXCR2 and CXCR4. Inhibition of these pathways along with anti-PD1 has shown to prevent immune evasion in preclinical models.

Combination with small molecule or large molecule targeted therapies has shown mixed results. Different growth factors and angiogenic factors are known to affect immune response leading to immune suppression. Immune therapy is often associated with reduced recruitment of blood vessels and activated cytotoxic

T-cells, anti-angiogenic therapy with similar effects could potentially act in synergistic manner. VEGF inhibitors in combination with immune checkpoint inhibitors have led to normalization of immune-suppressive TME and thereby reversing resistance. Based on positive clinical outcome, Bevacizumab in combination with atezolizumab and chemotherapy recently gained US-FDA approval in patients with metastatic NSCLC. Combining BRAF/MEK inhibitors with anti-CTLA4 led to increased toxicity while combination with anti-PD1 has shown promise with enhanced anti-tumor immunity and tolerability. Early preclinical data with inhibitors of PI3K, CDK4/6 in combination with checkpoint inhibitors have also shown promising results suggesting a potential treatment option.

2.9 Future Path

Cancer treatment modalities have shown a remarkable improvement with the introduction of immune checkpoint inhibitors. The development of immune checkpoint inhibitors has brought hope to patients with hard to treat tumors and opened new avenues in understanding of cancer immunology. As a consequence, there are several hundred clinical trials exploring anti-tumor effect in various tumor types. Despite these remarkable results, for the majority of those enrolled patients the benefits are short lived if at all it occurs followed by rapid resistance. High incidences of immune related adverse reactions make it ethically challenging to assign trials without a strong biological rationale. In some cases the curative effect though has been restricted to specific tumors due to tumor intrinsic or extrinsic resistance mechanism.

There are a several emerging checkpoint inhibitors that could act as monotherapy targets or as a part of combination therapy. Although the focus currently is on anti-CTLA-4 and anti-PD-1 inhibitors, some of these emerging checkpoint inhibitors are much more potent and could possibly offer a better patient outcome. Some of these next generation targets are VISTA, LAG-3, TIGIT, and TIM-3. One of the possible scenarios could be anti-CTLA-4 and anti-PD-1 inhibitors forming the primary activation pathway while the newer generation inhibitors which have overlapping effects could act in synergistic manner.

Overall, Immune checkpoint inhibitors are a powerful modulator of cell immunity forming backbone for all future cancer treatment modalities. However, to realize this would require an understanding of genomic, epigenomic, and cellular features that drive both tumor response and resistance. Having a clear understanding of immune pathway to outline whether a sequential or a combination approach would be best to prevent immune evasion. Although progress is made in identifying inflamed or mutated tumors that are likely to respond, ability to counteract cold and excluded tumors is still lacking. Further, current treatment modalities such as chemotherapy, targeted therapy, radiation that directly regulate TME, timing of immune checkpoint inhibitors to maximize immune response is critical. Patient selection in clinical trials is also critical since majority of those are heavily pretreated suggesting that early immune inhibition might be more beneficial. Further understanding of tumor biology and early induction of immune response might lead to early tumor effect followed by long-term remission. Overall, it is clear that Immune checkpoint inhibitors will be cornerstone for cancer therapy and in future could offer cure for devastating disease.

References

Chambers CA, Kuhns MS, Egen JG, Allison JP (2001) C TLA -4-mediated inhibition in regulation of T cell responses: mechanisms and manipulation in tumor immunotherapy. Annu Rev Immunol 19:565–594. https://doi.org/10.1146/annurev.immunol.19.1.565

Daud AI, Loo K, Pauli ML, Sanchez-Rodriguez R, Sandoval PM, Taravati K, Tsai K, Nosrati A, Nardo L, Alvarado MD, Algazi AP, Pampaloni MH, Lobach IV, Hwang J, Pierce RH, Gratz IK, Krummel MF, Rosenblum MD (2016) Tumor immune profiling predicts response to anti-PD-1 therapy in human melanoma. J Clin Investig 126(9):3447–3452. https://doi.org/10.1172/JCI87324

Freeman-Keller M, Kim Y, Cronin H, Richards A, Gibney G, Weber JS (2016) Nivolumab in resected and unresectable metastatic melanoma: Characteristics of immune-related adverse events and association with outcomes. Clin Cancer Res 22(4):886–894. https://doi.org/10.1158/1078-0432.CCR-15-1136

Hodi FS, Mihm MC, Soiffer RJ, Haluska FG, Butler M, Seiden MV, Davis T, Henry-Spires R, MacRae S, Willman A, Padera R, Jaklitsch MT, Shankar S, Chen TC, Korman A, Allison JP, Dranoff G (2003) Biologic activity of cytotoxic T lymphocyte-associated antigen 4 antibody blockade in previously vaccinated metastatic melanoma and ovarian carcinoma patients. Proc Natl Acad Sci USA 100(8):4712–4717. https://doi.org/10.1073/pnas.0830997100

Ishida Y, Agata Y, Shibahara K, Honjo T (1992) Induced expression of PD-1, a novel member of the immunoglobulin gene superfamily, upon programmed cell death. EMBO J 11 (11):3887–3895. https://doi.org/10.1002/j.1460-2075.1992.tb05481.x

Joseph G, Horak CE, Inzunza D, Cardona DM, Simon JS, Gupta AK, Sankar V, Park J-S, Kollia G, Taube JM, Anders R, Jure-Kunkel M, Novotny J Jr, Taylor CR, Zhang X, The JC (2013) Association of tumor PD-L1 expression and immune biomarkers with clinical activity in patients with non-small cell lung cancer (NSCLC) treated with nivolumab (Anti-PD-1; BMS-936558; ONO-4538). J Clin Oncol 31:2013. (Suppl; Abstr 3016)

Keir ME, Liang SC, Guleria I, Latchman YE, Qipo A, Albacker LA, Koulmanda M, Freeman GJ, Sayegh MH, Sharpe AH (2006) Tissue expression of PD-L1 mediates peripheral T cell tolerance. J Exp Med 203(4):883–895. https://doi.org/10.1084/jem.20051776

Leach DR, Krummel MF, Allison JP (1996) Enhancement of antitumor immunity by CTLA-4 blockade. Science 271:1734–1736. https://doi.org/10.1126/science.271.5256.1734

Ribas A, Camacho LH, Lopez-Berestein G, Pavlov D, Bulanhagui CA, Millham R, Comin-Anduix B, Reuben JM, Seja E, Parker CA, Sharma A, Glaspy JA, Gomez-Navarro J (2005) Antitumor activity in melanoma and anti-self responses in a phase I trial with the anti-cytotoxic T lymphocyte-associated antigen 4 monoclonal antibody CP-675,206. J Clin Oncol 23:8968–8977. https://doi.org/10.1200/JCO.2005.01.109

Walunas TL, Lenschow DJ, Bakker CY, Linsley PS, Freeman GJ, Green JM, Thompson CB, Bluestone JA (1994) CTLA-4 can function as a negative regulator of T cell activation. Immunity 1(5):405–413. https://doi.org/10.1016/1074-7613(94)90071-X

Vaccines as Immunotherapy

3

Pratik Ogale, Vandana S. Nikam, Manish Gautam, Sunil Gairola, and S. S. Jadhav

Abstract

Cancer is a disorder wherein normal cells get transformed and lose control over cell division potential leading to tumor formation. Various factors are responsible for causing cancer. Out of all known cancers, almost 12% of them caused because of viruses. Seven viruses, namely Epstein–Barr virus (EBV), hepatitis B virus (HBV), human papillomavirus (HPV), human T-cell leukemia virus (HTLV), hepatitis C virus (HCV), Kaposi's sarcoma-associated herpesvirus (KSHV), and Merkel cell polyomavirus (MCV) are associated with various types of cancers. This chapter provides an overview of types of cancer vaccines along with brief history, cancer causing viruses, and gives an idea regarding current research approaches towards oncoviral vaccine development. Currently, highly focused immunotherapeutic approaches against various cancers are monoclonal antibodies, small molecule inhibitors, cytokine therapy, and vaccines. Other therapeutic approaches like engineered T-cells, adoptive cell therapy are at the initial stage. As of now, the FDA has approved vaccines against HBV and HPV which causes hepatocellular carcinoma and cervical carcinoma, respectively. Extensive research is in the process of developing innovative vaccine strategies for cancer caused by EBV, KSHV, and MCV.

Keywords

Vaccine · Cancer immunotherapy · Immunomodulation · Tumor associated antigens · Oncovirus · Adoptive cell therapy

P. Ogale · M. Gautam · S. Gairola · S. S. Jadhav (✉)
Serum Institute of India Pvt. Limited, Pune, India
e-mail: ssj@seruminstitute.com

V. S. Nikam
Department of Pharmacology, STES's Smt. Kashibai Navale College of Pharmacy, S. P. Pune University, Pune, Maharashtra, India

S. P. Sawarkar et al. (eds.), *Immunotherapy – A Novel Facet of Modern Therapeutics*,
https://doi.org/10.1007/978-981-15-9038-2_3

3.1 Introduction

Cancer is a term to denote diseases characterized by abnormal growth of cells which can invade other tissues and other parts of body through the blood and lymph system. As per WHO estimates, cancer is the second leading cause of death globally, and is responsible for an estimated 9.6 million deaths in 2018. Globally, about one in six deaths is due to cancer. A wide range of human carcinogens are reported that includes occupational exposures, pharmaceuticals, X-ray, natural factors, and infectious agents (Blackadar 2016). Tobacco use is the most important risk factor for cancer and is responsible for approximately 22% of cancer deaths. Cancer causing infections, such as hepatitis (HB) and human papillomavirus (HPV), are responsible for up to 25% of cancer cases in low- and middle-income countries.[1] Other than HB and HPV, there are five other viruses that are reported to cause cancer (Parkin et al. 2002; Liao 2006).

Global cancer therapy market was valued at USD 136,254.35 million in 2018, and is estimated to be valued at USD 220,701.26 million in 2024, witnessing a CAGR of 8.37%. Cancer treatment includes the combination of localized therapies, such as surgery, radiation therapy, and/or systemic therapies such as chemotherapy, hormonal, immune, and targeted therapy. The treatment also includes supportive therapy that is used to reduce side effects and improve quality of life (e.g., medications to reduce nausea, protect against organ damage from chemotherapy or radiation, or stimulate blood cell production) (Damyanov et al. 2018).

Treatment (surgery, chemotherapy, radiotherapy) is based on rationale that cancer cells should be eliminated as they suppress natural immune responses. With advancements in technologies, it is now well established that various components of the immune system play pivotal roles in protecting humans from cancer. Approaches to exploit the human immune responses to treat cancer are based on principles of immune surveillance, wherein the immune system can destroy tumor cells during its initial stages. Several therapies based on immune surveillance were envisaged. Following several clinical failures, recent successes and regulatory approvals of autologous cellular immunotherapy, sipuleucel-T, for the treatment of prostate cancer, anti-cytotoxic T lymphocyte-associated protein 4 (CTLA-4) antibody, ipilimumab, anti-programmed cell death protein 1 (PD1) antibodies for melanoma, and success of HPV and HB vaccines, have revitalized the field of cancer immunotherapy.

The basic tenet of cancer immunotherapy is the use of host innate and adaptive immunity to defeat cancer with an anti-cancer response via the immune system. Cancer immunotherapy can be broadly classified into active and passive immunotherapy. The active immunotherapy is targeted at stimulating immune response through vaccination, non-specific immunomodulation or targeting specific antigen receptor. On the other hand, passive immunotherapy aims at administrating agents

[1]WHO Fact Sheets. https://www.who.int/news-room/fact-sheets/detail/cancer. Accessed on 10 Nov 2019.

that triggers anti-tumor response, e.g., monoclonal antibodies, lymphocytes (Naran et al. 2018). Cancer immunotherapy includes therapies including vaccination with tumor antigens, adoptive cellular therapy (ACT), oncolytic viruses (OVs), antibodies against targets that enhance T-cell activity. Immune therapy also includes strategies to neutralize immunosuppressor mechanisms such as CD25, CTLA-4, and PD1.

The book chapter provides an overview of cancer immunotherapy and discusses the recent advances in the therapy.

3.2 Targets of Cancer Immunotherapy

The tumor biology and immune system linkages are well identified by Paul Ehrlich way back in 1909 in his experiments of immunization with tumor cells (Ehrlich 1909). He suggested that tumor cells occur at high frequency in human, but kept under check and control by immune system. With advancement in technology, it is clear that many cancer antigens are differentially regulated compared to healthy cells and host T-cells can recognize them. A detailed mechanism of anti-tumor response is mediated by T-cells wherein the antigens are recognized with the help of major histocompatibility complex (MHC) class I or II molecules, priming of T-cells and activation of T-cells. In cancer, either the T-cell activation is suppressed or complete potentiation of T-cell activity is tethered or cancer cells escape immune surveillance system. In order to develop a highly specific therapy, it is necessary to first identify the best target that has a critical role in cancer development and its survival. A simple but effective approach for target identification is to compare the protein levels among cancer cells and normal healthy cells. Those proteins that are expressed specifically in tumors are potential targets of immunotherapy. Target antigens are broadly classified into tumor associated antigens (TAAs) and tumor specific antigens (TSAs). Other targets such as germ line antigens and virus-associated antigens are also potential targets for immunotherapy. Following paragraphs provide an overview of the target antigens.

3.2.1 Tumor Associated Antigens (TAAs)

TAAs are expressed on normal and cancer cells. They are considered as targets, as they are overexpressed in tumors in comparison to normal cells, e.g., carcinoembryonic antigen, PAP antigen for gastrointestinal cancer and prostate cancer, respectively (Tagliamonte et al. 2014; Arlen and Wood 2012). Whole cell cancer vaccines were pursued to target TAAs; however, clinical studies suggested them to be less precise. TAAs were also found to be associated with limitations like autoimmune toxicity (Ilyas and Yang 2015).

3.2.2 Tumor Specific Antigens (TSAs)

Tumor specific antigens are the proteins or peptides resulting from codon alteration at the gene level. TSAs also referred to as neoantigens, are present specifically on tumor cell and absent on normal ones. Thus, TSAs are considered ideal therapeutic targets recognized as non-self by the immune system (Schumacher et al. 2019). Also, TSAs can eliminate issues associated with immune tolerance and autoimmunity.

Researchers are currently focusing on the development of cancer vaccines using TSAs. The vaccines developed will be of personalized type. Studies with such personalized vaccines (RNA-based and peptide-based) have proved its safety and ability to generate an immune response. Various TSAs have already been identified from melanoma, hepatoma, lung, and renal cancers, and further research is still in continuation (Jiang et al. 2019).

3.2.3 Oncoviral Antigens

The research on oncoviruses started long back in 1911. The tumor viruses from mammalian cells were suggested to have a role in human cancer; however, the idea of viruses causing cancer did not receive much attention until the 1930s (Javier and Butel 2008). The first human virus discovered in the 1960s and 1970s was Epstein–Barr virus (EBV), also known as human herpesvirus 4 (HHV4) (Zur 2007). EBV first observed by electron microscopy from the cells of Burkitt's lymphoma, and this led the foundation for human tumor virology. EBV has a linear ds-DNA genome of about 164–184 Kbp consisting of 85 genes along with terminal and internal repeat regions (Baer et al. 1984). EBV-1 and EBV-2 are the two subtypes of this virus that differs at EBNA locus (Zur 2007).

Hepatitis B virus got its recognition in the 1970s for its key role in liver failure, cirrhosis, chronic liver infections, and hepatocellular carcinoma. It is a hepadenovirus with small, partially double-stranded circular DNA. HBV shows its symptoms after an incubatory period of 3–4 months (Zur 2007; Robinson et al. 1974).

In the same period, human papillomavirus (HPV) identified as an etiological agent of cervical carcinoma. In the 1980s, HPV16 and HPV18 directly identified from cervical cancer cells. HPV has a linear, 7–8 Kbp ds-DNA genome with at most 10 open reading frames (Boshart et al. 1984). HPV transmits through mucosal and skin contact. Sexual contact usually leads to anogenital HPV infections. Although, it is a major cause of cervical cancer, it is also associated with the cancer of anogenital tract, vulva, penis, oropharynx, vagina, and others (Zur 2007).

The first known human retrovirus discovered in the 1970s due to a high number of adult T-cell leukemia (ATL) cases that observed in south-west Japan. The viral genome sequence studies revealed its similarity with the human T-cell leukemia virus type-1 (HTLV-1) (Poiesz et al. 1980). The transmission of the virus occurs through breastfeeding, blood transfusion, and sexual contact (Taylor and Matsuoka 2005).

In the 1970s, non-A and non-B hepatitis viral infections observed which transformed from positive-strand RNA with ~10^4 nucleotides. These viruses later named as hepatitis C virus (HCV) (Choo et al. 1989). It is a flavivirus having a linear uncapped ss-RNA genome, with an open reading frame encoding polypeptides that are processed into ten structural and non-structural proteins using the host and viral proteases (Hoofnagle 2002).

In 1994, Kaposi's sarcoma associated herpesvirus (KSHV) was discovered by Chang et al. using a PCR based method of representational difference analysis. Kaposi's sarcoma is an exceptional skin disease, first explained by Moritz Kaposi in 1872 (Chang et al. 1994). The KSHV, formerly known as human herpesvirus-8 (HHV-8), described as an AIDS-defining disease in the 1980s. KSHV infection occurs in B-cells, macrophages, endothelial cells, keratinocytes, and α3β1 integrin receptor (Akula et al. 2002).

Merkel cell polyomavirus (MCV) is the most recent and only proven oncovirus among all other polyomaviruses identified. Its role in tumorigenesis in animals was already known. In 1971, two human polyomaviruses JC virus (JCV) and BK virus (BKV) were discovered, and more recently, eight new human polyomaviruses have been identified, and MCV is one of them (White et al. 2013). MCV was observed in Merkel cell carcinoma (MCC) patients which is a highly aggressive and very rare neuroectodermal tumor (Schrama et al. 2012). MCV has a circular DNA, about 5.4 Kbp, which get partially integrated into the host cell DNA. Approximately 80% of MCC cases showed the integration of viral DNA (Feng et al. 2007). Its site of replication is hypothesized to be Merkel cells of the skin as it is a causative agent of MCC. But, it has been a very difficult task to prove this. Because, no method has been developed to cultivate and maintain the viral genome infected or transfected cells longer than the primitive stage of low level virion formation (Arora et al. 2012) (Table 3.1 and Fig. 3.1).

3.3 Cancer Immunotherapeutics

Cancer immunotherapy can be broadly classified into active and passive immunotherapy. The active immunotherapy is targeted at stimulating immune response through vaccination, non-specific immunomodulation or targeting specific antigen receptor. On the other hand, passive immunotherapy aims at administrating agents that triggers anti-tumor response, e.g., monoclonal antibodies, lymphocytes (Naran et al. 2018).

Table 3.1 Cancer causing viruses and oncoproteins involved in tumorigenesis (modified from White et al. 2013)

Virus	Type and family	Integration	Oncoprotein(s)	Tropism(s)	Cancer
Human papillomaviruses 16 and 18	DNA; Papillomavirus	Episomal and Integrated	E6, E7	Keratinocytes	Cervical and penile carcinoma
Epstein–Barr virus	DNA; Herpesvirus	Episomal	EBNA-1. EBNA-2, LMP-1, LMP-2, EBRs, BARTs	B lymphocytes, epithelial cells	Burkit's lymphoma, nasopharyngeal carcinoma
Kaposi's sarcoma associated herpesvirus	DNA; Herpes virus	Episomal	LANA, vIL-6, vMIP-I, vFLIP, vBCL-2, v-cyclin-D, vGPCR, vIRF-1	B-cells, macrophages, keratinocytes, endothelial cells	Kaposi's sarcoma, primary effusion lymphoma, multicentric Castleman's disease
Hepatitis B virus	DNA; Hepadnavirus	Yes	HBx, HBsAg	Liver	Hepatocellular carcinoma
Hepatitis C virus	RNA; Flavivirus	No	Core protein, NS3	Liver	Hepatocellular carcinoma
Human adult T-cell leukemia virus type 1	RNA; Retrovirus	Yes	Tax, p12, p30p8, p13	T lymphocytes	T-cell leukemia
Merkel cell polyomavirus	DNA; Polyomavirus	Yes	T antigen	Merkel cells	Merkel cell carcinoma

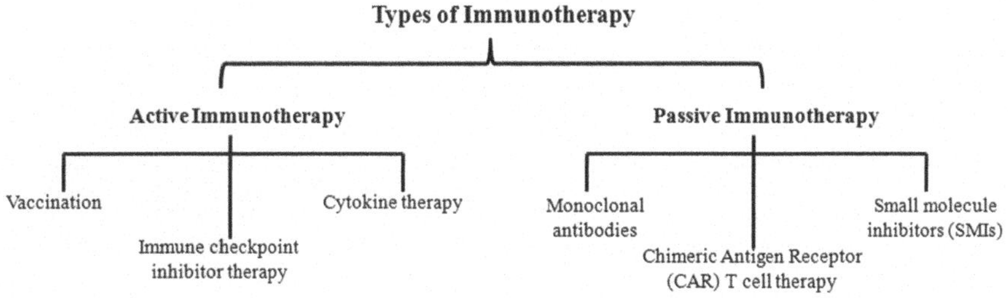

Fig. 3.1 Classification of immunotherapy

3.4 Active Immunotherapy

3.4.1 Vaccination

Vaccination is one of the breakthroughs in medicine due to which numerous infectious diseases are eradicated or prevented. As per the World Health Organization (WHO) definition, vaccines are biological preparations having killed or weakened microorganisms or their toxins or one of the surface proteins, and they help to develop acquired immunity. The studies highlighting the importance of immune surveillance in cancer and cancer cell modulation to immune response led to the design of the strategy boosting the immune system against cancer cells.

Vaccination is an *in vivo* mode of treatment that induces tumor-specific immune response. It can be either preventive vaccination or therapeutic vaccination. Cancer vaccines elicit or restore immune system capability to fight against cancer. Preventive vaccines, as their name suggests, do not allow cancer to occur, since it makes use of antigenic moieties from infectious agents which help immune system to recognize the invading particles. Till now, vaccines for HBV and HPV have received approval from the US FDA. On the other hand, therapeutic vaccines directly attack cancer cells by targeting immune system. In this approach, recombinant fusion protein conjugated with GM-CSF is widely used to activate specific T-cells. Due to this reason, the broadened immune response can be observed as additional tumor specific antigens might get attacked.

Administration of vaccine generally carried out using an adjuvant that elicits the activity of immune system. Dendritic cells (DCs) can be used as adjuvants due to their well-known ability to initiate and maintain the immune response.

The choice of antigen to design immunotherapy is challenging task for therapeutic vaccines compared to prophylactic vaccine due to established disease burden, lowering antigen expression by tumor cells, and escape from immune response. Thus, the prophylactic vaccines showed successful results, for example, hepatitis B vaccine and human papillomavirus causing liver and cervical cancer, respectively.

The general mechanism of activation of immune cells after vaccination is depicted in Fig. 3.2. Vaccination leads to activation of adaptive immunity that starts with entry of antigen into the antigen-presenting cells like dendritic cells,

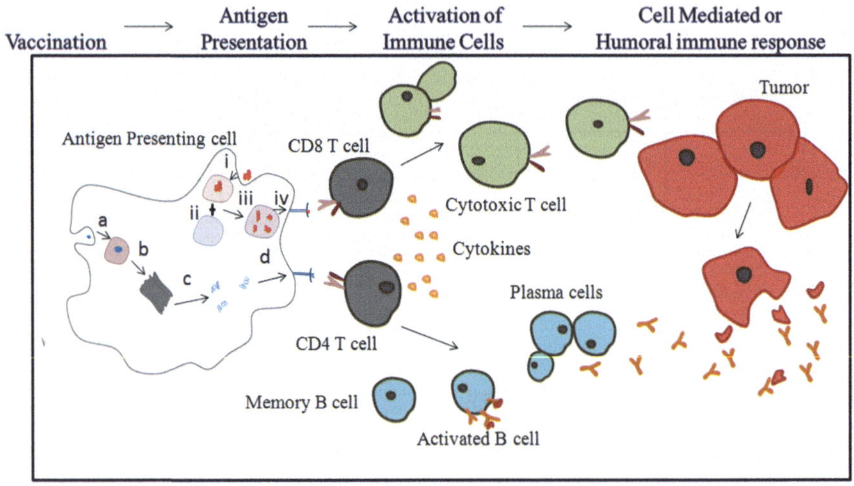

Antigen Presentation

Proteasomal degradation Pathway	Phagolysosomal Pathway
a) Lead molecule entry	i) Lead molecule entry
b) Phagosome formation	ii) Phagosome formation
c) Proteasomal degradation of antige	iii) Degradation of antigen in phagolysosomal complex
d) Presentation by CD4 T cell	iv) Presentation by CD8 T cell

Fig. 3.2 General mechanism of immune system after vaccination

macrophages, etc. After entering inside, antigen gets processed either by proteasomal degradation pathway or by phagolysosomal degradation pathway. The entered antigen gets cleaved by proteasomal complex and then gets presented by the cell using MHC class II molecules. CD4 cells then recognize the antigen and get bound to it, leading to activation of the humoral immune response. On the other hand, in phagolysosomal degradation, antigen gets phagocytosed and forms phagosome. The antigen enclosed phagosome comes in contact with lysosome containing various peptidases and acid hydrolases. Antigenic peptides then get cleaved inside the phagolysosome and are presented by MHC class I molecule to CD8 cells. This leads to initiation of priming of cell mediated immune response. In some cases, CD4 cells release cytokines that triggers activation of CD8 cells. After antigen recognition, CD4 cells activate B lymphocytes and form memory cells. This leads to quick activation of cell mediated or humoral immunity when the same antigen gets exposed again to the immune system (Fig. 3.2).

There are two groups of vaccines based on the live vectors used: bacterial vectors and viral vectors. The main advantage of live vectors is their ability to enter and replicate inside the host cell generating a potent immune response.

Live Bacterial Vector Vaccine

Most extensively studied live bacterial vectors are Listeria-based vectors (Pan et al. 1995a). *Listeria monocytogenes* (Lm) are gram-positive bacteria that have capability to mount potent innate and adaptive immunity in human. The immunogenic properties of Lm have been explored in vaccine technology (Pan et al. 1995b).

Live Viral Vector Vaccine

Viral vectors are receiving high preference due to their higher ability to infect and express virus-encoded antigen within infected cells. Various viral vectors like adenovirus, alphavirus, adeno-associated virus, vaccinia virus, fowlpox virus, and vesicular stomatitis virus have been tried to date for their efficient use in the synthesis of HPV vaccines (Kaufmann et al. 2002; Borysiewicz et al. 1996).

Peptide-Based Vaccine

Though peptide-based vaccines result in lesser immunogenicity, identification of a well-known peptide epitope as a candidate for vaccine development is difficult. Still, it has a better advantage over live vector-based vaccines in terms of stability, tolerability, and ease in production (Feltkamp et al. 1993). They are safe, stable and can be synthesized at a lesser cost, for example, Stimuvax for breast cancer and NSCLC, Oncophage for brain cancer and melanoma.

Nucleic Acid Based Vaccine

It can be either DNA or RNA replicon based, produced to encourage immune response against a specific antigen (Tewari and Monk 2014). The advantages of DNA based vaccine lie in its stability, safety, and cost-effectiveness. Another advantage is that it can be administered in patients multiple times with the same efficacy as it does not synthesize neutralizing antibodies (Vici et al. 2014). Several DNA vaccines are in clinical use, like mammaglobin-A is used against breast cancer, PAP for prostate cancer, VXM01 for pancreatic cancer.

Another type of nucleic acid vaccine is RNA based vaccine that can be administered either in the form of RNA or DNA. After administration, the molecules turn itself as RNA replicons within the inserted cells. Although the effectiveness of RNA replicons has been tested in pre-clinical trials, it has undergone very limited clinical testing till now.

3.5 Prophylactic and Therapeutic Vaccines in Clinical Uses or in Clinical Trials for Cancers Caused by Viruses

3.5.1 Cervical Cancer

At the initial stage of cervical cancer, no significant symptoms are visible, but in later stages of the disease, symptoms like pelvic pain, vaginal bleeding, or pain during sexual intercourse are common. In 90% of the cases, the causative agent for cervical cancer is HPV but smoking, weak immune system, birth control pills, starting sex at a young age and having many sexual partners also causes cervical cancer. These causes are not frequent in nature. Diagnosis for cervical cancer is carried out primarily by cervical screen followed by biopsy. Currently, Gardasil 9 is the only approved vaccine against HPV.

Table 3.2 Current approaches to develop cancer vaccines against cervical cancer (Eskander and Tewari 2015)

Type	Vaccine	Target
Live (Bacterial and viral) vector based vaccine	ADXS11–001 (bacterial) TA-HPV (viral)	HPV-16 E7 fusion protein HPV-16 E6 and E7 peptide
Peptide	HLA-A*201	HPV-16 E7 peptide
Protein	SGN-00101	Fusion protein of HPV-16 E7
Nucleic acid	ZYC101a VGV-3100a	HPV-16 E7 HLA-A2 restricted peptide Plasmid targeting HPV-16 and HPV-18 E6 and E7

Listeria, when used as a live bacterial vaccine, infects macrophages and also disrupts its phagosomal membrane to escape into the host cell cytoplasm. Listeria gains this ability of evasion due to the expression of listeriolysin O (LLO) (Wallecha et al. 2009; Shahabi et al. 2008). The presence of Listeria's in the cytoplasm and the phagolysosome, L-monocytogene-derived peptides get presented by MHC-I and MHC-II molecules giving rise to both CD4 and CD8 response. Listeria- based HPV vaccine was firstly reported to be clinically used in the year 2009 by Maciag et al. The study was carried out using a fusion of HPV-16 E7 antigen to nonhemolytic fragment of Lm protein LLO (Maciag et al. 2009).

Similarly, Lm-based vector which is genetically engineered live attenuated *L. monocytogenes* ADXS11–001 targets HPV transformed cells by secreting HPV-16E7 fusion protein. Further studies for improving the immunotherapeutic aspects are still in progress (Wallecha et al. 2012).

A peptide vaccine with E7 peptide stimulates cytotoxic T-cell response and functions as protective agent against HPV-16 E7 positive tumors (Feltkamp et al. 1993). Currently, the researchers are investigating the most effective adjuvant for vaccine development. The adjuvants under investigation are adenylyl cyclase from *B. pertussis*, exotoxin A from *P. aeruginosa*, mycobacterial heat shock protein, toll-like receptor agonists, and a penetrating peptide polyphemus protein.

ZYC101a is one of the nucleic acid vaccines in phase II/III encodes E6 and E7 peptide of HPV-16 and HPV-18 (Matijevic et al. 2011; Garcia et al. 2004). Another DNA based vaccine VGX-3100 targeting the same peptide chains as of ZYC101 has proven to be safe in phase I of clinical trial. It induces CD8 T-cell production and enhances cytolytic response (Bagarazzi et al. 2012) (Table 3.2).

3.5.2 Hepatocellular Carcinoma (HCC)

Primary liver cancer is considered as one of the common causes of deaths due to cancer, and hepatocellular carcinoma is a type of primary liver cancer (Torre et al. 2015; Qin 2012). Conventional therapies for HCC treatment include radioactive seed implantation, palliative liver resection, radiofrequency ablation (RFA), transarterial chemoembolization, and liver transplantation. Though these treatments are found to be effective, the major drawback for them is that none of them can completely address residual tumor cells leading to re-progression, and metastasis of cancer (Qin 2012). FDA has approved HEPLISAV-B vaccine against HBV for adult patients (age >18).

Currently, research is focused mainly on targeting tumor re-progression, relapse, and relocation (metastasis). On a similar line, studies are moving forward by making vaccines for HCC that focus antigenic substances to elicit a precise immune response against tumor cells. The vaccine utilizes whole cancer cells, dendritic cells (DCs), antigenic peptides, and nucleic acid (mostly DNA) (Xie et al. 2018). The tumor specific response becomes elevated when vaccines are based on HCC cell lysates or autologous HCC cells. GM-CSF (bi-shRNA/granulocyte- macrophage colony-stimulating factor) showed to have a satisfactory immune response in phase I clinical trial. In the follow-up studies, after 1043 days of vaccine administration, a low immune response and a decline in immunogenicity of cancer cells were observed (Nemunaitis et al. 2014).

GPC-3, NY-ESO-1, alpha-fetoprotein (AFP), human telomerase reverse transcriptase (hTERT), melanoma Ag A (MAGEA), and HCA 587 are the best peptide targets identified until now to treat HCC (Sun et al. 2015). Though AFP can be overexpressed on HCC cell surface, its immune response is observed to be limited as it acquires immune tolerance during immune system development. A recombinant AFP has been explored to overcome the limitation, and the study is still in progress (Zhang et al. 2008a).

Other antigen-presenting cells, dendritic cells (DCs), carry out absorption, antigen processing, and its presentation by maintaining high number of MHC molecules. They do induce T-cells by releasing interferon-gamma (IFN-γ) (Takakura et al. 2015). At the initial stage of vaccine preparation, cytokines (rhGM-CSF and rhIL-4) were used to activate DCs which on later stages get mature in the presence of TNF-α and are finally introduced to autologous antigens or tumor cells. When tested these DC vaccines in clinical phase II on patients already undergone primary treatment, DC vaccine is proved to be safe and can act as efficient adjuvant treatment (Lee et al. 2015).

3.5.3 Burkit's Lymphoma

Infectious mononucleosis (IM) is primarily caused by Epstein–Barr virus (EBV) which is also associated with various B-cells and epithelial tumorigenesis (Cohen 2015). The glycoproteins such as gp350, 9H, 9B, BMRF2, and 9 L frequently observed on virion surface. The gycoprotein-gp350 is widely studied glycoprotein that binds with CD21 of B lymphocyte leading to virus endocytosis (Chen et al. 2014). An amino acid sequence study revealed its conserved nature and is found to be 97% identical for EBV-1 and EBV-2 (Lees et al. 1993). Along with gp350, other glycoproteins serve as target for neutralizing antibodies. The B-cell infection gets inhibited by administrating monoclonal antibodies against gp42, whereas EBV infection to epithelial cells observed to get inhibited by administrating monoclonal antibodies against gH/gL or to BMRF2 (Backovic et al. 2007).

The glycoprotein, gp350, was the first ever vaccine developed against EBV associated malignancies for animals (Epstein et al. 1985). After consistent research, genetically modified gp350 was administered with ISCOMs (Immune Stimulating Complexes), muramyl dipeptide squalene, or alum protected cotton top tamarins from lymphoma. The glycoprotein, gp350, failed to produce neutralizing antibodies when isolated from adenovirus and vaccinia virus, but found to be effective as of natural gp350 (Wilson et al. 1996).

Rhesus lymphocryptovirus (LCV) shows most of the features as of EBV and hence is studied widely as a model for EBV infection and vaccination. Studies have shown identical functionality and 100% gene homology between EBV and LCV (Moghaddam et al. 1997).

When animals were tested for soluble gp350, the highest antibody titer against gp350 was observed, whereas similar testing showed specific antibodies against LCV and CD-4, CD-8 immune response on treatment with virus like replicon moieties with gp350, EBNA3A, and EBNA3B (Sashihara et al. 2011).

Like animals, gp350 is also the first human vaccine (in clinical trials) for EBV. Studies from clinical phase trials showed its effectiveness in production of neutralizing antibodies leading to reduction in IM but were not able to completely prevent EBV infection (Moutschen et al. 2007). Peptides that can elicit T-cell immune response to EBV associated proteins are another approach of vaccine development. EBNA-3A peptide conjugated with tetanus toxoid showed good amount of T-cell immune response after vaccination with no adverse effects (Elliott et al. 2008). Studies are now focusing on development of virus like particles lacking detectable EBV DNA. Mice studies for this resulted in neutralizing antibody production as well as T-cell immune response (Ruiss et al. 2011).

Therapeutic vaccines against EBV-related diseases primarily focus on induction of either B- or T-cell immunity. Studies showed an increased cellular response against non-Hodgkin lymphoma, Hodgkin lymphoma, and Burkitt's lymphoma. Clinical phase trials are in progress for therapeutic vaccines where T-cell immune response observed to develop when tested with patients having advanced as well as metastatic Burkitt's lymphoma (Bollard et al. 2014). Two kinds of therapeutic vaccines were tested consisting of a fusion of carboxy-terminal of EBNA-1 and LMP2 protein in modified vector. Carboxy-terminal of EBNA-3 found to elicit CD-4, while LMP2 protein elicits CD-8 immune response (Taylor et al. 2004). Injection of incubated EBV associated antigens or virus particles expressing EBV peptides with autologous DCs are one kind of recent approach for Burkitt's lymphoma treatment. For example, LMP2 protein incubated with DC resulted in CD-8 immune response when vaccinated to the patients having Burkitt's lymphoma (Lin et al. 2002).

3.5.4 Kaposi's Sarcoma

Kaposi's sarcoma is very rare, hence limited efforts have been taken to develop a specific therapy against the disease. Patients with low immunity or suffering from HIV-1 infection develop advanced sarcoma, and vaccination is a must for these patients.

Although there are some therapeutic approaches available, patients from resource-limited nations cannot use them due to its high cost. This is one of the major reasons for increased cases of Kaposi's sarcoma in sub-Saharan Africa (Uldrick and Whitby 2011).

Till now, no vaccine has reported in clinical trials for Kaposi's sarcoma (Chauhan et al. 2019). New trends for vaccine development include target finding against antigens such as glycoproteins that help entry of viruses inside the host cell. The targets are expected to elicit both cellular and humoral immune response by recognizing multiple TCRs. To enhance the immunogenicity and the duration of immune response, adjuvants can be applied to vaccine epitopes. The vaccine epitopes are preferably fully conserved, promiscuous, should reside at the site other than glycosylation and should not be same as of any human protein.

No human trials have performed for Kaposi's sarcoma, but primate studies showed successful results in terms of tumor cell mass reduction. In studies, MHV-68 (a gamma herpes virus similar to KSHV) was infected in mice to get an idea about a suitable mode of vaccination that will ultimately lead to the prevention of disease (Tibbetts et al. 2003).

Studies are now being carried out on reducing the viral load from patients already suffering from Kaposi's sarcoma. The goal behind this approach is to reduce the transmission of viral particles to other healthy individuals.

3.5.5 Merkel Cell Carcinoma (MCC)

Currently, surgery, chemotherapy, or radiation therapies are the only options available for treating MCC. With limited understanding of pathogenesis and an increasing number of MCC patients, the innovation in therapeutic strategies is warranted (Cassler et al. 2016).

Emerging treatments for MCC include immunotherapy and vaccination. When compared with MCV-positive cells, high mutation rate and increased tumor neoantigens have been observed for MCV-negative MCC (Goh et al. 2016). MCV-positive cells express viral proteins which act as good target for cellular and humoral immune response (Lyngaa et al. 2014). MCC cells express high number of PDL-1 (Programmed cell death ligand-1) on its surface. PD-1 (Programmed cell death receptor-1) present on T-cell binds with PDL-1 of MCC cell leading to tumor necrosis (Lipson et al. 2013).

Avelumab, a monoclonal antibody against PDL-1 is currently in phase II clinical trials and is tested for patients having advanced MCC (Cassler et al. 2016). Another anti-PD-1 monoclonal antibody, pembrolizumab received approval for small cell lung cancer and melanoma. The same antibody is being tested for MCC in clinical phase II studies that showed MCC cells' sensitivity towards PD-1 checkpoint inhibition (Patnaik et al. 2015; Nghiem et al. 2015). To circumvent metastatic melanoma, FDA has already approved ipilimumab which inhibits cytotoxic T lymphocyte antigen 4 (CTLA-4). The current phase II trial is estimating the efficiency of ipilimumab for MCC in combination with adjuvant therapy (Brummer et al. 2016). Similar approach for stimulation of immune system is being tested for MCC by conjugating the adjuvant with aldesleukin IL-2.

The effectiveness of such approach is not yet clear. More advanced study is being carried out in search of targeted molecules against PI3/AKT, mTOR, tyrosine kinases such as VEGFRs, PDGFRs, and c-KIT (Cassler et al. 2016).

Recent research focuses on the preparation of DNA vaccines that will have ability to stimulate antigen-specific CD-8 immune response for MCC. Zeng et al. and the group formulated a DNA vaccine and tested on the laboratory-derived LT B16 melanoma cell line to evaluate its potency. Large T antigen (LT) is necessary for MCV to proliferate within infected cells. The tested vaccine was found to be protective and resulted in LT specific CD-4 T-cell and natural killer cells mediated anti-tumor effects. CD-8 cells were observed to be activated but not specific for LT. To overcome this challenge, study is now focused on to find out the appropriate mode of vaccination to achieve better LT specific immune response. Also, as there is no vaccine reported to the date for MCC, other approaches like protein based, dendritic cell based, vector based vaccines are under vigorous study (Zeng et al. 2012).

Cytokine Therapy Against Cancer

Cytokines are either membrane linked or secreted proteins produced by the cells of both innate and adaptive immunity. They act as molecular coordinators between the immune cells of the body that triggers self-limited, highly specific immune response against its target. The efficacy of cytokine response depends on several factors such as cytokine concentration, number of cytokine receptors, and signalling cascade involved within the immune cells (Picaud et al. 2002).

There are almost seven different types of cytokine receptors, viz. Type-I, Type-II, immunoglobulin superfamily receptors, TNF receptors, G-protein coupled receptors (GPCRs), TGF- β receptors, and IL-17 receptors. Currently, cytokines that are under clinical studies are highly specific for Type-I and Type-II receptors. Type-I cytokine receptor binds with cytokines IL-21, IL-2, IL-7, IL-4, IL-15, and IL-9. The signalling pathway for all these cytokines involves Janus Kinases 1 and Janus Kinases 3 (JAK1 and JAK 3) and STATs. Another subgroup within Type-I receptor family is GM-CSF and IL-6 receptor. Both the receptors have common gp130 subunit that mediates the actual signalling between immune cells (Rochman et al. 2009). Type-II receptor family binds to IL10, IFN-α, IFN-β, and IFN- γ which signals through tyrosine kinases like JAK (Kotenko and Pestka 2000).

Cytokines such as IL-21, IL-15, IL-18, IL-7, IL-12 and GM-CSF have been tested for advanced cancers in clinical trials (Feldmann 2008). Currently, studies are being carried out using IL-10 and TGF-β to check for its anti-tumor effect. To date, only two cytokines have received the FDA approval for treating the cancer. One of them is bolus IL-2 (high dose) for metastatic melanoma and the other is IFN-γ for renal cell carcinoma (Lee and Margolin 2011).

Recent approach in vaccine development is to implement cytokines in dendritic cell based vaccines (DC) as they trigger T-cell activation. GM-CSF and IL-4 have been most commonly tested cytokines for increasing the concentration of antigen-specific monocytes and its maturation as antigen-presenting cells. Successful attempts of fusion of prostate cancer antigen to GM-CSF lead to the discovery of sipuleucel-T, the only cancer vaccine received FDA approval against prostate cancer (Trepiakas et al. 2009). Another approach under focus is transfection of the genes of cytokines such as IL-2 or GM-CSF to the tumor cells used in vaccination to activate the T-cells as well as to limit the effect of cytokines at the site of tumor (Lee and Margolin 2011).

Immune Checkpoint Inhibitor Therapy

Immune checkpoint pathways are the signalling cascades triggered by tumor cells to escape the activated anti-tumor response. Immune checkpoint inhibitors are thus emerged as milestone in immunotherapeutics. It plays a key role in inhibiting the immune checkpoint pathways that ultimately leads to eradication of tumor cells. Various targets have been identified so far and still the research is in progress. For the treatment of advanced melanoma, the US FDA has approved cytotoxic T lymphocyte antigen-4 targeting agent referred as ipilimumab (Robert et al. 2011). It activates effector T-cell proliferation and also prevents the inhibition of T lymphocytes (Hodi et al. 2010). Similarly, other approved immune checkpoint inhibitors pembrolizumab and nivolumab target programmed death-1 (PD-1) receptor and are used to treat patients with melanoma and non-small cell lung carcinoma (NSCLC) (Garon et al. 2015). Recent studies showed that PDL-1 blocking antibodies are found to be effective against almost ten different cancer types along with melanoma and NSCLC (Cheng et al. 2018). Currently, pembrolizumab and nivolumab are under phase IV trials for treatment against different malignancies.

Although the concept of immune checkpoint inhibitors found to be attractive, small fraction of patients actually gets benefited. Major reason behind this is tumor microenvironment which is differentiated into three classes with respect to presence of T-cells: immune desert, immune excluded, and immune inflamed. In the first class, T-cells are absent or unable to get primed and activated within tumor microenvironment, whereas gathered T-cells become unable to invade tumor microenvironment in case of immune excluded phenotype. In the case of immune inflamed phenotype, numerous specific or non-specific immune cells are observed to be present in the microenvironment. Reported data suggests the increasing number of immune-related adverse events (irAEs) like autoimmunity in the patients undergoing immune checkpoint inhibitor therapy for cancer (Feng et al. 2013).

The clinical trials have been carried out using combination of immune checkpoint inhibitors with conventional treatments like chemotherapies, radiation therapies, and also with other immunotherapeutic agents to overcome the problem associated with

Table 3.3 FDA approved Immune checkpoint inhibitors

Agent	Target	Cancer type
Ipilimumab	CTLA-4	Melanoma
Nivolumab	PD-1	Melanoma, lung
Pembrolizumab	PD-1	Melanoma
Atezolizumab	PD-L1	Bladder

immune checkpoint inhibitors. Combination of antibodies targeting CTLA-4 and PD-1 showed reduction in tumor when compared with ipilimumab administered patients (Wolchok et al. 2013).

Table 3.3 summarizes the US FDA approved immune checkpoint inhibitors along with their targets.

3.6 Passive Immunotherapy

The therapy works by attacking tumor cells, specifically with the use of drugs or some other molecules. There are various modes of targeted therapy. Small molecule inhibitors and monoclonal antibodies against precise targets are two kinds of approaches that are currently under focus.

3.6.1 Monoclonal Antibodies

Currently, FDA approved monoclonal antibodies (mAbs) are either of humanized, human, or chimeric type, developed with the help of recombinant DNA technology (Simpson and Caballero 2014). In the past few years, several mAbs have received US FDA approval and are in regular clinical use for cancers such as colon, breast, lymphomas, and others (Sathyanarayanan and Neelapu 2015). Although the monoclonal antibodies are effective, they can recognize and attack its target only if it is located outside the cell membrane. The first monoclonal antibody used in cancer immunotherapy was of murine type that showed reduced half-life and decreased efficacy when administered frequently to humans as the host produced antibodies against administered mouse monoclonal antibody (Liu 2014; Teillaud 2012).

The mechanism of action of mAbs is either by antibody dependent cell toxicity or through complement dependent toxicity. Antibody dependent cell toxicity involves binding of mAbs to the target antigen present on tumor cell. The complex of mAb and tumor cell is then recognized by Fc receptors present on immune cells like macrophages and natural killer cells, releasing cytotoxic proteins such as granzyme and perforins. Ultimately, cell dies by apoptosis. Monoclonal antibodies that follow ADCC are trastuzumab, pertuzumab, and cetuximab (Chung et al. 2014; Wang et al. 2015; Boyerinas et al. 2015). On the other hand, mAbs like alemtuzumab, cetuximab, and ofatumumab bind to extracellular tumor antigens and initiate intracellular signalling cascade leading to complement activation (Glassman and Balthasar 2014). Activated complement protein then binds to the monoclonal

Table 3.4 FDA approved monoclonal antibodies against cancer caused by viruses

Monoclonal antibody	Type	Treatment against
Trastuzumab	Humanized	Breast cancer
Alemtuzumab	Humanized	Chronic lymphocytic leukemia
Bevacizumab	Humanized	Metastatic colorectal, non-small cell lung, ovarian cancer, breast cancer
Cetuximab	Chimeric	Metastatic colorectal cancer, squamous cell cancer, non-small-cell lung cancer
Panitumumab	Human	Metastatic colorectal cancer
Ofatumumab	Human	Chronic lymphocytic leukemia
Denosumab	Human	Solid tumor bony metastases
Ipilimumab	Human	Metastatic melanoma
Pertuzumab	Humanized	Breast cancer
Nivolumab	Human	Melanoma, non-small cell lung cancer, renal cell carcinoma, Hodgkin's lymphoma
Pembrolizumab	Humanized	Melanoma, metastatic non-small cell lung cancer
Ramucirumab	Humanized	Gastric cancer, non-small cell lung cancer, breast cancer
Elotuzumab	Humanized	Multiple myeloma
Necitumumab	Human	Metastatic squamous non-small lung cancer
Avelumab	Human	Non-small cell lung cancer, ovarian and stomach cancers, renal cell carcinoma
Durvalumab	Human	Metastatic urothelial carcinoma

antibody and induces membrane attack complex formation. Membrane attack complex then performs cell lysis (Zhou et al. 2008).

There are two classes of therapeutic monoclonal antibodies. The class I mAbs, e.g., alemtuzumab and trastuzumab are self-acting and do not carry any drug or radioactive molecule with them. They are also termed as naked mAbs (Karlitepe et al. 2015; Oldham and Dillman 2008). Whereas class II mAbs like gemtuzumab ozogamicin bind with calicheamicin, a cytotoxic molecule and ibritumomab tiuxetan carries yttrium-90, a radioactive substance (Simpson and Caballero 2014; Scott et al. 2012). Some of the FDA approved monoclonal antibodies are used in cancer therapeutics (Table 3.4).

3.6.2 Small Molecule Inhibitors

Small molecule inhibitors can attack their target even if it is localized inside the cell. This becomes possible due to their very small size. They are usually taken in the form of pills. Small molecule inhibitors look for kinases and inhibit them. The inhibition is done by competing with ATP-binding site at tyrosine kinases. Due to such inhibitions, non-harmonized metabolic pathways get inactivated resulting into prevention of cancer. Major targets for these inhibitors are BCR-ABL, Akt, or mTOR (Gerber 2008).

BCR-ABL as a Target

The target is a fusion protein and is highly focused for drug development because of its presence in almost all cases of chronic myeloid leukemia (CML). Gleevec and Sprycel are the two drugs that are currently in use against CML in the USA. Although both these molecules target BCR-ABL complex, Sprycel shows more affinity for BCR-ABL than Gleevec (Zhang et al. 2009).

In CML, Gleevec acts specifically against inactivated form of tyrosine kinase of BCR-ABL complex. The molecule competitively inhibits ATP-binding site resulting into interference with tyrosine kinase activity (Hofmann et al. 2003). On the other hand, Sprycel is recommended to patients who have developed resistance for Gleevec. The advantage of using Sprycel is, apart from T315L mutant, it can recognize almost all variants of BCR-ABL. Studies have also shown its ability to target other tyrosine kinases of various protein families like Src, c-Kit, platelet-derived growth factor receptor (PDGFR), and EphA2. As Sprycel can target broad spectrum of tyrosine kinases, there is a least probability of development of resistance for it (Kiesel et al. 2009).

VEGFR (Vascular Endothelial Growth Factor Receptor) as a Target

VEGFRs present on numerous cell types protruding outside the cells. They are transmembrane tyrosine kinases which gets dimerized and activated by autophosphorylation only when the ligand binds them extracellularly. There are three VEGFRs, viz. VEGF receptor-1 (VEGFR-1; Flt-1), VEGFR-2 (KDR/Flk-1), and VEGFR-3 (Flt-4). Activated VEGFR-2 has an important role in increasing vascular permeability, migration, and proliferation. Because of these reasons, VEGFRs are now emerging as potential target and hence many of the small molecule inhibitors are currently under pre-clinical and clinical trials (Aprile et al. 2015; Pham et al. 2015).

Nexavar, a small molecule inhibitor, not only complexes with VEGFR-2 and VEGFR-3 but also targets both c-Kit, PDGFR-β (Wilhelm et al. 2004). Recently, the molecule has received approval from US-FDA against unresectable hepatocellular carcinoma and advanced renal cell carcinoma (RCC). Nexavar binds with kinase domain of VEGFR-2, VEGFR-3, c-Kit, and PDGFR-β and hence interfere with RAF/MEK/ERK pathway. Inhibition of these pathways is the main target for treatment of liver and renal cancers as they are observed to become non-regulated during tumor formation. The notable side effects of this treatment are improper wound healing, high blood pressure, rashes on skin, and thrombosis (Zhang et al. 2009).

The drug, Avastin, binds to VEGF and prevents binding to its receptor VEGFR leading to inhibition of other downstream signaling cascades PI3K, PLC-γ, and GRB2 (Baudino 2015).

EGFR and HER2 as a Target

Epidermal growth factor receptor (EGFR) involves in wide array of processes like growth of the cell, its proliferation, existence, tissue invasion, and cell movement. The mechanism of action of EGFR activation is somewhat similar to the activation of VEGFR. Researchers have observed abnormal expression pattern of EGFR and hence abnormal activity during human epithelial cancers. This finding leads the invention of novel molecules that can target EGFR (Baudino 2015).

A molecule named as Iressa precisely looks for ATP-binding site within EGFR and inhibits Ras-MEK pathway and cell growth. The molecule is approved for treating metastasis of tumor cells (Ma et al. 2015).

Likewise, overexpression of human epidermal growth factor-2 (HER2) is observed during breast cancers and the pattern is also concurrent with aggressive forms of tumors. Studies have proposed that there might be the presence of two different pathways for breast cancer development as overexpression of HER2 is related with sensitivity for anthracycline and resistance to endocrine therapy. The proposed two signalling cascade involves tyrosine kinase and hormone receptor cascade. One of these pathways is susceptive to chemotherapeutic drugs while another is to anti-estrogens. Due to this reasons, HER2 becomes a principle protein for targeted tumor remedy.

Tykerb is a small molecule inhibitor that shows greater affinity towards both EGFR and HER2 than Iressa. It competes with ATP for binding at ATP-binding site and hence inhibits further signalling cascade. This leads to inactivation of downstream signalling molecules like c-*myc*, c-*jun,* and c-*fos* (Medina and Goodin 2008; Dai et al. 2008). Herceptin (MAb against breast cancer) when becomes ineffective, Tykerb is administered. Currently, FDA has approved administration of Tykerb in combination with Xeloda (an oral chemotherapeutic agent) against metastatic breast cancer with higher expression of HER2 (Bedard et al. 2009; Le and Hay 2009).

mTOR as a Target

mTOR is one of the most important protein of signalling as it acts as a connecting link for many pathways, also for tumor suppressor phosphates or tensin homolog (PTEN). It has been proved that when PTEN loses its activity there is gain in function of AKT/PKB and mTOR leading to higher protein expression and entry of cell into cell division cycle (Wouters and Koritzinsky 2008).

When cancer of bladder, brain, breast, lung, prostate, melanoma, renal cell, and thyroid occur, PI3K/Akt/mTOR signalling gets hyper-activated than basal level that ultimately leads to overexpression of cyclin D (Martin and Hall 2005; Costa 2007). Torisel is a small molecule inhibitor recommended for renal cell carcinoma as it specifically looks for mTOR (Hudes et al. 2007). Research is still in progress for the use of mTOR as a target in cancer therapeutics in broader way.

The US-FDA approved small molecule inhibitors and monoclonal antibodies against cancers caused by viruses are summarized in Table 3.5.

Table 3.5 FDA approved targeted therapeutic drugs against cancers caused by viruses

FDA approved monoclonal antibody (mAb) or small molecule inhibitor (SMI)	FDA approved indication
Avelumab (Bavencio) (mAb)	Merkel cell carcinoma
Bevacizumab (Avastin)	Cervical cancer
Nivolumab (Opdivo) Regorafenib (Stivarga) Sorafenib (Nexavar)	Hepatocellular carcinoma
Tositumomab (Bexxar) Ibritumomab tiuxetan (Zevalin)	Non-Hodgkin's lymphoma
Romidepsin (Istodax) Belinostat (Beleodaq)	Peripheral T-cell lymphoma
Ramucirumab (Cyramza) Nivolumab (Opdivo)	Colorectal cancer

3.6.3 Chimeric Antigen Receptor (CAR) T-Cell Therapy

Though the concept of CAR T cell therapy was proposed back in the 1980s, the positive outcome of clinical studies helped to gain attention towards it in recent times. Major disadvantage of other immunotherapies is the recognition and elicitation of immune response against self-antigens. To circumvent this, researchers are now focusing on synthesis of genetically modified T-cells with chimeric antigen receptors. Ideally, these receptors should function like normal T-cells with enhanced specificity towards predefined antigens in non-MHC restricted approach. The antigen binding region of CAR has variable light and heavy chains similar to an antibody, and is linked with a spacer of ~15 peptide residues (Mullaney and Pallavicini 2001). Three generations of CARs have been introduced till date of which development of third generation CARs is still underway (Abate-Daga and Davila 2016; Stone and Kranz 2013). The advantage of CARs is their high degree of antigen specificity than normal TCRs. It can also recognize other moieties such as glycolipids, peptidoglycans, and small peptide residues that generally escape antigen recognition (Schmidt-Wolf et al. 1991).

The process of CAR T-cell production starts with extraction of cells from patients by leukapheresis and separation of T lymphocytes from leukapheresis isolates. Though the T-cell separation is carried out, it has been observed that these isolated T-cells are contaminated with other immune cells like myeloid cells, natural killer cells, and also malignant cells (Stroncek et al. 2016). It is now at the priority for scientists to achieve highest purification of T lymphocytes. To develop these T-cells as CAR cells, retroviruses and lentiviruses are used as vectors to achieve permanent transgene expression. Lentiviral vectors are found to act as safe vectors when expressed in human T-cells (Naldini et al. 1996).

Various regulatory bodies and also US FDA have approved administration of Kymriah (tisagenlecleucel-T) against acute lymphoid leukemia in pediatric and young adult patients (age 3–25). The same is under the review of FDA for other malignancies like B-cell lymphoma, non-Hodgkin lymphoma, follicular lymphoma, chronic lymphoid leukemia, and against multiple myeloma. FDA has given approval for axicabtagene ciloleucel against aggressive B-cell non-Hodgkin lymphoma. Both the therapies have used retroviruses as vectors (Yip and Webster 2018).

Although the therapy has proven to be effective, adverse effects have been also observed in some patients. These patients developed enhanced proliferation and clonal expansion of injected CAR T-cells (Ruella et al. 2018). Studies are under progress to enhance the efficacy as well as long lasting functionality in vivo and to develop strategies to reduce immunosenescence and other adverse effects.

3.7 Vaccine Against Cancers: Non-infectious targets

The prostate cancer vaccine is the best example of this class. Prostate cancer is the second cause of cancer death among males after lung cancer (Siegel et al. 2016). The FDA-approved autologous vaccine, sipuleucel-T, targets prostatic acid phosphatase (PAP), the levels of which are observed to be elevated during cancer progression. The vaccine is administered by separating out the peripheral blood cells like monocytes, T and B lymphocytes from the same patient's blood by leukapheresis (Weiner et al. 2016). Combination studies have been done for metastatic castrate-resistant prostate cancer using sipuleucel-T and cyclophosphamide. DCVAC/PCa is also an autologous vaccine which has undergone phase III clinical trials. It is a dendritic cell based vaccine (Podrazil et al. 2015). Another approach using vaccinia and fowlpox virus leads to the development of viral vector based vaccine, Prostvac-VF. The vaccinia virus acts as an immunologic priming agent, whereas fowlpox acts as boosting agent. GVAX is an allogeneic genetically engineered GM-CSF bearing whole cell based vaccine. GM-CSF triggers activation of APCs and also induces anti-tumor response (Silvestri et al. 2016). Recently, DNA based vaccines against prostate cancer have undergone phase I of clinical trial. The immunologic activity for this has been determined but efficacy for the same is not yet estimated (Colluru et al. 2016).

With such advancements, immunotherapies have revolutionized treatments for cancer with a lot more scope to improve the modes of therapy for better patient outcome (Mohindra 2018).

3.8 Challenges and Future of Cancer Therapy

Cancer expresses unique and typical foreign antigens, as observed in diseases related to bacteria, viruses, and fungi. Antigen-specific approaches have challenges for mass immunizations as these antigens result from tumor-specific mutations and hence cannot be used for a large population. Such approaches have shown promise in personalized therapy by acting as potential targets for activating an innate immune response that leads to the release of chemical stimuli and activation of antigen-presenting cells. However, even in personalized therapy approaches, challenges related to the suppression of DC activation and maturation leading, to a condition called tumor immunosuppression. Also, it has been observed and practiced that the antigens overexpressed in tumors compared to normal cells were targeted for vaccine development. The major drawback with such an approach is that these so-called antigens are self and can escape from the innate immune system (Butterfield 2015; Bowen et al. 2018).

To overcome the effects of traditionally designed vaccines, strategies are now changing with various advanced strategies such as T-cell transfer therapies that can recognize the antigen specifically and help in tumor regression. For this treatment to be called as effective, various factors like identification of target antigen, ex-vivo expansion of T-cells, its transport at tumor site, and durability need to be considered first. Otherwise, as observed by Bai et al. group, T-cells become less efficient at tumor area than expected but show excellent efficiency at sites other than the tumor space (Bai et al. 2008).

It is envisaged that novel targetted delivery systems will play a key role in therapy by allowing focussed targetting of tumor sites and thereby minimizing side effects. Very recent approach in delivery systems is application of nanostructure based agents that can be applied with other immunotherapeutic techniques. The major focus in developing such agents is to deliver the molecules (can be drugs or biomolecules) specifically at tumor site reducing the side effects and drug resistance (Zhang et al. 2008b). Nanoscale based agents can also be used for diagnosis by detecting cancer cells and their biomarkers. Nanoscale based approach has various advantages over other treatment options. It is possible to synthesize them in specific sizes with the ability of enhanced permeation and retention within tumor cells. Also, it can help improving the half-life of chemotherapeutic molecules carried along with them. Currently, numerous nanostructure based agents are under clinical studies.

Another approach in eradication of tumors is photodynamic therapy. The therapy uses photosensitizer. After exposing to visible or near infrared light, the photosensitizer gets activated and generates oxygen free radicals leading to cell death due to oxidation of biomolecules. Currently this treatment is applied only for skin cancer as its efficacy for other cancer types is not yet determined (Chatterjee et al. 2008; Fayter et al. 2010).

Hyperthermia is another treatment that uses heat (temperature over 43 °C) to inhibit tumor cell proliferation. The heat is generated by using ultrasounds, microwaves, and radiofrequency to focus the tumor cells (Griffin et al. 2010). It is currently being applied as an option for conventional tumor treatments. Studies on use of nanotechnology to improve heat delivery are underway. Prostate carcinoma in application with hyperthermia using magnetic nanoparticles underwent clinical phase studies and found to be effective in recurrent cancer. Patients are uncomfortable with a high magnetic field strength used to deliver heat. The direct administration of intratumor injection leads to uneven heat distribution in the tumor. The research is in progress to overcome the limitations associated with hyperthermia treatment (Johannsen et al. 2010; van der Horst et al. 2018).

3.9 Summary

Cancer is a disease where some cells from body lose control over programmed cell death and form tumors and get metastasized to other sites. From the past many years and at present most of the cancers are treated with surgery, chemotherapy, and radiotherapy, but these methods cannot eliminate resistant tumor cells or tumor metastases present in the body and thus relapse cancer. To have more focused therapy for cancer, antigens that can be targeted are of two kinds. Tumor associated antigens are the ones that are present on both tumors as well as normal cells. TAAs express at very low levels in normal cells. The tumor-specific antigens are cancer cell-specific and are a result of mutations like codon alteration that ultimately leads to abnormal cellular function. Viral antigens are also emerged as key molecules in developing anti-cancer immunotherapy. Along with other factors, various viruses cause several cancers. The first virus to be discovered as cancer causing agent was Epstein–Barr virus. The process of discovering new agents is still in progress. MCV is the recently discovered virus to cause Merkel cell carcinoma.

The novel approach of cancer immunotherapeutics is in vigorous research and can be classified into passive and active immunotherapy. Active immunotherapy elicits anti-tumor response by vaccination, non-specific immunomodulation or targeting specific antigen receptor, while passive immunotherapy acts by administrating molecules that can trigger anti-tumor response.

Current research focuses on developing new therapeutic strategies against viral cancers. The active immunotherapeutic strategies involve vaccination, cytokine therapy, and immune checkpoint inhibitor therapy. A new approach of vaccine development is emerging as an effective therapeutics to prevent cancers. Till now, Gardasil-9 and HEPLISAV-B are the only two vaccines received approval from the US-FDA against cancer caused by HBV and HPV. Studies are being carried out to develop vaccine strategies against cancers caused by viruses. It has been noted that the research is highly limited for the innovative vaccine development against human T-cell adult leukemia. The current vaccine strategies include vaccine development by using whole cells or lysates, DNA based and protein or peptide-based vaccines. Large numbers of research groups are working on to develop vaccines by using live bacterial or virus vectors. Most of the findings are still in it naive stages though some have undergone to clinical phase II trials proving their effectiveness and safety.

Cytokine therapy is another type of active immunotherapy. Cytokines are the protein molecules synthesized by both the cells of innate and adaptive immunity. They are either membrane linked or secreted proteins and act as molecular coordinators between the immune cells. This induces self-limited and specific immune response against its target. Currently, cytokines that are under clinical studies are highly specific for Type-I and Type-II receptors.

Immune checkpoint inhibitors play a key role in inhibiting the immune checkpoint pathways that leads to eradication of tumor cells. Many targets have been found so far and still the research is in progress. Previous data shows increased number of events for immune-related adverse events (irAEs) like autoimmunity in patients undergoing immune checkpoint inhibitor therapy.

The discovery of small molecule inhibitors, as well as targeted response by monoclonal antibodies, is another approach in cancer immunotherapy. Various mAbs have been produced till date and have received approval from the US FDA. The major drawback with mAbs is that they can recognize the targets only if it is present outside the cells. On the other hand, small molecule inhibitors, due to their small size can enter inside the cell and can act on intracellular targets. CAR-T-cell therapy is another approach of synthesizing unique tumor antigen receptor bearing T lymphocytes ex-vivo and injecting them within patient suffering from cancer.

Although the field of immunotherapy, particularly vaccine development is not new; in case of cancer vaccines, it has many challenges to face. The primary reason behind this is antigen variation, not only within cancer type but also within patients. It is a bottleneck for researchers to find the targets for vaccine development and to apply them for a large population. Though recent advancements in technologies have come up with new strategies as of nanoparticle based immunotherapy, it has long way to go. The continuous efforts in identifying new targets and corresponding immunotherapy-based strategies to combat cancer are warranted.

References

Abate-Daga D, Davila ML (2016) CAR models: next-generation CAR modifications for enhanced T-cell function. Molecular Ther Oncol 3:16014

Akula SM, Pramod NP, Wang FZ, Chandran B (2002) Integrin α3β1 (CD 49c/29) is a cellular receptor for Kaposi's sarcoma-associated herpesvirus (KSHV/HHV-8) entry into the target cells. Cell 108(3):407–419

Aprile G, Ongaro E, Del Re M, Lutrino SE, Bonotto M, Ferrari L, Rihawi K, Cardellino GG, Pella N, Danesi R, Fasola G (2015) Angiogenic inhibitors in gastric cancers and gastroesophageal junction carcinomas: a critical insight. Crit Rev Oncol Hematol 95(2):165–178

Arlen PM, Wood LV (2012) Prostate cancer vaccines: moving therapeutic vaccination forward in the post-Provenge™ era. Expert Rev Vaccines 11(3):287–302

Arora R, Chang Y, Moore PS (2012) MCV and Merkel cell carcinoma: a molecular success story. Curr Opin Virol 2(4):489–498

Backovic M, Jardetzky TS, Longnecker R (2007) Hydrophobic residues that form putative fusion loops of Epstein-Barr virus glycoprotein B are critical for fusion activity. J Virol 81 (17):9596–9600

Baer R, Bankier AT, Biggin MD, Deininger PL, Farrell PJ, Gibson TJ, Hatfull G, Hudson GS, Satchwell SC, Seguin C, Tuffnell PS (1984) DNA sequence and expression of the B95-8 Epstein-Barr virus genome. Nature 310(5974):207

Bagarazzi ML, Yan J, Morrow MP, Shen X, Parker RL, Lee JC, Giffear M, Pankhong P, Khan AS, Broderick KE, Knott C (2012) Immunotherapy against HPV16/18 generates potent TH1 and cytotoxic cellular immune responses. Sci Transl Med 4(155):155

Bai A, Higham E, Eisen HN, Wittrup KD, Chen J (2008) Rapid tolerization of virus-activated tumor-specific CD8+ T cells in prostate tumors of TRAMP mice. Proc Natl Acad Sci 105 (35):13003–13008

Baudino T (2015) Targeted cancer therapy: the next generation of cancer treatment. Curr Drug Discov Technol 12(1):3–20

Bedard PL, de Azambuja E, Cardoso F (2009) Beyond trastuzumab: overcoming resistance to targeted HER-2 therapy in breast cancer. Curr Cancer Drug Targets 9(2):148–162

Blackadar CB (2016) Historical review of the causes of cancer. World J Clin Oncol 7(1):54

Bollard CM, Gottschalk S, Torrano V, Diouf O, Ku S, Hazrat Y, Carrum G, Ramos C, Fayad L, Shpall EJ, Pro B (2014) Sustained complete responses in patients with lymphoma receiving autologous cytotoxic T lymphocytes targeting Epstein-Barr virus latent membrane proteins. J Clin Oncol 32(8):798

Borysiewicz LK, Fiander A, Nimako M, Man S, Wilkinson GW, Westmoreland D, Evans AS, Adams M, Stacey SN, Boursnell ME, Rutherford E (1996) A recombinant vaccinia virus encoding human papillomavirus types 16 and 18, E6 and E7 proteins as immunotherapy for cervical cancer. Lancet 347(9014):1523–1527

Boshart M, Gissmann L, Ikenberg H, Kleinheinz A, Scheurlen W, zur Hausen H (1984) A new type of papillomavirus DNA, its presence in genital cancer biopsies and in cell lines derived from cervical cancer. EMBO J 3(5):1151–1157

Bowen WS, Svrivastava AK, Batra L, Barsoumian H, Shirwan H (2018) Current challenges for cancer vaccine adjuvant development. Expert Rev Vaccines 17(3):207–215

Boyerinas B, Jochems C, Fantini M, Heery CR, Gulley JL, Tsang KY, Schlom J (2015) Antibody-dependent cellular cytotoxicity activity of a novel anti–PD-L1 antibody avelumab (MSB0010718C) on human tumor cells. Cancer Immunol Res 3(10):1148–1157

Brummer GC, Bowen AR, Bowen GM (2016) Merkel cell carcinoma: current issues regarding diagnosis, management, and emerging treatment strategies. Am J Clin Dermatol 17(1):49–62

Butterfield LH (2015) Cancer vaccines. BMJ 350:h988

Cassler NM, Merrill D, Bichakjian CK, Brownell I (2016) Merkel cell carcinoma therapeutic update. Curr Treat Options Oncol 17(7):36

Chang Y, Cesarman E, Pessin MS, Lee F, Culpepper J, Knowles DM, Moore PS (1994) Identification of herpesvirus-like DNA sequences in AIDS-associated Kaposi's sarcoma. Science 266 (5192):1865–1869

Chatterjee DK, Fong LS, Zhang Y (2008) Nanoparticles in photodynamic therapy: an emerging paradigm. Adv Drug Deliv Rev 60(15):1627–1637

Chauhan V, Rungta T, Goyal K, Singh MP (2019) Designing a multi-epitope based vaccine to combat Kaposi Sarcoma utilizing immunoinformatics approach. Sci Rep 9(1):2517

Chen J, Zhang X, Jardetzky TS, Longnecker R (2014) The Epstein-Barr virus (EBV) glycoprotein B cytoplasmic C-terminal tail domain regulates the energy requirement for EBV-induced membrane fusion. J Virol 88(20):11686–11695

Cheng W, Fu D, Xu F, Zhang Z (2018) Unwrapping the genomic characteristics of urothelial bladder cancer and successes with immune checkpoint blockade therapy. Oncogene 7(1):2

Choo QL, Kuo G, Weiner AJ, Overby LR, Bradley DW, Houghton M (1989) Isolation of a cDNA clone derived from a blood-borne non-A, non-B viral hepatitis genome. Science 244 (4902):359–362

Chung S, Lin YL, Reed C, Ng C, Cheng ZJ, Malavasi F, Yang J, Quarmby V, Song A (2014) Characterization of in vitro antibody-dependent cell-mediated cytotoxicity activity of therapeutic antibodies—impact of effector cells. J Immunol Methods 407:63–75

Cohen JI (2015) Epstein–Barr virus vaccines. Clin Trans Immunol 4(1):e32

Colluru VT, Johnson LE, Olson BM, McNeel DG (2016) Preclinical and clinical development of DNA vaccines for prostate cancer. Urol Oncol: Sem Orig Invest 34(4):193–204

Costa LJ (2007) Aspects of mTOR biology and the use of mTOR inhibitors in non-Hodgkin's lymphoma. Cancer Treat Rev 33(1):78–84

Dai CL, Tiwari AK, Wu CP, Su XD, Wang SR, Liu DG, Ashby CR, Huang Y, Robey RW, Liang YJ, Chen LM (2008) Lapatinib (Tykerb, GW572016) reverses multidrug resistance in cancer cells by inhibiting the activity of ATP-binding cassette subfamily B member 1 and G member 2. Cancer Res 68(19):7905–7914

Damyanov CA, Maslev IK, Pavlov VS, Avramov L (2018) Conventional treatment of cancer realities and problems. Annals Comp Alt Med 1(1):1–9

Ehrlich P (1909) Uber den jetzigen Stand der Karzinomforschung. Ned Tijdschr Gneneesk 53:273–290

Elliott SL, Suhrbier A, Miles JJ, Lawrence G, Pye SJ, Le TT, Rosenstengel A, Nguyen T, Allworth A, Burrows SR, Cox J (2008) Phase I trial of a CD8+ T-cell peptide epitope-based vaccine for infectious mononucleosis. J Virol 82(3):1448–1457

Epstein MA, Morgan AJ, Finerty S, Randle BJ, Kirkwood JK (1985) Protection of cottontop tamarins against Epstein–Barr virus-induced malignant lymphoma by a prototype subunit vaccine. Nature 318(6043):287

Eskander RN, Tewari KS (2015) Immunotherapy: an evolving paradigm in the treatment of advanced cervical cancer. Clin Ther 37(1):20–38

Fayter D, Corbett M, Heirs M, Fox D, Eastwood A (2010) A systematic review of photodynamic therapy in the treatment of pre-cancerous skin conditions, Barrett's oesophagus and cancers of the biliary tract, brain, head and neck, lung, oesophagus and skin. In NIHR Health Technology Assessment programme: Executive Summaries. NIHR Journals Library

Feldmann M (2008) Many cytokines are very useful therapeutic targets in disease. J Clin Invest 118 (11):3533–3536

Feltkamp MC, Smits HL, Vierboom MP, Minnaar RP, De Jongh BM, Drijfhout JW, Schegget JT, Melief CJ, Kast WM (1993) Vaccination with cytotoxic T lymphocyte epitope-containing peptide protects against a tumor induced by human papillomavirus type 16-transformed cells. Eur J Immunol 23(9):2242–2249

Feng H, Taylor JL, Benos PV, Newton R, Waddell K, Lucas SB, Chang Y, Moore PS (2007) Human transcriptome subtraction by using short sequence tags to search for tumor viruses in conjunctival carcinoma. J Virol 81(20):11332–11340

Feng Y, Roy A, Masson E, Chen TT, Humphrey R, Weber JS (2013) Exposure–response relationships of the efficacy and safety of ipilimumab in patients with advanced melanoma. Clin Cancer Res 19(14):3977–3986

Garcia F, Petry KU, Muderspach L, Gold MA, Braly P, Crum CP, Magill M, Silverman M, Urban RG, Hedley ML, Beach KJ (2004) ZYC101a for treatment of high-grade cervical intraepithelial neoplasia: a randomized controlled trial. Obstet Gynecol 103(2):317–326

Garon EB, Rizvi NA, Hui R, Leighl N, Balmanoukian AS, Eder JP, Patnaik A, Aggarwal C, Gubens M, Horn L, Carcereny E (2015) Pembrolizumab for the treatment of non–small-cell lung cancer. N Engl J Med 372(21):2018–2028

Gerber DE (2008) Targeted therapies: a new generation of cancer treatments. Am Fam Physician 1:77

Glassman PM, Balthasar JP (2014) Mechanistic considerations for the use of monoclonal antibodies for cancer therapy. Cancer Biol Med 11(1):20

Goh G, Walradt T, Markarov V, Blom A, Riaz N, Doumani R, Stafstrom K, Moshiri A, Yelistratova L, Levinsohn J, Chan TA (2016) Mutational landscape of MCPyV-positive and MCPyV-negative Merkel cell carcinomas with implications for immunotherapy. Oncotarget 7 (3):3403

Griffin RJ, Dings RP, Jamshidi-Parsian A, Song CW (2010) Mild temperature hyperthermia and radiation therapy: role of tumour vascular thermotolerance and relevant physiological factors. Int J Hyperthermia 26(3):256–263

Hodi FS, O'Day SJ, McDermott DF, Weber RW, Sosman JA, Haanen JB, Gonzalez R, Robert C, Schadendorf D, Hassel JC, Akerley W (2010) Improved survival with ipilimumab in patients with metastatic melanoma. N Engl J Med 363(8):711–723

Hofmann WK, Komor M, Wassmann B, Jones LC, Gschaidmeier H, Hoelzer D, Koeffler HP, Ottmann OG (2003) Presence of the BCR-ABL mutation Glu255Lys prior to STI571 (imatinib) treatment in patients with Ph+ acute lymphoblastic leukemia. Blood 102(2):659–661

Hoofnagle JH (2002) Course and outcome of hepatitis C. Hepatology 36(S1):S21–S29

Hudes G, Carducci M, Tomczak P, Dutcher J, Figlin R, Kapoor A, Staroslawska E, Sosman J, McDermott D, Bodrogi I, Kovacevic Z (2007) Temsirolimus, interferon alfa, or both for advanced renal-cell carcinoma. N Engl J Med 356(22):2271–2281

Ilyas S, Yang JC (2015) Landscape of tumor antigens in T cell immunotherapy. J Immunol 195 (11):5117–5122

Javier RT, Butel JS (2008) The history of tumor virology. Cancer Res 68(19):7693–7706
Jiang T, Shi T, Zhang H, Hu J, Song Y, Wei J, Ren S, Zhou C (2019) Tumor neoantigens: from basic research to clinical applications. J Hematol Oncol 12(1):93
Johannsen M, Thiesen B, Wust P, Jordan A (2010) Magnetic nanoparticle hyperthermia for prostate cancer. Int J Hyperthermia 26(8):790–795
Karlitepe A, Ozalp O, Avci CB (2015) New approaches for cancer immunotherapy. Tumor Biol 36 (6):4075–4078
Kaufmann AM, Stern PL, Rankin EM, Sommer H, Nuessler V, Schneider A, Adams M, Onon TS, Bauknecht T, Wagner U, Kroon K (2002) Safety and immunogenicity of TA-HPV, a recombinant vaccinia virus expressing modified human papillomavirus (HPV)-16 and HPV-18 E6 and E7 genes, in women with progressive cervical cancer. Clin Cancer Res 8(12):3676–3685
Kiesel H, Müller AM, Schmitt-Graeff A, Veelken H (2009) Dramatic and durable efficacy of imatinib in an advanced angiosarcoma without detectable KIT and PDGFRA mutations. Cancer Biol Ther 8(4):319–321
Kotenko SV, Pestka S (2000) Jak-Stat signal transduction pathway through the eyes of cytokine class II receptor complexes. Oncogene 19(21):2557
Le QA, Hay JW (2009) Cost-effectiveness analysis of lapatinib in HER-2–positive advanced breast cancer. Cancer 115(3):489–498
Lee S, Margolin K (2011) Cytokines in cancer immunotherapy. Cancer 3(4):3856–3893
Lee JH, Lee Y, Lee M, Heo MK, Song JS, Kim KH, Lee H, Yi NJ, Lee KW, Suh KS, Bae YS (2015) A phase I/IIa study of adjuvant immunotherapy with tumour antigen-pulsed dendritic cells in patients with hepatocellular carcinoma. Br J Cancer 113(12):1666
Lees JF, Arrand JE, Pepper SD, Stewart JP, Mackett M, Arrand JR (1993) The Epstein-Barr virus candidate vaccine antigen gp340/220 is highly conserved between virus types A and B. Virology 195(2):578–586
Liao JB (2006) Cancer issue: viruses and human cancer. Yale J Biol Med 79(3–4):115
Lin CL, Lo WF, Lee TH, Ren Y, Hwang SL, Cheng YF, Chen CL, Chang YS, Lee SP, Rickinson AB, Tam PK (2002) Immunization with Epstein-Barr Virus (EBV) peptide-pulsed dendritic cells induces functional CD8+ T-cell immunity and may lead to tumor regression in patients with EBV-positive nasopharyngeal carcinoma. Cancer Res 62(23):6952–6958
Lipson EJ, Vincent JG, Loyo M, Kagohara LT, Luber BS, Wang H, Xu H, Nayar SK, Wang TS, Sidransky D, Anders RA (2013) PD-L1 expression in the Merkel cell carcinoma microenvironment: association with inflammation, Merkel cell polyomavirus, and overall survival. Cancer Immunol Res 1(1):54–63
Liu JK (2014) The history of monoclonal antibody development–progress, remaining challenges and future innovations. Annals Med Surgery 3(4):113–116
Lyngaa R, Pedersen NW, Schrama D, Thrue CA, Ibrani D, Met Ö, Thor Straten P, Nghiem P, Becker JC, Hadrup SR (2014) T-cell responses to oncogenic Merkel cell polyomavirus proteins distinguish patients with Merkel cell carcinoma from healthy donors. Clin Cancer Res 20 (7):1768–1778
Ma L, Wang DD, Huang Y, Yan H, Wong MP, Lee VH (2015) EGFR Mutant Structural Database: computationally predicted 3D structures and the corresponding binding free energies with gefitinib and erlotinib. BMC Bioinformatics 16(1):85
Maciag PC, Radulovic S, Rothman J (2009) The first clinical use of a live-attenuated Listeria monocytogenes vaccine: a Phase I safety study of Lm-LLO-E7 in patients with advanced carcinoma of the cervix. Vaccine 27(30):3975–3983
Martin DE, Hall MN (2005) The expanding TOR signaling network. Curr Opin Cell Biol 17 (2):158–166
Matijevic M, Hedley ML, Urban RG, Chicz RM, Lajoie C, Luby TM (2011) Immunization with a poly (lactide co-glycolide) encapsulated plasmid DNA expressing antigenic regions of HPV 16 and 18 results in an increase in the precursor frequency of T cells that respond to epitopes from HPV 16, 18, 6 and 11. Cell Immunol 270(1):62–69

Medina PJ, Goodin S (2008) Lapatinib: a dual inhibitor of human epidermal growth factor receptor tyrosine kinases. Clin Ther 30(8):1426–1447

Moghaddam A, Rosenzweig M, Lee-Parritz D, Annis B, Johnson RP, Wang F (1997) An animal model for acute and persistent Epstein-Barr virus infection. Science 276(5321):2030–2033

Mohindra N (2018) Current state of immunotherapy: chipping away at the tip of the iceberg. J Cancer 1:1–2

Moutschen M, Léonard P, Sokal EM, Smets F, Haumont M, Mazzu P, Bollen A, Denamur F, Peeters P, Dubin G, Denis M (2007) Phase I/II studies to evaluate safety and immunogenicity of a recombinant gp350 Epstein–Barr virus vaccine in healthy adults. Vaccine 25(24):4697–4705

Mullaney BP, Pallavicini MG (2001) Protein-protein interactions in hematology and phage display. Exp Hematol 29(10):1136–1146

Naldini L, Blömer U, Gallay P, Ory D, Mulligan R, Gage FH, Verma IM, Trono D (1996) In vivo gene delivery and stable transduction of nondividing cells by a lentiviral vector. Science 272 (5259):263–267

Naran K, Nundalall T, Chetty S, Barth S (2018) Principles of immunotherapy: implications for treatment strategies in cancer and infectious diseases. Front Microbiol 9:3158

Nemunaitis J, Barve M, Orr D, Kuhn J, Magee M, Lamont J, Bedell C, Wallraven G, Pappen BO, Roth A, Horvath S (2014) Summary of bi-shRNAfurin/GM-CSF augmented autologous tumor cell immunotherapy (FANG™) in advanced cancer of the liver. Oncology 87(1):21–29

Nghiem P, Bhatia S, Daud A, Friedlander P, Kluger H, Kohrt H, Kudchadkar R, Lipson E, Lundgren L, Margolin K, Reddy S (2015) 22LBA Activity of PD-1 blockade with pembrolizumab as first systemic therapy in patients with advanced Merkel cell carcinoma. Eur J Cancer 51:S720–S721

Oldham RK, Dillman RO (2008) Monoclonal antibodies in cancer therapy: 25 years of progress. J Clin Oncol 26(11):1774–1777

Pan ZK, Ikonomidis G, Pardoll D, Paterson Y (1995a) Regression of established tumors in mice mediated by the oral administration of a recombinant Listeria monocytogenes vaccine. Cancer Res 55(21):4776–4779

Pan ZK, Ikonomidis G, Lazenby A, Pardoll D, Paterson Y (1995b) A recombinant Listeria monocytogenes vaccine expressing a model tumour antigen protects mice against lethal tumour cell challenge and causes regression of established tumours. Nat Med 1(5):471

Parkin DM, Whelan SL, Ferlay J, Teppo L, Thomas DB (2002) Cancer incidence in five continents. IARC Sci Publ VIII:155

Patnaik A, Kang SP, Rasco D, Papadopoulos KP, Elassaiss-Schaap J, Beeram M, Drengler R, Chen C, Smith L, Espino G, Gergich K (2015) Phase I study of pembrolizumab (MK-3475; anti–PD-1 monoclonal antibody) in patients with advanced solid tumors. Clin Cancer Res 21 (19):4286–4293

Pham K, Luo D, Siemann DW, Law BK, Reynolds BA, Hothi P, Foltz G, Harrison JK (2015) VEGFR inhibitors upregulate CXCR4 in VEGF receptor-expressing glioblastoma in a TGFβR signaling-dependent manner. Cancer Lett 360(1):60–67

Picaud S, Bardot B, De Maeyer E, Seif I (2002) Enhanced tumor development in mice lacking a functional type I interferon receptor. J Interferon Cytokine Res 22(4):457–462

Podrazil M, Horvath R, Becht E, Rozkova D, Bilkova P, Sochorova K, Hromadkova H, Kayserova J, Vavrova K, Lastovicka J, Vrabcova P (2015) Phase I/II clinical trial of dendritic-cell based immunotherapy (DCVAC/PCa) combined with chemotherapy in patients with metastatic, castration-resistant prostate cancer. Oncotarget 6(20):18192

Poiesz BJ, Ruscetti FW, Gazdar AF, Bunn PA, Minna JD, Gallo RC (1980) Detection and isolation of type C retrovirus particles from fresh and cultured lymphocytes of a patient with cutaneous T-cell lymphoma. Proc Natl Acad Sci 77(12):7415–7419

Qin LX (2012) Inflammatory immune responses in tumor microenvironment and metastasis of hepatocellular carcinoma. Cancer Microenviron 5(3):203–209

Robert C, Thomas L, Bondarenko I, O'Day S, Weber J, Garbe C, Lebbe C, Baurain JF, Testori A, Grob JJ, Davidson N (2011) Ipilimumab plus dacarbazine for previously untreated metastatic melanoma. N Engl J Med 364(26):2517–2526

Robinson WS, Clayton DA, Greenman RL (1974) DNA of a human hepatitis B virus candidate. J Virol 14(2):384–391

Rochman Y, Spolski R, Leonard WJ (2009) New insights into the regulation of T cells by γ c family cytokines. Nat Rev Immunol 9(7):480

Ruella M, Xu J, Barrett DM, Fraietta JA, Reich TJ, Ambrose DE, Klichinsky M, Shestova O, Patel PR, Kulikovskaya I, Nazimuddin F (2018) Induction of resistance to chimeric antigen receptor T cell therapy by transduction of a single leukemic B cell. Nat Med 24(10):1499

Ruiss R, Jochum S, Wanner G, Reisbach G, Hammerschmidt W, Zeidler R (2011) A virus-like particle-based Epstein-Barr virus vaccine. J Virol 85(24):13105–13113

Sashihara J, Hoshino Y, Bowman JJ, Krogmann T, Burbelo PD, Coffield VM, Kamrud K, Cohen JI (2011) Soluble rhesus lymphocryptovirus gp350 protects against infection and reduces viral loads in animals that become infected with virus after challenge. PLoS Pathog 7(10):e1002308

Sathyanarayanan V, Neelapu SS (2015) Cancer immunotherapy: strategies for personalization and combinatorial approaches. Mol Oncol 9(10):2043–2053

Schmidt-Wolf IG, Negrin RS, Kiem HP, Blume KG, Weissman IL (1991) Use of a SCID mouse/human lymphoma model to evaluate cytokine-induced killer cells with potent antitumor cell activity. J Exp Med 174(1):139–149

Schrama D, Ugurel S, Becker JC (2012) Merkel cell carcinoma: recent insights and new treatment options. Curr Opin Oncol 24(2):141–149

Schumacher TN, Scheper W, Kvistborg P (2019) Cancer neoantigens. Annu Rev Immunol 37:173–200

Scott AM, Allison JP, Wolchok JD (2012) Monoclonal antibodies in cancer therapy. Cancer Immunity Arch 12(1):14

Shahabi V, Reyes-Reyes M, Wallecha A, Rivera S, Paterson Y, Maciag P (2008) Development of a Listeria monocytogenes based vaccine against prostate cancer. Cancer Immunol Immunother 57 (9):1301–1313

Siegel RL, Miller KD, Jemal A (2016) Cancer statistics, 2016. CA Cancer J Clin 66(1):7–30

Silvestri I, Cattarino S, Giantulli S, Nazzari C, Collalti G, Sciarra A (2016) A perspective of immunotherapy for prostate cancer. Cancer 8(7):64

Simpson A, Caballero O (2014) Monoclonal antibodies for the therapy of cancer. BioMed Central 8 (4):6

Stone JD, Kranz D (2013) Role of T cell receptor affinity in the efficacy and specificity of adoptive T cell therapies. Front Immunol 4:244

Stroncek DF, Ren J, Lee DW, Tran M, Frodigh SE, Sabatino M, Khuu H, Merchant MS, Mackall CL (2016) Myeloid cells in peripheral blood mononuclear cell concentrates inhibit the expansion of chimeric antigen receptor T cells. Cytotherapy 18(7):893–901

Sun TY, Yan W, Yang CM, Zhang LF, Tang HL, Chen Y, Hu HX, Wei X (2015) Clinical research on dendritic cell vaccines to prevent postoperative recurrence and metastasis of liver cancer. Genet Mol Res 14(4):16222–16232

Tagliamonte M, Petrizzo A, Tornesello ML, Buonaguro FM, Buonaguro L (2014) Antigen-specific vaccines for cancer treatment. Hum Vaccin Immunother 10(11):3332–3346

Takakura K, Kajihara M, Ito Z, Ohkusa T, Gong J, Koido S (2015) Dendritic-tumor fusion cells in cancer immunotherapy. Discov Med 19(104):169–174

Taylor GP, Matsuoka M (2005) Natural history of adult T-cell leukemia/lymphoma and approaches to therapy. Oncogene 24(39):6047

Taylor GS, Haigh TA, Gudgeon NH, Phelps RJ, Lee SP, Steven NM, Rickinson AB (2004) Dual stimulation of Epstein-Barr Virus (EBV)-specific CD4+-and CD8+-T-cell responses by a chimeric antigen construct: potential therapeutic vaccine for EBV-positive nasopharyngeal carcinoma. J Virol 78(2):768–778

Teillaud JL (2012) From whole monoclonal antibodies to single domain antibodies: think small. In: Single domain antibodies. Humana Press, Totowa, pp 3–13

Tewari KS, Monk BJ (2014) New strategies in advanced cervical cancer: from angiogenesis blockade to immunotherapy. Clin Cancer Res 20(21):5349–5358

Tibbetts SA, Loh J, Van Berkel V, McClellan JS, Jacoby MA, Kapadia SB, Speck SH, Virgin HW (2003) Establishment and maintenance of gammaherpesvirus latency are independent of infective dose and route of infection. J Virol 77(13):7696–7701

Torre LA, Bray F, Siegel RL, Ferlay J, Lortet-Tieulent J, Jemal A (2015) Global cancer statistics, 2012. CA Cancer J Clin 65(2):87–108

Trepiakas R, Pedersen AE, Met Ö, Svane IM (2009) Addition of interferon-alpha to a standard maturation cocktail induces CD38 up-regulation and increases dendritic cell function. Vaccine 27(16):2213–2219

Uldrick TS, Whitby D (2011) Update on KSHV epidemiology, Kaposi Sarcoma pathogenesis, and treatment of Kaposi Sarcoma. Cancer Lett 305(2):150–162

van der Horst A, Versteijne E, Besselink MG, Daams JG, Bulle EB, Bijlsma MF, Wilmink JW, van Delden OM, van Hooft JE, Franken NA, van Laarhoven HW (2018) The clinical benefit of hyperthermia in pancreatic cancer: a systematic review. Int J Hyperthermia 34(7):969–979

Vici P, Mariani L, Pizzuti L, Sergi D, Di Lauro L, Vizza E, Tomao F, Tomao S, Cavallotti C, Paolini F, Venuti A (2014) Immunologic treatments for precancerous lesions and uterine cervical cancer. J Exp Clin Cancer Res 33(1):29

Wallecha A, Carroll KD, Maciag PC, Rivera S, Shahabi V, Paterson Y (2009) Multiple effector mechanisms induced by recombinant Listeria monocytogenes anticancer immunotherapeutics. Adv Appl Microbiol 66:1–27

Wallecha A, French C, Petit R, Singh R, Amin A, Rothman J (2012) Lm-LLO-based immunotherapies and HPV-associated disease. J Oncol 2012:542851

Wang W, Erbe AK, Hank JA, Morris ZS, Sondel PM (2015) NK cell-mediated antibody-dependent cellular cytotoxicity in cancer immunotherapy. Front Immunol 6:368

Weiner AB, Matulewicz RS, Eggener SE, Schaeffer EM (2016) Increasing incidence of metastatic prostate cancer in the United States (2004–2013). Prostate Cancer Prostatic Dis 19(4):395

White MK, Gordon J, Khalili K (2013) The rapidly expanding family of human polyomaviruses: recent developments in understanding their life cycle and role in human pathology. PLoS Pathog 9(3):e1003206

Wilhelm SM, Carter C, Tang L, Wilkie D, McNabola A, Rong H, Chen C, Zhang X, Vincent P, McHugh M, Cao Y (2004) BAY 43-9006 exhibits broad spectrum oral antitumor activity and targets the RAF/MEK/ERK pathway and receptor tyrosine kinases involved in tumor progression and angiogenesis. Cancer Res 64(19):7099–7109

Wilson AD, Shooshstari M, Finerty S, Watkins P, Morgan AJ (1996) Virus-specific cytotoxic T cell responses are associated with immunity of the cottontop tamarin to Epstein–Barr virus (EBV). Clin Exp Immunol 103(2):199–205

Wolchok JD, Kluger H, Callahan MK, Postow MA, Rizvi NA, Lesokhin AM, Segal NH, Ariyan CE, Gordon RA, Reed K, Burke MM (2013) Nivolumab plus ipilimumab in advanced melanoma. N Engl J Med 369(2):122–133

Wouters BG, Koritzinsky M (2008) Hypoxia signalling through mTOR and the unfolded protein response in cancer. Nat Rev Cancer 8(11):851

Xie Y, Xiang Y, Sheng J, Zhang D, Yao X, Yang Y, Zhang X (2018) Immunotherapy for hepatocellular carcinoma: current advances and future expectations. J Immunol Res 2018:8740976

Yip A, Webster RM (2018) The market for chimeric antigen receptor T cell therapies. Nat Rev Drug Discov 17(3):161–162

Zeng Q, Gomez BP, Viscidi RP, Peng S, He L, Ma B, Wu TC, Hung CF (2012) Development of a DNA vaccine targeting Merkel cell polyomavirus. Vaccine 30(7):1322–1329

Zhang W, Liu J, Wu Y, Xiao F, Wang Y, Wang R, Yang H, Wang G, Yang J, Deng H, Li J (2008a) Immunotherapy of hepatocellular carcinoma with a vaccine based on xenogeneic homologous α fetoprotein in mice. Biochem Biophys Res Commun 376(1):10–14

Zhang L, Gu FX, Chan JM, Wang AZ, Langer RS, Farokhzad OC (2008b) Nanoparticles in medicine: therapeutic applications and developments. Clin Pharmacol Therapeut 83(5):761–769

Zhang J, Yang PL, Gray NS (2009) Targeting cancer with small molecule kinase inhibitors. Nat Rev Cancer 9(1):28

Zhou X, Hu W, Qin X (2008) The role of complement in the mechanism of action of rituximab for B-cell lymphoma: implications for therapy. Oncologist 13(9):954–966

Zur HH (2007) Infections causing human cancer. Wiley, New York

Immunotherapy for Autoimmune Diseases 4

Aniket Mali, Apurva Sawant, Anagha Mahadik, and Sujit Nair

Abstract

Various pathologies caused by a dysregulated immune system characterized by chronic inflammation leading to pain or permanent damage to tissue are grouped under an umbrella term "autoimmune disorders." Immunotherapy is a field of immunology that facilitates discovery of therapies for diseases by means of stimulation, augmentation, or suppression of an immunoresponse. Several emerging and promising next-generation immunotherapy modalities for autoimmune diseases such as checkpoint based immunotherapy, antigen-specific immunotherapies, anti-cytokine therapy, anti-T-cell therapy, anti-B-cell therapy and biologics and their combination therapy, etc., have evolved and initiated the new era of immunotherapy for autoimmune diseases in the recent past. We discuss these modalities in detail along with comprehensive tables that elucidate the specific therapies. Further, we delineate current immunotherapeutics in clinical trials for autoimmune diseases and discuss the "financial toxicity" of current immunotherapies in autoimmune diseases.

A. Mali
National Centre for Cell Science, Pune, India

A. Sawant
Department of Pharmaceutics, SVKM's Dr. Bhanuben Nanavati College of Pharmacy, University of Mumbai, Mumbai, India

A. Mahadik
Shobhaben Pratapbhai Patel School of Pharmacy & Technology Management, SVKM's NMIMS University, Mumbai, India

S. Nair (✉)
SVKM's Dr. Bhanuben Nanavati College of Pharmacy, University of Mumbai, Mumbai, India
e-mail: sujit108@gmail.com

S. P. Sawarkar et al. (eds.), *Immunotherapy – A Novel Facet of Modern Therapeutics*,
https://doi.org/10.1007/978-981-15-9038-2_4

Keywords

Immunotherapy · Autoimmune diseases · Checkpoint immunotherapy · Antigen-specific immunotherapy · Anti-cytokine therapy · Anti-T-cell therapy · Anti-B-cell therapy · Biologics · Clinical trials · Financial toxicity

4.1 The Immune System: An Overview

The human immune system is a complicated set of cellular and molecular mechanisms involving different proteins and biochemicals, which protects the human body against infectious pathogens, cancer cells, and alien substances without attacking the endogenous molecules. The protection of the host body against all types of infections by specifically recognizing and eliminating foreign agents is the prime function of the immune system (Viswanath 2013). In general, the immune system has two lines of defense: innate immunity and adaptive immunity. The first immunological, antigen-independent (non-specific) mechanism for combating an intruding pathogen is innate immunity. It is a rapid immune response which occurs within minutes or hours after attack. The innate immune response has no immunologic memory and, therefore, is unable to recognize or "memorize" the same pathogen when the body gets exposed to it in the future. Various cells are employed in the innate immune response such as phagocytes (macrophages and neutrophils), dendritic cells, mast cells, basophils, eosinophils, natural killer (NK) cells, and lymphocytes (T-cells) as well as complement system (Warrington et al. 2011). Both types of phagocytes act by a similar mechanism of engulfing the microbes. Besides this similar function, neutrophils release their specific granules which assist in the elimination of pathogenic microbes and macrophages also play an important role in antigen presentation to T-cells. Dendritic cells act as important messengers between innate and adaptive immunity by their ability to phagocytose and function as antigen-presenting cells (APCs). Both mast cells (which reside in the connective tissue surrounding blood vessels) and basophils (which reside in the circulation) are involved in the initiation of acute inflammatory responses, such as those seen in allergy and asthma. Eosinophils are granulocytes that possess phagocytic properties and play an important role in the destruction of large parasites which are difficult to phagocytose. NK cells also known as large granular lymphocytes (LGLs) play a major role in the rejection of tumors and the destruction of cells infected by viruses which is achieved through the release of perforins and granzymes from NK-cell granules which induce apoptosis (programmed cell death) (Stone et al. 2010). On the other hand, adaptive immunity is antigen-dependent and antigen-specific due to which it involves a lag time between exposure to the antigen and maximal response. When infection is established due to the inability of innate immunity to effectively eliminate infectious agents, adaptive immunity develops. The main characteristic feature of adaptive immunity is its ability to memorize the initial immunologic response which endows the host to generate a more rapid and efficient immune response upon subsequent exposure to the antigen. The most important functions of the adaptive immune response are the detection of specific "non-self" antigens in the presence of "self" antigens; the activation of pathogen-specific immunologic effector pathways that eradicate specific pathogens or pathogen-infected cells; and the development of an immunologic memory that can promptly eliminate a specific pathogen when subsequent infections occur in future (Warrington et al. 2011). Adaptive responses are of two types: cell-mediated immunity, conducted by T-cells and facilitated by APCs; and humoral immunity (antibody-mediated immunity), mediated by antibodies produced by B-cells. The T lymphocytes account for 60–80% of total lymphocytes and have a very high lifetime. They primarily eradicate the intracellular pathogens by activating macrophages and kill virally infected cells by recognizing the primary structure of an antigen. T helper (Th) lymphocytes represent 2/3 of total lymphocytes and secrete interleukins (messenger molecules that assist the communications between immune system cells). Depending on the type of cytokines secreted, two types of Th cells are distinguished: Th1 cells which produce interleukin-2, IFN-γ, and TNF-α and trigger inflammatory reactions; and Th2 cells which produce interleukins 3, 4, and 5. In humoral immunity, activation of B lymphocytes results into the synthesis of antigen-specific immunoglobulins (antibodies) by plasma cells and development of immunological memory by memory B-cells (Grigore and Inform 2017). In conclusion, the tightly regulated interplay between T-cells, B-cells, and APCs play a significant role(s) in the development of adaptive immunity in concurrence with innate immunity to eradicate infectious agents. Thus, defects in either system can lead to immune pathological disorders such as hypersensitivity reactions, autoimmune diseases, and immunodeficiencies (Warrington et al. 2011).

4.2 Autoimmunity and Immune Tolerance

In simple words, the defect in the host's immune system that results in loss of normal immune homeostasis and produces an abnormal response to its own tissues is referred to as autoimmunity. The presence of self-reactive T-cells, auto-antibodies, and inflammation are the hallmarks of the autoimmunity (Warrington et al. 2011). The Nobel Prize-winning hypothesis of the "forbidden clone" by Macfarlane Burnet led to a better understanding of not only autoimmunity but also of lymphoid cell development, thymic education, apoptosis, and deletion of autoreactive cells and mechanisms of autoimmunity that led to clinical disease (Wang et al. 2015). Thus, autoimmunity is considered to be an interruption in the process of antigenic detection and elimination. Body's cells may undergo antigenic variation as a result of physical, chemical, or biological influences. Such "neo-antigens" (altered antigens) may elicit an immune response that destroys body's own cells (Ganapathy et al. 2017). Thus, autoimmunity may also be defined as the genesis of immune system reactivity via autoantibodies or T-cell responses to self-structures (Viswanath 2013).

The concept of immune tolerance was defined as an ability of the immune system to prevent itself from targeting self-molecules, cells, or tissues (Wang et al. 2014). During maturation of the immune system, it becomes "tolerant" to self by eliminating the immune cells that react against self-tissues. To understand immune tolerance, it is important to realize the key concepts such as central tolerance, peripheral anergy, T regulatory cells (Tregs), and the homeostasis produced by cytokines and chemokines and their cognate receptors. During central tolerance in the thymus, developing lymphocytes go through positive selection in the cortex prior to maturing and entering the circulation while lymphocytes with impending reactivity against self-peptides are negatively selected and deleted in the thymic medulla. After leaving the thymus, mature T-cells undergo secondary selection (peripheral tolerance) by which the majority of self-reactive T-cells are deleted or rendered anergic. Further in the process of clonal deletion or clonal anergy, immature B-cells expressing surface immunoglobulin M (IgM) capable of recognizing ubiquitous self cell surface antigens are eliminated. Even though mature B-cells are under the control of peripheral tolerance, with the help of process known as clonal deletion or clonal anergy, autoreactive B-cells can escape deletion (Wang et al. 2015; Salinas et al. 2013). It is important to note that, in normal individuals also, potentially self-reacting lymphocytes can still "leak out" in small numbers into the periphery, even under the strict surveillance of central and peripheral tolerance. Thus, depending upon the existence of self-reactive T and B lymphocytes and their abilities to produce autoantibodies, autoimmunity can be classified as "physiological" and "pathological" autoimmunity (Avrameas and Selmi 2013). Physiological autoimmunity is generally staged without evidence of clinical disease where natural autoantibodies help to maintain normal immune homeostasis by eliminating self and foreign antigens. On the other hand, pathological autoimmunity is a stage that develops when immune tolerance is broken and autoantibodies and self-reactive lymphocytes become involved in inflammation which further lead to development of autoimmune diseases (Wang et al. 2015).

4.3 Autoimmune Diseases

The breach of immune tolerance, i.e. the failure to differentiate self from non-self, leading to the development of autoimmunity is the basis for autoimmune diseases. Various pathologies caused by a dysregulated immune system characterized by chronic inflammation leading to pain or permanent damage to tissue are grouped under an umbrella term "autoimmune disorders." In simple words, autoimmune diseases are the variety of diseases arising due to the irregular functioning of the immune system, that leads to the generation of immune system reactivity (Viswanath 2013). Indeed, autoimmune diseases are multi-etiological entities that develop due to disturbed immunoregulatory processes, as well as environmental and genetic abnormalities. Heredity accounts for about 30% of the risk of developing an autoimmune disease, while non-inherited, environmental factors account for the remaining 70% risk (Viswanath 2013; Nagy et al. 2015). There is a permanent failure of one or several tolerance mechanisms in autoimmune diseases due to the cumulative effect of various specific human leukocyte antigen (HLA), non-HLA genes, and environmental factors and/or derailed immune regulatory processes. This leads to the development of self-reactive B- and T-cell clones, which cause damage to tissues or organs (Ermann and Fathman 2001). There are nearly 100 distinct autoimmune diseases, some of which are organ-specific such as primary biliary cirrhosis (PBC) and some of which reflect a variety of immunological dysfunction involving multiple organs such as systemic lupus erythematosus (SLE) (Wang et al. 2015; Yu et al. 2014). Thus, clinically autoimmune diseases can be classified as organ-specific (e.g., Type 1 diabetes mellitus) or systemic (e.g., systemic lupus erythematosus). The common types of autoimmune diseases such as Addison's disease, autoimmune hepatitis, celiac disease, Type 1 diabetes, Grave's disease (overactive thyroid), Guillain–Barre syndrome, Hashimoto's disease, hemolytic anemia, inflammatory myopathies, idiopathic thrombocytopenic purpura, multiple sclerosis, myasthenia gravis, psoriasis, primary biliary cirrhosis, rheumatoid arthritis, Sjogren's syndrome, systemic lupus erythematosus, and vitiligo with their pathogenesis aspects are described elsewhere (Viswanath 2013; Wang et al. 2015).

A simple hypothesis for autoimmune diseases is that polymorphisms in various genes result in imperfect regulation or reduced threshold for lymphocyte activation, and environmental factors commence or enhance activation of self-reactive lymphocytes that have escaped control and are composed to react against self-constituents (Rosenblum et al. 2015). Autoimmune disorders are a group of diseases wherein structural/functional damage to cells/tissues/organs/organ systems is caused by the action of immunologically competent cells/antibodies against normal body constituents. To initiate autoimmunity several endocrine, genetic, and environmental factors interact together on immune system by the following mechanisms (Ganapathy et al. 2017; Wucherpfennig 2001):

1. *Cytolysis of the target cells*: Due to the release of tissue-specific autoantibodies via complement;
2. *Immune complex deposition*: Due to the binding of auto-antibody to soluble mediators;
3. *Phagocytosis, cytotoxicity, and antibody-mediated cellular immunity*: Due to the auto-antibody-mediated attack on immune system;
4. *Molecular Mimicry*: Auto-antibody against foreign antigen and auto-antigen epitopes which mimic foreign antigen (cross reactive antigen) leading to tissue damage;
5. *Stimulation/obstruction of the target structure*: Due to the action on cell surface structures by autoantibodies.

Thus, the stimulation and maintenance of immune tolerance signify major therapeutic goals in autoimmunity-caused autoimmune diseases (Janikashvili et al. 2016).

4.4 Immunotherapy for Autoimmune Diseases: General Considerations

Immunotherapy is a field of immunology that facilitates discovery of therapies for diseases by means of stimulation, augmentation, or suppression of an immunoresponse. Simply, immunotherapy is a type of therapy which uses substances made by a body or in a laboratory to stimulate or suppress the immune system in order to improve or restore normal functions of the immune system, so that the body can effectively fight against cancer, infection, and other diseases. Immunotherapies capable of initiating or boosting the immune response are referred to as "activating immunotherapies," while those capable of repressing the immune response are referred to as "suppressive immunotherapies" (Wraith 2017). More recently, the potential of immunotherapy to enhance or repress immune responses has been globally recognized and appreciated particularly in two areas of immunotherapy, i.e. suppressing immunotherapies for autoimmune diseases and activating immunotherapies for cancer. In the twenty-first century, the "immunotherapy revolution" started, with the approval of ipilimumab for melanoma. These cancer treatment approaches were based on activating immunotherapies which inhibit the inhibitors of the immune system releasing the brakes on the immune system. Subsequently, different immune checkpoint inhibitors, vaccines, and co-stimulatory agonists have been discovered and commercialized for a number of cancer types (Wraith 2017; De Miguel-Luken et al. 2017; Chen and Mellman 2013). On the other hand, the increased understanding of mechanisms of autoimmunity in recent years has paved the way to new promising therapeutic strategies for treating autoimmune diseases. This leads to the development of new types of immunotherapeutics that are capable of effectively and selectively targeting the self-reacting immune cells, cytokines, and other mediators of the immune response and are now available as cutting-edge therapies for autoimmune disease patients (Ostrov 2015).

The foremost challenge in the treatment of autoimmune diseases is to selectively suppress the autoimmune disease without affecting the control of rest of the functional immune system over cancers and infectious diseases. Hence, development of novel treatments with increasing specificity for the particular autoimmune disease is important with no or decreased risk of potential side effects (Wraith 2017). Historically in the 1980s, intra-venous immunoglobulin (IVIG) became a standard approach in managing autoimmune disorders, after the serendipitous discovery of polyclonal IgG immunoglobulin for the treatment of autoimmune thrombocytopenia (Imbach et al. 1981). After subsequent trials on many autoimmune disorders, greater than 70% of the IVIG prescribed in the United States by 2014 was for autoimmune and inflammatory diseases rather than for immunodeficiency (Ballow 2014). In recent years, further advancements in the research with various kinds of innovative work identifying new receptors, signaling pathways, monoclonal antibodies (MABs) and with the development of hybridomas and molecular cloning led to the discovery of new biologic agents directing the new treatments for autoimmune diseases (Ostrov 2015).

The current treatment strategies of autoimmune diseases include two major approaches, first is a "conservative approach" where a symptomatic or replacement therapy is given and second is an "aggressive approach" where immunosuppressive or immune modulation therapy is preferred. For instance, autoimmune thyroid disease is mainly managed either by reducing the production of thyroxin at the stage of hyper functioning of the thyroid gland or by hormone replacement therapy when the gland is damaged. On the other hand, in systemic diseases like SLE which targets vital organs like kidney, the primary treatment is immunosuppressive therapy in order to prevent more organ damage. Generally in autoimmune disease, 60–70% response is observed for immunosuppression with gradual decrease in the response to the drug used. Although in few cases there is a long-lasting remission of autoimmune diseases, some of the autoimmune diseases go for clinical remission to relapse after sometime (Chandrashekara 2012). It is important to note that currently available steroid and non-steroid immunosuppressive medicines for autoimmune diseases also have limited efficacy and we have been dependent on non-specific immunosuppressive therapies for quite some time (Wraith 2017). Hence, there is an immense need to develop new approaches and ways to modulate the immune system for developing new therapeutic strategies of immunotherapy for different autoimmune diseases. In the last few decades, significant advancements have occurred in the approaches of immunosuppressive therapy for autoimmune diseases. Compared to the initial immunosuppressive drugs which were non-specific and interfering with larger pathways and cells, current immunosuppressive drugs are more target-specific with profound immunosuppression effect, increased remission rate, and reduced toxicity on other collateral systems (Feldmann and Steinman 2005; Böhm et al. 2006). In recent years, several new types and therapeutic strategies of immunotherapy for autoimmune diseases have evolved. James P. Allison and Tasuku Honjo were awarded the 2018 Nobel Prize in Physiology or Medicine for the discovery of cancer therapy by inhibition of negative immunoregulation of CTLA4 and PD1 immune checkpoints. Immune checkpoint therapy has led to tremendous progress in clinical development and revolutionized cancer treatment. This seminal discovery has fundamentally improved the outcomes for many people with advanced cancer (Smyth and Teng 2018). Similarly, several emerging and promising next-generation immunotherapy modalities for autoimmune diseases such as checkpoint based immunotherapy, antigen-specific immunotherapies, anti-cytokine therapy, anti-T-cell therapy, anti-B-cell therapy and biologics and their combination therapy, etc., have evolved and initiated the new era of immunotherapy for autoimmune diseases over a past few decades. These modalities are being discussed in upcoming sections of this chapter.

4.5 Checkpoint-Based Immunotherapy for Autoimmune Diseases

As mentioned earlier, the important characteristic of an autoimmune disease is the induction of B-cell and T-cell autoreactivity directed against self proteins, i.e. autoantigens. In other words, self-tolerance is the unresponsiveness of the immune system to self-antigens, and dysregulation of immune homeostasis along with self-tolerance leads to autoimmunity, resulting in harmful inflammation in and destruction of autoantibodies generated by B-cells and self-tissues mediated by autoreactive T-cells (Zhang and Vignali 2016). During the T-cell development process, the majority of T-cells which are specific for self-antigens are erased or deleted in a process of thymic elimination to set up a focal tolerance prior to the entry of T-cells into the periphery (Hogquist et al. 2005). However, it is a known fact that, potentially self-reacting lymphocytes can still "leak out" in small numbers into the periphery, even under the strict surveillance of central and peripheral tolerance due to the incomplete thymic deletion process. Thus, in order to circumvent the attack on normal host cells by remaining self-specific T-cells, additional mechanisms are required. These mechanisms include inhibition of proliferation of self-antigen specific T-cells by development of regulatory T-cells (Tregs) and regulation of T-cell activation and their functions by development of checkpoint pathways. The peripheral tolerance mechanisms play a significant role in checking autoimmune diseases (He et al. 2017). According to the two-signal model, activation of native T-cells requires two signaling processes: stimulation by major histocompatibility complex (MHC)–peptide molecules of T-cell receptor (TCR), and co-stimulation on antigen-presenting cells (APCs) via co-stimulatory receptors and their corresponding ligands (Zhang and Vignali 2016). APCs express B7–1 (CD80) or B7–2 (CD86), the co-stimulatory molecules, which initiate the subsequent signals. T-cell co-stimulatory receptor CD28 recognizes these co-stimulating molecules and thus an engagement of both TCR and CD28 on same T-cells triggers their multiplication by prompting an initiating signal to the T-cells, leading to a T-cell response to a self-antigen (in autoimmunity) or a foreign antigen (He et al. 2017). On the other hand, the cytotoxic T lymphocyte associated antigen-4 (CTLA-4) which is a T-cell receptor inhibitor, has a more prominent affinity for CD86 and CD80 ligands than the stimulatory receptor CD28. Consequently, CTLA-4 competes with CD28 for CD80 and CD86 (co-stimulating molecules) thereby serving as a checkpoint for T-cell response further leading to hyporesponsiveness or T-cell anergy (Linsley et al. 1994). Similarly, programmed death-1 (PD-1) has also been recognized as an immune checkpoint on T-cells or other immune cells. Along with its cognate ligands PD-L1 or PD-L2, PD-1 plays an important role in the process of peripheral tolerance to protect normal host tissue against self-reactive or specific T-cells by two mechanisms: blocking the escape of self-reactive T-cells into the periphery and promoting Treg development and function (Francisco et al. 2010; Fife and Pauken 2011). Thus, the immune checkpoint pathways play a crucial role in maintaining health by modulating harmony between protective T-cell response and T-cell tolerance.

Despite the fact that advancements in checkpoint-based immunotherapies for autoimmune diseases are relatively slow compared to that for cancer, this field has attracted a great deal of research interest. In both cases the aims for checkpoint-based immunotherapies are different where activation of T-cells is the prime aim in treating the cancer and chronic infections, whereas blocking the activation of self-specific or self-reactive T-cells is the prime goal in the treatment of autoimmune diseases (He et al. 2017). CTLA4-Ig (Abatacept) is an FDA-approved drug used to treat diseases like juvenile idiopathic arthritis and rheumatoid arthritis, and is currently being tested for other autoimmune diseases in several clinical studies. Abatacept is a soluble recombinant human fusion protein that is characterized by an extracellular domain of human CTLA-4 which is linked to a modified Fc domain of human IgG1. This agent binds to the co-stimulatory molecules B7–1/B7–2 present on APCs and mimics the action of the native CTLA-4. This results in the downregulation of autoreactive effector T-cell responses due to competitive inhibition of the crucial CD28:B7–1/B7–2 co-stimulatory signaling pathway(s) required for T-cell activation (Ruperto et al. 2008). The use of an Fc-chimeric version of PD-L1 in an in vitro model has demonstrated collapse of self-reactive T-cells on administration of a PD-1 agonist (McKinney et al. 2015). It is also demonstrated that the de novo generation of Tregs from naïve CD4 T-cells is amplified by PD-L1 (Francisco et al. 2009). These discoveries imply that it is possible to achieve dual benefits by utilizing the therapeutic capability of Tregs and concurrently reducing the augmentation proliferation and the role of activated self-reactive T-cells. Thus, it is important to understand and study the different molecular mechanisms of checkpoint-based immunotherapeutic agents particularly on Tregs and autoreactive activated T-cells in order to overcome the several autoimmune diseases by this novel approach. However, it is also critical to understand the fact that immunotherapies that repress activation or induce collapse of autoreactive T-cells can possibly trigger global immunosuppression. This may cause damaged immune function against infected or newly mutated cells or decreased immune control of malignancy and chronic infections. Hence, the use of checkpoint-based immunotherapies for autoimmune diseases remains challenging, where there is a strong need to improve specificity of these agents in order to minimize the immune-related adverse effects (irAE) (He et al. 2017).

4.6 Auto-Antigen Specific Immunotherapies for Autoimmune Diseases

Antigen specificity is considered as a fundamental mechanism of adaptive immunity. An alternate appealing approach to avoid global immune-suppression that can affect overall control of the immune system during the treatment of autoimmune diseases by immunotherapy is believed to be auto-antigen specific immunotherapies (ASIs). Hypothetically, minimal damage would be caused to the overall defensive immunity against foreign antigens acquired from microbial pathogens and cancer cells by restricting the induction of T-cell exhaustion only to the activated autoreactive (autoantigen-specific) T-cells that have escaped thymic deletion (He et al. 2017; Bluestone and Bour-Jordan 2012). It is a known fact that T-cell remains anergic (unresponsive) to the resultant antigenic challenges if it receives only antigen-specific stimulation through its TCR without the subsequent stimulation signal via its co-stimulating receptor (Chen and Flies 2013) .In concurrence with this idea, ASI has been investigated in mouse models and has been reported to reverse or prevent autoimmune diseases, thus demonstrating that inducing tolerance to a finite number of autoantigens or epitopes is adequate to elicit therapeutic benefits (Macleod and Anderton 2015). Thus, the goal of ongoing research in immune tolerance is the development of autoantigen-specific immunotherapeutic treatments such as ASI that allow for the specific blockade of the harmful effects of self-reactive immune-cell function while retaining the ability of the immune system to clear non-self antigens (Miller et al. 2007).

Antigen-specific tolerance can be induced by introducing an antigen under tolerogenic conditions rather than immunogenic conditions. In reality, antigen introduced orally or in soluble form appears to decrease, and not potentiate, subsequent immune response to the antigen. Thus, antigen-specific tolerance forms the basis for the use of allergen extract-based immunotherapy to treat allergies, and has been suggested as a potential means to treat autoimmune diseases (Smilek et al. 2014). Although ASI for autoimmune disease has the potential to control the disease much like allergen-specific immunotherapy, there are basic differences between both, including that allergic diseases consist of helper T-cell; Th2 dominant responses, whereas autoimmune diseases consist of Th1and Th17 dominant responses. If we understand the pathophysiology and identify the autoantigens involved in particular autoimmune diseases, it is possible to manipulate autoantigen-related pathways to induce immune tolerance against self-antigens. Based on this concept, several considerable efforts have been made to use ASI approach to modify the immune response in autoimmune diseases. Several studies in animal models that stimulate chronic inflammatory conditions have found that controlled administration of autoantigens can provide protection from autoimmune disease (Hirsch and Ponda 2014). Table 4.1 summarizes several ASI studies reported for the treatment of different autoimmune diseases.

From Table 4.1, it is clear that T1D is one of the most researched autoimmune diseases due to the availability of well-defined autoantigens and NOD mouse models. Although ASI for T1D has shown promising results in animal models and Phase I trials, few have shown efficacy in Phase II studies, raising concern that ASI for T1D therapy may not be a viable option. In the case of MS, some animal studies of EAE reported limited efficacy of oral/nasal administration of soluble myelin peptides in prevention of EAE but not in treatment of EAE after onset. Unfortunately, human clinical trials with oral bovine MBP have not been successful, and the oral route appears limited in inducing tolerance in ongoing disease. ASIs for RA have been limited by the lack of systematic knowledge of the pathogenesis of autoimmunity; whereas more studies are necessary to study the putative role of ASI for celiac disease, SLE, and other autoimmune diseases (Hirsch and Ponda 2014).

Table 4.1 Reported ASI studies in the treatment of autoimmune diseases

Autoimmune disease	Auto-antigen targeted	Type of study	ASI treatment approach	Findings	Ref
Type 1 diabetes (T1D)	Insulin, proinsulin, insulin peptides	Animal study in NOD mice model	Administration of insulin, proinsulin, and insulin peptides by IV, SC, IN, and oral routes	Protection/suppression of T1D in NOD mice	Chatenoud (2010)
	Insulin	A multicenter RCT	Oral insulin (2.5 mg or 7.5 mg daily)	No difference observed between placebo and treatment groups	Chaillous et al. (2000)
	Insulin	DPT-1: Double-blind trial	Oral insulin (5 mg daily)	No difference observed between placebo and treatment groups	Pozzilli et al. (2000)
	Insulin	A double-blind crossover study	Intranasal insulin (1.6 mg)	Intranasal insulin was safe and caused immune changes that were consistent with insulin mucosal tolerance	Harrison et al. (2004)
	Insulin	A RCT, non-blinded	SC insulin (0.25 unit/kg daily)	In persons at high risk for diabetes, insulin does not delay or prevent type 1 diabetes	Diabetes Prevention Trial--Type 1 Diabetes Study Group (2002)
	Insulin	A double-blind, RCT	Intranasal insulin (1 unit/kg daily)	No difference observed between placebo and treatment groups	Näntö-Salonen et al. (2008)
	Insulin peptide	A randomized, four-arm, placebo-controlled, dose-ranging phase 2 trial	SC injections of the altered peptide ligand, NBI-6024 at 1, 0.5, or 0.1 mg repeated doses	No effects	Walter et al. (2009)
	GAD65	Animal study in NOD mice model	IP injection of 100 μg GAD65 in 50% IFA	Prevents and inhibits autoimmune damage of β-cells	Tian et al. (1996)
	GAD65	Animal study in NOD mice model	IV administration	Suppression of the diabetogenic response is by	Tisch et al. (1998)

(continued)

Table 4.1 (continued)

Autoimmune disease	Auto-antigen targeted	Type of study	ASI treatment approach	Findings	Ref
				induction of GAD65-specific CD4+ regulatory T-cells	
	GAD65	Animal study in NOD mice model	IN administration using CTB as a vaccine carrier	Using CTB as a vaccine carrier, tandem GAD65 peptides can prevent T1D in NOD mice at the late stage of disease	Wang et al. (2009)
	GAD65	Animal study in NOD mice model	Oral administration of GAD65 CTB	Significant decrease in inflammation of pancreatic islets and diabetes	Gong et al. (2010)
	GAD65	Randomized, double-blind, placebo-controlled phase II study	SC injections of 20 μg of recombinant human GAD65 (Diamyd®)	Significant efficacy was reported in preservation of β-cell function	Ludvigsson et al. (2008)
	GAD65	A randomized double-blind trial, phase II	SC injections of 20 μg GAD-alum (Diamyd®)	No difference in loss of insulin secretion	Wherrett et al. (2011)
	GAD65	Phase III study	SC injections of 20 μg GAD-alum (Diamyd®)	There was no significant difference in β-cell function (assessed by C-peptide levels) between treatment and placebo after follow-up of 15 months	Ludvigsson et al. (2012)
	HSP60	Animal study in NOD mice model	IN vaccination of DiaPep277, a synthetic peptide derived from human HSP60	Prevents T1D development	Fierabracci (2011)
	HSP60	A phase III, double-blind, randomized, parallel group, placebo-controlled, study	SC injections of DiaPep277 or placebo	DiaPep277 safely contributes to preservation of β-cell function and to improved glycemic control in patients with type 1 diabetes	Raz et al. (2014)

Multiple sclerosis (MS)	MBP	EAE animal model for MS	IN administration of encephalitogenic peptide	IN dose of peptide profoundly inhibited EAE when administered prior to disease induction	Metzler and Wraith (1993)
	MBP	Novel DA rat system	IN administration of MBP	Prevented induced EAE	Bai et al. (1998)
	PLP	EAE animal model for MS	Treatment of oligomerized T-cell epitope of myelin proteolipid protein (PLP139–151)	Disease protection was shown to be antigen-specific and effective in both prevention and treatment of ongoing EAE	Puentes et al. (2013)
	MBP	Double-blind human clinical trial	Daily capsules of bovine myelin or a control protein	Reduction in T-cells that are reactive with myelin basic protein	Weiner et al. (1993)
	MBP	Phase I clinical trial	IV administration of synthetic peptides containing epitope P85 VVHFFKNIVTP96	IV administration of this peptide induced tolerance	Warren et al. (1997)
	APL of MBP	Phase II clinical trial	SC administration of 50 mg to 5 mg CGP77116 (APL) weekly	APL resulted in disease exacerbation	Bielekova et al. (2000)
	APL of MBP	Double-blind, randomized, placebo-controlled phase II trial	SC administration of 50 mg to 5 mg NBI 5788 (APL) weekly	Hypersensitivity reactions (9% incidence)	Kappos et al. (2000)
	MBP MOG PLP	1 year, placebo-controlled cohort study	TD treatment with a mixture of 3 myelin peptides: MBP85–99, PLP139–151, and MOG35–55	TD treatment caused immunologic tolerance to myelin antigens	Juryńczyk et al. (2010)
	MBP MOG PLP	1 year double-blind, placebo-controlled cohort study	Skin patch comprising a mixture of three myelin peptides: MOG35–55, MBP85–99, PLP139–155	The patch was safe and well tolerated and significantly reduced the measures of disease activity	Walczak et al. (2013)

(continued)

Table 4.1 (continued)

Autoimmune disease	Auto-antigen targeted	Type of study	ASI treatment approach	Findings	Ref
Rheumatoid arthritis (RA)	CII	Animal study in polyarthritis model (mice)	ID injection of 300 ug of CII in emulsified form	Decreased incidence of collagen-induced arthritis	Nagler-Anderson et al. (1986)
	CII	Animal study in WA/KIR rats	Oral administration of CII by gavage, before disease onset	Reduced (IgG)-collagen specific antibodies; developed a T-cell population that could transfer tolerance to naïve mice	Thompson et al. (1993)
	CII	Animal study in inbred female Lewis rats	Oral administration of 20 mg mBSA/ml of distilled water	No therapeutic effect after oral administration	Yoshino (1995)
	CII	Animal study in male DBA/1 mice	Single oral administration of PLGA-CII (nanoparticles)	PLGA may serve as a powerful vehicle to promote the tolerance effect of oral CII	Kim et al. (2002)
	CII	A double-blind, randomized clinical trial	A daily oral dose of 0.5 mg triple helical CII	A significant increase in disease activity	Hauselmann (1998)
	dnaJP1	A pilot clinical trial	Oral administration of dnaJP1 for 6 months	Upregulated T-cell production of IL-4/IL-10 and downregulated production of IL-2, IFNγ, TNFα	Prakken et al. (2004)
	dnaJP1	A double-blind, placebo-controlled, pilot phase II trial	Oral doses of 25 mg of dnaJP1 or placebo daily for 6 months	Induced IL-10 production and inhibited TNFα	Koffeman et al. (2009)
	Hsp60	Animal study in female inbred Lewis rats and BALB/c and BALB/c Thy1.1 mice	ID, SC treatment by APL-1 designed from a novel epitope of Hsp60	Enhanced Tregs and inhibited disease progression	Dominguez et al. (2011)
	Hsp60	Clinical study	APL-1 treatment to PBMCs derived from RA patients	APL-1 upregulated Tregs and caused apoptosis of activated CD4+ cells, thus regulating inflammatory immune responses	Barberá et al. (2013)

Systemic lupus erythematosus (SLE)	Nucleosomal histones	Animal study in (SWR x NZB)F1 mouse model of lupus	IV administration of peptides from histone region	Treatment delayed the onset of severe lupus nephritis, prolonged survival, and even halted the progression of renal disease	Kaliyaperumal et al. (1999)
	snRNP	Animal study in lupus-prone MRL/lpr mice	IV administration of phosphorylated peptide, P140 in saline	Decreased proteinuria and anti-DNA antibody production, and significantly prolonged survival of treated mice	Monneaux et al. (2003)
	snRNP	An open-label, dose-escalation phase II clinical study	SC administrations of a clinical batch of P140 (IPP-201101) peptide at 2-week intervals	Safe and tolerable; primary endpoint of reduction in anti-dsDNA antibody titers achieved	Muller et al. (2008)

NOD non-obese diabetic, IV intra-venous, SC subcutaneous, IN intranasal, RCT randomized controlled trial, DPT-1 diabetes prevention trial of type I diabetes, GAD65 glutamic acid decarboxylase 65-kilodalton isoform, CTB cholera toxin B subunit, IP intraperitoneal, HSP60 heat-shock protein 60, MBP myelin basic protein, MOG myelin oligodendrocyte glycoprotein, PLP proteolipid protein, APL altered peptide ligands, TD transdermal, CII type II collagen, ID intradermal, IgG immunoglobulin G, mBSA methylated bovine albumin, Hsp60 heat-shock protein 60, PBMCs peripheral blood mononuclear cells, snRNP ribonucleoproteins

4.7 Anti-Cytokine (Anti-IL-1, Anti-IL-6, Anti-TNF Agents) Therapy for Autoimmune Diseases

During the development of the normal immune response, cytokines not only regulate a broad range of physiological processes but are also involved in the pathogenesis of autoimmune diseases. Autoimmune pathogenesis can be triggered when immune system cells recognize self tissue as foreign and the balance between pro- and anti-inflammatory cytokines is disturbed. Thus, anti-cytokine treatment alone and/or in combination with varied classes of immune suppressive molecules is highly effective, where some cytokines have been successfully identified as potential targets for the therapy of inflammatory/autoimmune diseases. Many anti-cytokine therapeutics are currently being used clinically, and many biologicals are in the pipeline (Astrakhantseva et al. 2014). There are three main categories of anti-cytokine agents which are popular for the treatment of autoimmune diseases: anti-interleukin 1 (IL-1), anti-interleukin 6 (IL-6), and anti-tumor necrosis factor (TNF) agents.

In 1975, TNF-α was found to be a specific product of macrophages and lymphocytes that induced breakdown of specific types of cells which also include tumor cells (Carswell et al. 1975). TNF-α is present on cell surfaces of lymphocytes and macrophages as a transmembrane protein. Cleavage of this protein leads to release of soluble TNFα. There are two TNF receptors that regulate the function of this mediator—TNFR1 and TNFR2. The TNFR1 receptor is membrane bound and upon stimulation by TNF-α releases other cytokines such as IL-2 and interferon (IFN), while the soluble TNFR2 receptor is present in the extracellular milieu where it serves to deactivate soluble TNF and blunts its inflammatory activity (Ostrov 2015). Various clinical trials of TNF inhibitors have revealed that it is possible to abrogate immune system activation, control inflammation, mitigate damage to joints, and sometimes cause stable remission in patients after discontinuing anti-TNF therapy merely by inhibiting a single cytokine (Huang et al. 2012; Verazza et al. 2013; Regueiro et al. 2014). Infliximab is the first TNF inhibitor that was found to be effective for patients with RA and Crohn's disease (unresponsive to conventional therapy). Later, it was also demonstrated to be efficacious in the treatment of psoriasis and ankylosing spondyloarthritis. Currently, various autoimmune diseases are effectively being treated with the use of TNF inhibitors, and five TNF inhibitors are approved in the majority of developed countries (Astrakhantseva et al. 2014).

IL-1, the first identified cytokine, was called the "endogenous pyrogen" because of its main action of inducing fever. IL-1α and IL-1β are the active products of this cytokine. Inactive IL-1β is cleaved to the active form by the inflammasome complex leading to the signs of inflammation (Ostrov 2015). IL-1 cytokines are a key factor in regulating the immune response and developing inflammation by controlling the expression of numerous effector proteins like chemokines, cytoplasmic metalloproteinases, cytokines, etc. (Dinarello 1996). Dysregulated IL-1α/β synthesis/secretion may result in grave pathologies. Upregulated IL-1α/β due to activation of its synthesis/secretion by triggering inflammasomes is generally responsible for many "classic" chronic inflammatory diseases. Several chronic inflammatory diseases including cryopyrin-associated periodic syndromes (CAPS), gout, multiple sclerosis, hypertension, type-2 diabetes, etc., are associated particularly with the increased level or production of IL-1β. These chronic inflammatory diseases are actively treated by IL-1 inhibitors like anakinra, rilonacept, and canakinumab which inhibit signal transduction pathways via IL-1/IL-1receptor (IL-1R) and thus crosstalk with the cycle of inflammation (Astrakhantseva et al. 2014).

Another key cytokine which along with IL-1 and TNF contributes to inflammation in autoimmune diseases is IL-6. It stimulates B-cell antibody production, elevates inflammatory serum markers (especially C-reactive protein), and promotes Th17 cell maturation (Ostrov 2015). Dimerization of gp130 is initiated by the receptor binding of IL-6 which leads to the activation of JAK tyrosine kinases. Activated JAKs further phosphorylate and activate STAT transcription factors, e.g. STAT3 for IL-6 receptor; hence, dysregulation of this cytokine network may result in autoimmune diseases, chronic and acute inflammations, and neoplastic disorders (Astrakhantseva et al. 2014; Heinrich et al. 1998). The uncontrolled production of IL-6 may cause various chronic inflammatory and autoimmune diseases by shifting the balance to the side of Th17/Th1 side with Treg reduction (Kimura and Kishimoto 2010). Tocilizumab, an inhibitor of IL-6, was the initial molecule clinically approved in its class. It alters common IL-6-receptor (IL-6R) complex functioning, and inhibits downstream activation of adhesion molecules, osteoclasts, and maturation of both B- and T-cells (Rosman et al. 2013). This agent is approved by the FDA to treat RA, adult-onset Still's disease (AOSD), systemic juvenile idiopathic arthritis (JIA), and polyarticular JIA. Tocilizumab has also undergone recent promising trials in SLE and Crohn's Disease (CrD) (Ostrov 2015). The clinical success of tocilizumab led to the development of other inhibitors of IL-6 such as sirukumab, sarilumab, olokizumab, and clazakizumab, which are now in the second phase of clinical trial (Tanaka and Mola 2014). Table 4.2 summarizes the recent anti-cytokine immunotherapeutics with their characteristic features and therapeutic applications. In summary, besides the few limitations of anti-cytokine immunotherapy due to its ability to affect basic protective bodily functions through specific cytokines, its use in the clinical setting for autoimmune diseases and chronic inflammatory diseases is, indeed, revolutionary. Based on enhanced understanding of the molecular mechanisms of cytokine-associated pathologies, it is already being actively used in several countries and will most certainly become a trail-blazing trend in clinical medicine in the future (Astrakhantseva et al. 2014).

4.8 Anti-T-Cell Therapy for Autoimmune Diseases

Emerging knowledge from the current developments in the field of immunotherapy has revealed that peripheral tolerance mechanisms that fail in autoimmunity are implicated in progressive malignancies and chronic infections. Thus, pathways targeted for therapeutic intervention in autoimmune diseases can be modulated in the opposite sense in malignancy and infectious disease (Bucktrout et al. 2018). Major therapeutic strategies of anti-T-cell therapy for autoimmune diseases are

Table 4.2 List of recent anti-cytokine immunotherapeutics used for various autoimmune diseases

Cytokine target	Drug	Structure	Half life (approx.)	Therapeutic applications
Soluble and membrane bound TNF-α	Infliximab	Mouse/human chimeric IgG1 mAb	14 days	CrD, UC, RA, AS, PsA, Ps
	Etanercept	TNFR2 dimer bound to fc-fragment of human IgG1	70 hours	RA, JIA, PsA, AS, Ps
	Adalimumab	Human IgG1 mAb	14 days	RA, JIA, PsA, CrD, AS
	Golimumab	Human IgG1 mAb	14 days	RA, PsA, AS
	Certolizumab pegol	PEGylated fab-fragment of human IgG1 mAb	14 days	CrD, RA, PsA, AS
IL-1α/β	Anakinra	Recombinant human IL-1 receptor antagonist	4–6 hours	RA, CAPS
	Rilonacept	Dimer of IL-1R1 and IL-1RAcP bound to fc-fragment of IgG1	7.5 days	CAPS, Muckle–Wells syndrome
	Canakinumab	Human IgG1 mAb	26 days	CAPS, Muckle–Wells syndrome
IL-6R	Tocilizumab	Human IgG1 mAb	11–13 days	RA, Castleman disease, JIA clinical
	Sarilumab	Human IgG1 mAb	8–10 days	RA, AS
	Sirukumab	Human IgG1 mAb	15–19 days	RA, SLE
sIL-6R	Sgp130Fc	Human gp130 extracellular domain bound to IgG1Fc-fragment	72 hours	RA
IL-6	Olokizumab	Human IgG1 mAb	31.5 days (SD 12.4 days)	CrD, RA
	Clazakizumab	Human IgG1 mAb	30 days	RA

AS ankylosing spondyloarthritis, CAPS cryopyrin-associated periodic syndromes, CrD Crohn's disease, IL-6R IL-6 receptor, JIA juvenile idiopathic arthritis, mAb monoclonal antibody, PsA psoriatic arthritis, Ps psoriasis, RA rheumatoid arthritis, sIL-6R soluble IL-6 receptor, UC ulcerative colitis

immunomodulation of T-cell co-stimulation, migration, and inflammation. The rationale behind the development of such therapies is that such immunotherapeutics could selectively target pathogenic T-cells during autoimmune conditions. It has been observed that CD28, a co-stimulatory molecule, expresses on T-cells; and that interaction of CD28 with B7–1 or B7–2 is necessary for T-cell activation as well as effector function. Thus, inhibiting CD28/B7 interactions in the TCR signaling axis may lead to tolerance resulting from deletion of T-cells and/or anergy, which is important to re-establish tolerance in autoimmune diseases (Bluestone and

Bour-Jordan 2019). In vivo studies showed that abrogation of CD28 signaling by means of a fusion protein of CTLA4Ig was efficacious in ameliorating many autoimmune diseases such as MS or SLE (Scalapino and Daikh 2008). Two fusion proteins (CTLA4Ig) composed of the Fc region of the immunoglobulin IgG1 fused to the extracellular domain of CTLA-4, viz., Abatacept and Belatacept (the higher-affinity Belatacept which is a second-generation variant) demonstrated more than 50% response rate in patients with psoriasis and RA who were refractory to anti-TNF-α therapy; these are approved by FDA for RA/JIA (Genovese et al. 2005). However, abatacept could not reduce disease flares during Phase II studies on SLE patients treated with oral corticosteroids, and was found ineffective in a Phase III trial of CrD (Merrill et al. 2010). The lack of response to abatacept in patients with CrD may be due to limited co-stimulatory activity by intestinal T-cells which do not express CD28 (Mayer et al. 2012). In a recent multi-center trial in newly diagnosed T1D patients, treatment with abatacept for 2 years was well tolerated and delayed the reduction in B-cell function compared with placebo (Orban et al. 2011). Many clinical studies of abatacept or belatacept are currently ongoing in SLE, MS, or T1D. Current therapy with CTLA4Ig was not found to lead to extensive immunosuppression or augmented infection rates, which is clearly beneficial in the therapy of autoimmune disease (Bluestone and Bour-Jordan 2019). Alefacept is a fusion protein of LFA-3-Ig that inhibits the interaction of lymphocyte function-associated antigen 3 (LFA-3/CD58) on the APCs with CD2, a co-stimulatory molecule on T-cells. It was reported to be efficacious in decreasing lesions in a psoriasis Phase III trial and is currently FDA-approved for psoriasis (Sugiyama et al. 2008). It has been reported that interactions between CD154 and CD40, the former on T-cells and the latter on APCs, are critical for stimulating autoreactive T-cells, activating APCs and producing autoantibodies; hence therapeutic abrogation of this pathway appeared promising in the 1990s. Unfortunately, clinical studies of anti-CD154 monoclonal antibodies were initiated in many autoimmune diseases such as SLE, CrD, MS, and psoriasis, but had to be discontinued due to occurrence of many thromboembolic events (Bluestone and Bour-Jordan 2019).

It is a well-known fact that key factors for the inducing or maintaining tolerance include T-cell trafficking and lymph node occupancy. Hence, immunotherapies which target T-cell migration in autoimmune disease may be used in combination therapy during immunosuppression to enable homing to lymph nodes of autoreactive T-cells and their tolerization (Bluestone and Bour-Jordan 2019; Ochando et al. 2005). Based on this theory, two most successful drugs developed for cell trafficking blockade are anti-integrin mAbsnatalizumab and efalizumab. These drugs are indicated in relapsing-remitting multiple sclerosis (RRMS), CrD, and psoriasis (Dubertret et al. 2006; Derfuss et al. 2013). Natalizumab, an FDA-approved mAb for CrDor RRMS, for patients suffering from severe disease or disease which is non-responsive to other standard-of-care, needs to carry an additional label warning. Further, efalizumab has been voluntarily withdrawn by the manufacturer due to the association of progressive multifocal leukoencephalopathy (PML) cases with the treatment (Hartung 2009). Another drug fingolimod capable of regulating T-cell trafficking by crosstalk with members of the sphingosine pathway was also found to improve the rate of relapse and progression to disability within 1–2 years in a Phase III clinical trial of RRMS patients and thereby turned out to become the first oral therapeutic approved by the FDA for MS (Kappos et al. 2010; Cohen et al. 2010). Due to the adverse effects associated with fingolimod therapy, second-generation molecules targeting the sphingosine signaling pathway are in development with potentially decreased side effects, and clinical trials are in progress for many autoimmune diseases (Bluestone and Bour-Jordan 2019). Along with these T-cell immunomodulation strategies, immunotherapies targeting proinflammatory cytokines (discussed in the previous section) or other inflammatory mediators in autoimmune diseases are also found to be useful, not just for improvement in clinical parameters but also for their ability to restore tolerance.

4.9 Anti-B-Cell Therapy for Autoimmune Diseases

The B-cell humoral response leading to production of autoantibodies and immune complexes contributes to manifestations of autoimmune diseases such as SLE and Sjogren's syndrome. The B-cell dysfunction contributes to the development of autoimmune phenomena via anomalies in B-cell's mechanisms such as antigen presentation, cytokine release, and T-cell activation (Ostrov 2015). From the last 10–15 years, B-cells are the recognized therapeutic targets for the treatment of autoimmune diseases. Presently, several promising and very efficient drugs specifically targeting plasma cells or B-cells are either in clinical use or under development for the treatment of several autoimmune diseases. These B-cell-directed therapies have proven to be therapeutically effective not only in classic B-cell/autoantibody-driven disorders, such as antibody/immune-complex-mediated SLE, autoimmune blistering skin diseases, or myasthenia gravis, but also in diseases that are believed to be mainly driven by T-cells, most importantly MS or RA (Hofmann et al. 2018). B-cells were recognized for their function as immune response enhancers in autoimmunity, as a result of their ability to generate autoantibody-producing plasma cells, and elicit $CD4^{+}$T-cell responses by antigen presentation. Such B-cells are typically classified as effector B-cells. Recently, studies indicated a potential role(s) of B-cells as negative sensors of immune response in autoimmunity, implicating interleukin 10 (IL-10) regulatory B-cell compartment (Breg). Thus, the abrogation of autoreactive effector B-cells in consonance with enhancement of autoantigen-driven Bregs, with immune surveillance maintenance, may be a key strategy to target B-cells. Anti-B-cell immunotherapies employ drugs directed against B-cell surface markers such as CD20/CD22, activating factors such as BAFF/TACI, and cytokines such as IL-6/TNFα/IFNα to target these B-cells (Musette and Bouaziz 2018). Table 4.3 summarizes the recent strategies to target B-cells with their drugs and implications in autoimmune diseases.

From the literature, it is clear that a great advancement has been made in decreasing resident and circulating B-cells in inflamed tissues or secondary lymphoid organs. The future of immunotherapy may require specific targeting,

Table 4.3 Recent anti-B-cell immunotherapy strategies with respective drugs and their implications

Anti-B-cell immunotherapy strategy	Specific target	Structure	Drugs under development	Implications in autoimmune diseases	Ref.
Drugs directed against B-cell surface markers	CD20	Type I anti-CD20 humanized/human mAbs	Rituximab,	RA, pemphigus, GPA, and MPA	Guillevin et al. (2014) and Joly et al. (2017)
			Ofatumumab	RA, MS	Pers and Jorgensen (2016)
			Ocrelizumab	RRMS	Hauser et al. (2017)
		Type II anti-CD20 humanized mAb	Obinutuzumab	–	Musette and Bouaziz (2018)
	CD19	A human IgG1 anti-CD19 mAb	Inebilizumab	MS	Agius et al. (2015)
	CD19 & CD3	A recombinant mouse derived bispecific mAb	Blinatumomab	–	Velasquez et al. (2018)
	CD22	Humanized mAb	Epratuzumab	SLE	Wallace et al. (2013)
Drugs directed against B-cell activating factors	BAFF	Human anti-BAFF mAb	Belimumab	SLE	Blair and Duggan (2018)
	BAFF	Human IgG4 mAb	Tabalumab	SLE	Hofmann et al. (2018)
	APRIL	Recombinant fusion protein combined with the extracellular ligand binding portion of TACI	Atacicept	RA, SLE	Hofmann et al. (2018)
Direct targeting of plasma cells	Plasma cells	Small molecule proteasome inhibitor	Bortezomib	Sjögren's syndrome, refractory SLE, TTP	Hofmann et al. (2018)
Inhibition of B-cell receptor signaling	BTK	Pharmacological BTK inhibitor	Ibrutinib	SLE, GVHD	Kil et al. (2012) and Miklos et al. (2017)

mAbs monoclonal antibodies, RA rheumatoid arthritis, GPA granulomatosis with polyangiitis, MPA microscopic polyangiitis, RRMS relapsing-remitting multiple sclerosis, SLE systemic lupus erythematosus, APRIL a proliferation-inducing ligand, TACI transmembrane activator and calcium-modulating cyclophilin ligand interactor, TTP thrombotic thrombocytopenic purpura, BTK Bruton's tyrosine kinase, GVHD graft versus host disease particularly of B-cell pathogenic effector functions, and augmentation of their regulatory role(s) without altering B-cell-dependent immune surveillance. Indeed, better patient management will be possible with better targeted therapeutics against specific B-cell populations and their functions (Musette and Bouaziz 2018).

4.10 Current Immunotherapeutics in Clinical Trials for Autoimmune Diseases

Several clinical trials are being performed on immunotherapeutics for autoimmune diseases as seen in Table 4.4. The majority of these molecules are being tested for safety and efficacy in psoriasis and rheumatoid arthritis. In both these diseases, one notes a larger proportion of immunotherapeutics in advanced clinical testing including Phases II and III. There is also considerable work that is ongoing with several immunotherapeutics in clinical trials for SLE, many of which are in Phase I or II. Interestingly, there are fewer immunotherapeutics in clinical trials for IBD, Crohn's disease, and MS; however, the few that are being investigated are mostly in Phase III. As seen in Table 4.4, there are a huge number of immunotherapeutics in clinical trials for various autoimmune diseases, and the number is likely to increase in the near future given the promise of immunotherapy in these diseases.

4.11 Financial Toxicity of Immunotherapies in Autoimmune Diseases

In recent years, the cost of immunotherapies in general and indeed for autoimmune diseases is reaching great proportions which brings into picture the financial burden which the patients and their families suffer from in different parts of the world despite having access to health insurance (Nipp et al. 2018; Zaprutko et al. 2017). The spectrum from diagnosis to treatment and post treatment care involves huge investments, financial as well as emotional, by the patients and their families who are afflicted by autoimmune diseases like CrD, inflammatory bowel syndrome (IBD), Grave's disease, MS, psoriasis, RA, Sjogren's syndrome, SLE, etc. (Lerner et al. 2015).

Inflammatory Bowel Syndrome, Crohn's Disease

Zheng et al. and Ylisaukko-Oja et al. discussed in their studies how along with cost of biologics, the secondary costs for outpatient visits, hospitalization, telephone consultation, laboratory visit, surgery, imaging, endoscopy added to the total cost of the treatments of inflammatory bowel syndrome (IBD) (Zheng et al. 2017; Ylisaukko-Oja et al. 2019). According to Berns et al., in previous years a patient with CrD was burdened with surgical intervention and hospitalization charges only but with modern anti-TNF-α biologic era the treatment itself accounted for 64% of the total cost, which was around $22,663 for infliximab in the United States. However such burden can be brought under control in future by the use of

Table 4.4 Current immunotherapeutics in clinical trials for autoimmune diseases

Sr no.	Autoimmune disease	Intervention	Number of patients	Recruitment status	Clinical trial registry identifier	Phase	Ref.
1	Crohn's disease	Guselkumab, Ustekinumab	2000	Recruiting	NCT03466411	II/III	Clinicaltrials.gov. A Study of the Efficacy and Safety of Guselkumab in Participants With Moderately to Severely Active Crohn's Disease (GALAXI). - Study Results - Clin [online] Available https://clinicaltrials.gov/ct2/show/NCT03466411
2	Crohn's disease	Risankizumab	940	Recruiting	NCT03105128	III	Clinicaltrials.gov. A Study of the Efficacy and Safety of Risankizumab in Subjects With Moderately to Severely Active Crohn's Disease. - Study Results - Clin [online] Available https://clinicaltrials.gov/ct2/show/NCT03105128
3	Crohn's disease	Ustekinumab	600	Recruiting	NCT03942120	IV	Clinicaltrials.gov. Post-Marketing Surveillance for Crohn's Disease Participants Treated With Stelara (Ustekinumab) (STELARA CD PMS). - Study Results - Clin [online] Available https://clinicaltrials.gov/ct2/show/NCT03942120

(continued)

Table 4.4 (continued)

Sr no.	Autoimmune disease	Intervention	Number of patients	Recruitment status	Clinical trial registry identifier	Phase	Ref.
4	Crohn's disease	Ustekinumab, Adalimumab	350	Recruiting	NCT03464136	III	Clinicaltrials.gov. Safety and Efficacy of Adalimumab Versus Ustekinumab for One Year (SEAVUE). - Stuy Results - Clin [online] Available https://clinicaltrials.gov/ct2/show/NCT03464136
5	Inflammatory bowel disease	Infliximab	60	Recruiting	NCT03445624		Clinicaltrials.gov. Abdominal Ultrasound With Doppler and Peripheral Hemogram in Assesment Inflammatory Bowel Disease. - Study Results - Clin [online] Available https://clinicaltrials.gov/ct2/show/NCT03445624
6	Inflammatory bowel disease	Infliximab	194	Recruiting	NCT02994836	IV	Clinicaltrials.gov. GIS-SUSANTI-TNF-2015 (Anti-TNF Discontinuation). - Study Results - Clin [online] Available https://clinicaltrials.gov/ct2/show/NCT02994836
7	Inflammatory bowel disease	Vedolizumab	2500	Recruiting	NCT03375424	III	Clinicaltrials.gov. Vedolizumab Study With Inflammatory Bowel Disease Patients in Germany (VEDOibdI). - Study Results - Clin [online] Available https://clinicaltrials.gov/ct2/show/NCT03375424

8	Grave's ophthalmopathy	RVT-1401	8	Recruiting	NCT03922321	II	Clinicaltrials.gov. Study of RVT-1401 for the Treatment of Patients With Moderate to Severe Active Graves' Ophthalmopathy (GO). - Study Results - Clin [online] Available https://clinicaltrials.gov/ct2/show/NCT03922321
9	Grave's orbitopathy	Teprotumumab	83	Active, not recruiting	NCT03298867	III	Clinicaltrials.gov. Treatment of Graves' Orbitopathy (Thyroid Eye Disease) to Reduce Proptosis With Teprotumumab Infusions in a Randomized, Placebo-Controlled, Clinical Study (OPTIC). - Study Results - Clin [online] Available https://clinicaltrials.gov/ct2/show/NCT03298867
10	Multiple sclerosis	ATA188	60	Recruiting	NCT03283826	I	Clinicaltrials.gov. Phase 1 Study to Evaluate the Safety of ATA188 in Subjects With Progressive and Relapsing-Remitting Multiple Sclerosis. - Study Results - Clin [online] Available https://clinicaltrials.gov/ct2/show/NCT03283826

(continued)

Table 4.4 (continued)

Sr no.	Autoimmune disease	Intervention	Number of patients	Recruitment status	Clinical trial registry identifier	Phase	Ref.
11	Multiple sclerosis	Elezanumab	165	Recruiting	NCT03737851	II	Clinicaltrials.gov. A Study to Assess the Safety and Efficacy of Elezanumab When Added to Standard of Care in Relapsing Forms of Multiple Sclerosis. - Study Results - Clin [online] Available https://clinicaltrials.gov/ct2/show/NCT03737851
12	Multiple sclerosis	Ocrelizumab	1500	Recruiting	NCT03599245	III	Clinicaltrials.gov. This is an Extension Study of the Roche P-trial to Investigate Safety and Effectiveness of a Single Ocrelizumab Dose in Participants With Multiple Sclerosis (MS). - Study Results - Clin [online] Available https://clinicaltrials.gov/ct2/show/NCT03599245
13	Multiple sclerosis	Ofatumumab	929	Active, not recruiting	NCT02792218	III	Clinicaltrials.gov. Efficacy and Safety of Ofatumumab Compared to Teriflunomide in Patients With Relapsing Multiple Sclerosis (ASCLEPIOS I). - Study Results - Clin [online] Available https://clinicaltrials.gov/ct2/show/NCT02792218

14	Multiple sclerosis	Rituximab	200	Recruting	NCT03979456	III	Clinicaltrials.gov. RItuximab Long-Term DOSE Trial in Multiple Sclerosis - RIDOSE-MS (RIDOSE-MS). - Study Results - Clin [online] Available https//clinicaltrials.gov/ct2/show/NCT03979456
15	Multiple sclerosis	Ublituximab	48	Active, not recruiting	NCT02738775	II	Clinicaltrials.gov. Phase 2 Study of Ublituximab in Patients With Relapsing Forms of Multiple Sclerosis. - Study Results - Clin [online] Available https://clinicaltrials.gov/ct2/show/NCT02738775
16	Plaque Psoriasis	Adalimumab	400	Recruiting	NCT03849404	III	Clinicaltrials.gov. Efficacy, Safety, and Immunogenicity of AVT02 With Moderate-to-Severe Chronic Plaque Psoriasis. - Study Results - Clin [online] Available https://clinicaltrials.gov/ct2/show/NCT03849404
17	Psoriasis	Adalimumab	56	Completed	NCT03217734	II	Clinicaltrials.gov. MAP Study: Methotrexate and Adalimumab in Psoriasis (MAP). - Study Results - Clin [online] Available https://clinicaltrials.gov/ct2/show/NCT03217734

(continued)

Table 4.4 (continued)

Sr no.	Autoimmune disease	Intervention	Number of patients	Recruitment status	Clinical trial registry identifier	Phase	Ref.
18	Generalized Pustular Psoriasis (GPP)	ANB019	10	Recruiting	NCT03619902	II	Clinicaltrials.gov. A Study to Evaluate the Efficacy and Safety of ANB019 in Subjects With Generalized Pustular Psoriasis (GPP). - Study Results - Clin [online] Available https://clinicaltrials.gov/ct2/show/NCT03619902
19	Chronic Plaque Psoriasis	Bimekizumab	43	Completed	NCT03230292	II	Clinicaltrials.gov. This is a Study to Evaluate the Long-term Safety, Tolerability and Efficacy of Bimekizumab in Adult Patients With Chronic Plaque Psoriasis. - Study Results - Clin [online] Available https://clinicaltrials.gov/ct2/show/NCT03230292
20	Moderate to Severe Chronic Plaque Psoriasis	Bimekizumab	1355	Recruiting	NCT03598790	III	Clinicaltrials.gov. A Study to Evaluate the Efficacy and Safety of Bimekizumab in Adult Subjects With Moderate to Severe Chronic Plaque Psoriasis (BE BRIGHT). - Study Results - Clin [online] Available https://clinicaltrials.gov/ct2/show/NCT03598790

21	Moderate to Severe Plaque Psoriasis	Bimekizumab, Adalimumab	480	Active, not recruiting	NCT03412747	III	Clinicaltrials.gov. A study to evaluate the efficacy and safety of bimekizumab in adult subjects with moderate to severe chronic plaque psoriasis (BE SURE). - Study results - Clin [online] Available https://clinicaltrials.gov/ct2/show/NCT03412747
22	Moderate to Severe Chronic Plaque Psoriasis	Bimekizumab, Secukinumab	743	Active, not recruiting	NCT03536884	III	Clinicaltrials.gov. A study to evaluate the efficacy and safety of bimekizumab compared to an active comparator in adult subjects with moderate to severe chronic plaque psoriasis (BE RADIANT). - Study results - Clin [online] Available https://clinicaltrials.gov/ct2/show/NCT03536884
23	Moderate to Severe Plaque Psoriasis	Brodalumab	39	Completed	NCT03403036	IV	Clinicaltrials.gov. Brodalumab in subjects with moderate to severe plaque psoriasis who have failed IL-17A therapies. - Study results - Clin [online] Available https://clinicaltrials.gov/ct2/show/NCT03403036

(continued)

Table 4.4 (continued)

Sr no.	Autoimmune disease	Intervention	Number of patients	Recruitment status	Clinical trial registry identifier	Phase	Ref.
24	Psoriasis Vulgaris	Brodalumab	210	Completed	NCT03331835	IV	Clinicaltrials.gov. A trial comparing the efficacy of subcutaneous injections of brodalumab to oral administrations of fumaric acid esters in adults with moderate to severe plaque psoriasis. - Study results - Clin [online] Available https://clinicaltrials.gov/ct2/show/NCT03331835
25	Plaque Psoriasis	Etanercept, Ixekizumab	201	Active, not recruiting	NCT03073200	III	Clinicaltrials.gov. Study of ixekizumab (ly2439821) in children 6 to less than 18 years with moderate-to-severe plaque psoriasis (Ixora-peds). - Study results - Clin [online] Available https://clinicaltrials.gov/ct2/show/NCT03073200
26	Psoriasis	Guselkumab	888	Recruiting	NCT03818035	III	Clinicaltrials.gov. A study to evaluate further therapeutic strategies with guselkumab in participants with moderate-to-severe plaque-type psoriasis (GUIDE). - Study results - Clin [online] Available https://clinicaltrials.gov/ct2/show/NCT03818035

27	Chronic Plaque Psoriasis	Guselkumab, Etanercept	125	Recruiting	NCT03451851	III	Clinicaltrials.gov. No a study to evaluate the efficacy, safety, and pharmacokinetics of subcutaneously administered guselkumab for the treatment of chronic plaque psoriasis in pediatric participants (PROTOSTAR). - Study results - Clin [online] Available https://clinicaltrials.gov/ct2/show/NCT03451851
28	Plaque Psoriasis	HLX03, Adalimumab	216	Recruiting	NCT03316781	III	Clinicaltrials.gov. Study of efficacy and safety of HLX03 in subjects with moderate to severe plaque psoriasis. - Study results - Clin [online] Available https://clinicaltrials.gov/ct2/show/NCT03316781
29	Psoriasis	Humira®, BI 695501	259	Recruiting	NCT03210259	III	Clinicaltrials.gov. The VOLTAIRE-X trial looks at the effect of switching between Humira® and BI 695501 in patients with plaque psoriasis. - Study results - Clin [online] Available https://clinicaltrials.gov/ct2/show/NCT03210259

(continued)

Table 4.4 (continued)

Sr no.	Autoimmune disease	Intervention	Number of patients	Recruitment status	Clinical trial registry identifier	Phase	Ref.
30	Generalized Pustular Psoriasis and Erythrodermic Psoriasis	Ixekizumab	12	Recruiting	NCT03942042	IV	Clinicaltrials.gov. A study of ixekizumab (LY2439821) in participants in Japan With generalized pustular psoriasis and erythrodermic psoriasis. - Study results - Clin [online] Available https:// clinicaltrials.gov/ct2/show/ NCT03942042
31	Psoriasis	Mirikizumab	1816	Recruting	NCT03556202	III	Clinicaltrials.gov. A Long-term Study to Evaluate Safety and Maintenance of Treatment Effect of LY3074828 in Participants With Moderate-to-Severe Plaque Psoriasis (OASIS-3). - Study Results - Clin [online] Available https//clinicaltrials. gov/ct2/show/NCT03556202
32	Psoriasis	Mirikizumab	530	Active, not recruiting	NCT03482011	III	Clinicaltrials.gov. A study to evaluate the efficacy and safety of mirikizumab (LY3074828) in participants with moderate-to-severe plaque psoriasis (OASIS-1). - Study results - Clin [online] Available https:// clinicaltrials.gov/ct2/show/ NCT03482011

33	Psoriasis	Mirikizumab	26	Recruiting	NCT03718884	I	Clinicaltrials.gov. A study of mirikizumab in participants with plaque psoriasis. - Study results - Clin [online] Available https://clinicaltrials.gov/ct2/show/NCT03718884
34	Psoriatic Arthritis	Risankizumab	964	Recruiting	NCT03675308	III	Clinicaltrials.gov. A study comparing risankizumab to placebo in subjects with active psoriatic arthritis (PsA) who have a history of inadequate response to or intolerance to at least one disease modifying anti-rheumatic drug (DMARD) therapy. - Study results - Clin [online] Available https://clinicaltrials.gov/ct2/show/NCT03675308
35	Psoriatic Arthritis (PsA)	Risankizumab	444	Recruiting	NCT03671148	III	Clinicaltrials.gov. A study comparing risankizumab to placebo in subjects with active psoriatic arthritis including those who have a history of inadequate response or intolerance to biologic therapy (IES). - Study results - Clin [online] Available https://clinicaltrials.gov/ct2/show/NCT03671148

(continued)

Table 4.4 (continued)

Sr no.	Autoimmune disease	Intervention	Number of patients	Recruitment status	Clinical trial registry identifier	Phase	Ref.
36	Psoriasis	Risankizumab	50	Active, not recruiting	NCT03518047	III	Clinicaltrials.gov. Risankizumab therapy versus placebo for subjects with psoriasis in the Russian Federation (IMMPRESS). - Study results - Clin [online] Available https://clinicaltrials.gov/ct2/show/NCT03518047
37	Psoriasis	Risankizumab	2171	Active, not recruiting	NCT03047395	III	Clinicaltrials.gov. A Study to Assess the Safety and Efficacy of Risankizumab for Maintenance in Moderate to Severe Plaque Type Psoriasis (LIMMITLESS). - Study Results - Clin [online] Available https//clinicaltrials.gov/ct2/show/NCT03047395
38	Psoriasis	Risankizumab, Secukinumab	327	Active, not recruiting	NCT03478787	III	Clinicaltrials.gov. Risankizumab versus secukinumab for subjects with moderate to severe plaque psoriasis. - Study results - Clin [online] Available https://clinicaltrials.gov/ct2/show/NCT03478787

39	Psoriasis	Secukinumab	24	Recruiting	NCT03131570	II	Clinicaltrials.gov. Safety and efficacy of secukinumab in mild psoriasis. - Study results - Clin [online] Available https://clinicaltrials.gov/ct2/show/NCT03131570
40	Plaque Psoriasis	Secukinumab	122	Recruiting	NCT03589885	III	Clinicaltrials.gov. Study of efficacy and safety of secukinumab 2 mL auto-injector (300 mg) in subjects with moderate to severe plaque psoriasis (MATURE). - Study results - Clin [online] Available https://clinicaltrials.gov/ct2/show/NCT03589885
41	Plaque Psoriasis	Secukinumab	196	Recruiting	NCT03020199	IV	Clinicaltrials.gov. Study of the efficacy of early intervention with secukinumab 300 mg s.c. compared to narrow-band UVB in patients with new-onset moderate to severe plaque psoriasis (STEPin). - Study results - Clin [online] Available https://clinicaltrials.gov/ct2/show/NCT03020199

(continued)

Table 4.4 (continued)

Sr no.	Autoimmune disease	Intervention	Number of patients	Recruitment status	Clinical trial registry identifier	Phase	Ref.
42	Generalized Pustular Psoriasis	Spesolimab	171	Recruiting	NCT03886246	III	Clinicaltrials.gov. A 5-year study to test BI 655130 in patients with generalized pustular psoriasis who took part in previous studies with BI 655130. - Study results - Clin [online] Available https://clinicaltrials.gov/ct2/show/NCT03886246
43	Psoriasis	Ustekinumab	75	Recruiting	NCT03218488	I	Clinicaltrials.gov. A safety study of ustekinumab in the treatment of pediatric participants aged 12 years and older with moderate to severe plaque psoriasis. - Study results - Clin [online] Available https://clinicaltrials.gov/ct2/show/NCT03218488
44	Rheumatoid arthritis	Abatacept	25	Recruiting	NCT03652961	IV	Clinicaltrials.gov. Rheumatoid arthritis memory B cells and abatacept (RAMBA) (RAMBA). - Study results - Clin [online] Available https://clinicaltrials.gov/ct2/show/NCT03652961

45	Rheumatoid Arthritis	ABBV-3373, Adalimumab	48	Recruiting	NCT03823391	II	Clinicaltrials.gov. A study to evaluate the safety, tolerability, pharmacokinetics, and efficacy of ABBV-3373 in participants with moderate to severe rheumatoid arthritis. - Study results - Clin [online] Available https://clinicaltrials.gov/ct2/show/NCT03823391
46	Rheumatoid Arthritis	AMG 570	34	Active, not recruiting	NCT03156023	I	Clinicaltrials.gov. Multiple ascending doses of AMG 570 in subjects with rheumatoid arthritis. - Study results - Clin [online] Available https://clinicaltrials.gov/ct2/show/NCT03156023
47	Rheumatoid Arthritis	BAT1806, Actemra	621	Recruiting	NCT03830203	III	Clinicaltrials.gov. Comparative study of BAT1806 to RoActemra® in rheumatoid arthritis patients with inadequate response to methotrexate. - Study results - Clin [online] Available https://clinicaltrials.gov/ct2/show/NCT03830203

(continued)

Table 4.4 (continued)

Sr no.	Autoimmune disease	Intervention	Number of patients	Recruitment status	Clinical trial registry identifier	Phase	Ref.
48	Rheumatoid Arthritis	CT-P17, Humira	564	Active, not recruiting	NCT03789292	III	Clinicaltrials.gov. A study to compare efficacy and safety of CT-P17 with humira in patients with active rheumatoid arthritis. - Study results - Clin [online] Available https://clinicaltrials.gov/ct2/show/NCT03789292
49	Rheumatoid Arthritis	Etanercept	50	Completed	NCT03193957	II	Clinicaltrials.gov. An multicentre clinical study to evaluate the usability and safety of the autoinjector and pre-filled syringe of SB4 in subjects with rheumatoid arthritis. - Study results - Clin [online] Available https://clinicaltrials.gov/ct2/show/NCT03193957
50	Rheumatoid Arthritis	Infliximab	300	Recruiting	NCT03885037	III	Clinicaltrials.gov. Infliximab biosimilar for intravenous drip infusion 100 mg “Pfizer” drug use investigation (rheumatoid arthritis). - Study results - Clin [online] Available https://clinicaltrials.gov/ct2/show/NCT03885037

51	Rheumatoid Arthritis	Olokizumab	1880	Recruiting	NCT03120949	III	Clinicaltrials.gov. Evaluation of the long term safety, tolerability and efficacy of two dosing regimens of olokizumab (OKZ), in subjects with rheumatoid arthritis (RA) who previously completed 24 weeks of blinded treatment in one of the core studies - CREDO 1, 2 or 3. (CRED. - Study results - Clin [online] Available https://clinicaltrials.gov/ct2/show/NCT03120949
52	Rheumatoid Arthritis	Sarilumab	84	Active, not recruiting	NCT03449758	IV	Clinicaltrials.gov. Effect of sarilumab on patient-reported outcomes in patients with active rheumatoid arthritis (SariPRO). - Study results - Clin [online] Available https://clinicaltrials.gov/ct2/show/NCT03449758
53	Rheumatoid Arthritis	Tocilizumab, Etanercept, Infliximab, Adalimumab, Golimumab, Certolizumab pegol	208	Recruiting	NCT03100253	IV	Clinicaltrials.gov. Rheumatoid arthritis treatment after first anti-TNF investigation (RAFTING). - Study results - Clin [online] Available https://clinicaltrials.gov/ct2/show/NCT03100253

(continued)

Table 4.4 (continued)

Sr no.	Autoimmune disease	Intervention	Number of patients	Recruitment status	Clinical trial registry identifier	Phase	Ref.
54	Rheumatoid Arthritis	VAY736	50	Recruiting	NCT03574545	I	Clinicaltrials.gov. Study comparing two VAY736 drug products in patients with rheumatoid arthritis. - Study results - Clin [online] Available https://clinicaltrials.gov/ct2/show/NCT03574545
55	Sjogren's syndrome	RSLV-132	28	Completed	NCT03247686	II	Clinicaltrials.gov. A study of RSLV-132 in subjects with primary Sjogren's syndrome (RSLV-132). - Study results - Clin [online] Available https://clinicaltrials.gov/ct2/show/NCT03247686
56	Primary Sjogren Syndrome	VAY736	195	Recruiting	NCT02962895	II	Clinicaltrials.gov. Safety and efficacy of VAY736 in patients with primary Sjogren's syndrome (pSS). - Study results - Clin [online] Available https://clinicaltrials.gov/ct2/show/NCT02962895
57	Systemic lupus erythematosus	Aldesleukin	16	Recruiting	NCT03312335	II	Clinicaltrials.gov. Low-dose interleukin-2 for treatment of systemic lupus erythematosus (Charact-IL-2). - Study results - Clin [online] Available https://clinicaltrials.gov/ct2/show/NCT03312335

58	Systemic Lupus Erythematosus	AMG 592	29	Recruiting	NCT03451422	I	Clinicaltrials.gov. Safety and efficacy of AMG 592 in subjects with active systemic lupus erythematosus. - Study results - Clin [online] Available https://clinicaltrials.gov/ct2/show/NCT03451422
59	Systemic Lupus Erythematosus	Anifrolumab	36	Completed	NCT02962960	II	Clinicaltrials.gov. A study to characterize the pharmacokinetics, pharmacodynamics, and safety of anifrolumab in adult type I interferon test high systemic lupus erythematosus subject with active skin manifestations. - Study results - Clin [online] Available https://clinicaltrials.gov/ct2/show/NCT02962960
60	Systemic Lupus Erythematosus	Belimumab + Rituximab	292	Active, not recruiting	NCT03312907	III	Clinicaltrials.gov. A study to evaluate the efficacy and safety of belimumab administered in combination with rituximab to adult subjects with systemic lupus erythematosus (SLE) - BLISS-BELIEVE. - Study results - Clin [online] Available https://clinicaltrials.gov/ct2/show/NCT03312907

(continued)

Table 4.4 (continued)

Sr no.	Autoimmune disease	Intervention	Number of patients	Recruitment status	Clinical trial registry identifier	Phase	Ref.
61	Systemic Lupus Erythematosus	Belimumab	20	Completed	NCT02880852	I	Clinicaltrials.gov. Belimumab phase i study in Chinese subjects with systemic lupus erythematosus. - Study results - Clin [online] Available https://clinicaltrials.gov/ct2/show/NCT02880852
62	Systemic Lupus Erythematosus	Belimumab	30	Recruiting	NCT03747159	II	Clinicaltrials.gov. Synergetic B-cell immunomodulation in SLE - 2nd study. (SynBioSe-2). - Study results - Clin [online] Available https://clinicaltrials.gov/ct2/show/NCT03747159
63	Systemic Lupus Erythematosus	Benlysta	600	Recruiting	NCT03370263	III	Clinicaltrials.gov. BENLYSTA® special drug use investigation. - Study results - Clin [online] Available https://clinicaltrials.gov/ct2/show/NCT03370263
64	Systemic Lupus Erythematosus	BOS161721	143	Recruiting	NCT03371251	I, II	Clinicaltrials.gov. Study of BOS161721 in systemic lupus erythematosus (sle) patients on a background of limited standard of care. - Study results - Clin [online] Available https://clinicaltrials.gov/ct2/show/NCT03371251

65	Systemic Lupus Erythematosus	ILT-101	100	Completed	NCT02955615	II	Clinicaltrials.gov. ILT-101 in patients with active moderate to severe systemic lupus erythematosus (SLE) (LUPIL-2). - Study results - Clin [online] Available https://clinicaltrials.gov/ct2/show/NCT02955615
66	Systemic Lupus Erythematosus	NKTR-358	48	Recruiting	NCT03556007	I	Clinicaltrials.gov. A study of NKTR-358 in participants with systemic lupus erythematosis (SLE). - Study results - Clin [online] Available https://clinicaltrials.gov/ct2/show/NCT03556007
67	Systemic Lupus Erythematosus Arthritis	Rituximab	30	Recruiting	NCT03054259	II	Clinicaltrials.gov. Rituximab objective outcome measures trial in SLE (ROOTS). - Study results - Clin [online] Available https://clinicaltrials.gov/ct2/show/NCT03054259

(continued)

Table 4.4 (continued)

Sr no.	Autoimmune disease	Intervention	Number of patients	Recruitment status	Clinical trial registry identifier	Phase	Ref.
68	Systemic Lupus Erythematosus	TAK-079	24	Recruiting	NCT03724916	I	Clinicaltrials.gov. A study to evaluate the safety, pharmacokinetics (PK), and pharmacodynamics (PD) of TAK-079 in combination with standard background therapy in participants with moderate to severe systemic lupus erythematosus (SLE). - Study results - Clin [online] Available https://clinicaltrials.gov/ct2/show/NCT03724916
69	Systemic Lupus Erythematosus	Ustekinumab	516	Recruiting	NCT03517722	III	Clinicaltrials.gov. A study of ustekinumab in participants with active systemic lupus erythematosus. - Study results - Clin [online] Available https://clinicaltrials.gov/ct2/show/NCT03517722
70	Systemic Lupus Erythematosus	VAY736, CFZ533	120	Recruiting	NCT03656562	II	Clinicaltrials.gov. Study the efficacy and safety of VAY736 and CFZ533 in SLE patients. - Study results - Clin [online] Available https://clinicaltrials.gov/ct2/show/NCT03656562

Multiple diseases							
71	Crohn's disease, Inflammatory bowel disease, Multiple sclerosis, Rheumatoid arthritis, Sjogren's syndrome, Systemic lupus erythematosus	Nivolumab	264	Recruiting	NCT03816345	I	Clinicaltrials.gov. Nivolumab in treating patients with autoimmune disorders or advanced, metastatic, or unresectable cancer. - Study results - Clin [online] Available https://clinicaltrials.gov/ct2/show/NCT03816345
72	Systemic lupus erythematosus, Rheumatoid arthritis	PF-06835375	112	Recruiting	NCT03334851	I	Clinicaltrials.gov. Safety and tolerability study Of PF-06835375 in subjects with seropositive systemic lupus erythematosus or rheumatoid arthritis. - Study results - Clin [online] Available https://clinicaltrials.gov/ct2/show/NCT03334851
73	Systemic lupus erythematosus, Sjogren's syndrome	VIB7734	32	Recruiting	NCT03817424	I	Clinicaltrials.gov. A study to evaluate VIB7734 in participants with systemic lupus erythematosus (SLE), cutaneous lupus erythematosus (CLE), Sjogren's syndrome, systemic sclerosis, polymyositis, and dermatomyositis. - Study results - Clin [online] Available https://clinicaltrials.gov/ct2/show/NCT03817424

biosimilars which have equivalent efficacy and safety profile as soon as the patent for branded biologics expire, which will no doubt have a positive economic impact with immense cost savings for the patients (Berns and Hommes 2016). When Severs et al. through COIN study compared the cost burden of biologic drugs and biosimilars it was found that in a Dutch IBD population of 85,400 patients [equaling 507 patients per 100,000 inhabitants, 55% UC and 45% CD patients], there was a cost saving of €493 million with biosimilars which is almost a reduction of 28% in the total healthcare costs which also covered the costs of IBD-specific hospitalizations, outpatient clinic visits, and surgeries of the inhabitants (Severs et al. 2016).

Multiple Sclerosis

Hartung et al. reported that the first generation of disease modifying therapies (DMTs) for MS in 1993 costs the patient \$8000 to \$11,000, but with the entry of new DMTs the prices increased up to \$60,000 in the year 2013 because they are prescription drugs which were affected by medical inflation in the United States (Hartung et al. 2015). Hartung et al. and Chen et al. found out that this amount went up to \$70,000 in the United states in the year 2017, which was due to an increase in the costs of patient care facilities (Chen et al. 2017; Hartung 2017). Similarly, the annual costs for MS in European countries like Spain and France were around €30,050 and €38,100, respectively, on a per patient basis in the year 2017 (Fernández et al. 2017; Kobelt et al. 2017).

Psoriasis

The economic burden associated with psoriasis is significant and it increases even further as the disease progresses from moderate to severe (Al Sawah et al. 2017; Augustin et al. 2017). The United States in the year 2013 estimated an overall expense of \$11,498 was paid by an individual patient of psoriasis throughout his treatment (Brezinski et al. 2015; Vanderpuye-Orgle et al. 2015). Meanwhile in European countries like the United Kingdom, France, Germany, Spain, and Italy, the cost of the treatment was in the range of US\$2077–13,132 (Augustin et al. 2017; Burgos-Pol et al. 2016).

Rheumatoid Arthritis

RA, a systemic autoimmune disorder accounts for economic burden in the range of €2.0 billion per year in European nations which included direct costs of the biologics and indirect costs of the various services and maintenance for the early rapidly progressing RA (ERPRA) patients (Mennini et al. 2017; Cross et al. 2014). A similar approach of using biosimilars of these biologics will result in decreased economic burden in the future (Gulácsi et al. 2015).

Systemic Lupus Erythematosus

SLE is a multisystem autoimmune disease that could potentially lead to serious organ complications and even death with its incidence being as low as 0.3–31.5 cases per 100,000 individuals every year (Carter et al. 2016; Sebastiani et al. 2016). The annual cost for hospitalization was US\$51,808.41 per patient in Rochester city of New York state for an approximate stay of 8.5 days in the hospital (Anandarajah et al. 2017). Meacock et al. concluded that in a randomized population in the United States, the mean annual treatment cost was in the range of US$2239–$35,540 until the year 2010 (Meacock et al. 2013).

4.12 Conclusion and Future Perspectives

There has been great progress and renewed interest in immunotherapies for autoimmune diseases. As discussed, various modalities ranging from immune checkpoint blockade to anti-T-cell therapy or anti-B-cell therapy, amongst others are in current use. This has seen the emergence of very many novel immunotherapies which we have delineated earlier. Of interest, there are a sizeable number of immunotherapies for autoimmune diseases in early and advanced stages of clinical trials and recent trends indicate that their number is only going to increase. Although this is laudable to provide access to better care to suffering patients, one needs to also balance the financial toxicity of these immunotherapies which can oftentimes defeat the very purpose of providing healthcare to those who need it most. It is recommended that all stakeholders from discovery scientists to clinicians to health management organizations and insurance providers as also lay members of the public be brought on the same page to appreciate all these "facets" of immunotherapies to make it a successful and go-to healthcare therapy for autoimmune diseases in the future that will benefit a greater proportion of patients.

References

Agius M, Klodowska-Duda G, Maciejowski M, Potemkowski A, Sweeny S, Li J et al (2015) Safety and tolerability of MEDI-551 in patients with relapsing forms of multiple sclerosis: results from a phase 1 randomised, placebo-controlled, escalating intravenous and subcutaneous dose study. Mult Scler 23(11):235–236

Al Sawah S, Foster SA, Goldblum OM, Malatestinic WN, Zhu B, Shi N et al (2017) Healthcare costs in psoriasis and psoriasis sub-groups over time following psoriasis diagnosis. J Med Econ 20(9):982–990

Anandarajah AP, Luc M, Ritchlin CT (2017) Hospitalization of patients with systemic lupus erythematosus is a major cause of direct and indirect healthcare costs. Lupus 26(7):756–761

Astrakhantseva IV, Efimov GA, Drutskaya MS, Kruglov AANS (2014) Modern anti-cytokine therapy of autoimmune diseases. Biochemist 79(12):1308–1321

Augustin M, Vietri J, Tian H, Gilloteau I (2017) Incremental burden of cardiovascular comorbidity and psoriatic arthritis among adults with moderate-to-severe psoriasis in five European countries. J Eur Acad Dermatol Venereol 31(8):1316–1323

Avrameas S, Selmi C (2013) Natural autoantibodies in the physiology and pathophysiology of the immune system. J Autoimmun 41:46–49

Bai XF, Li HL, Shi FD, Liu JQ, Xiao BG, Van Der Meide PH et al (1998) Complexities of applying nasal tolerance induction as a therapy for ongoing relapsing experimental autoimmune encephalomyelitis (EAE) in DA rats. Clin Exp Immunol 111(1):205–210

Ballow M (2014) Mechanisms of immune regulation by IVIG. Curr Opin Allergy Clin Immunol 14 (6):509–515

Barberá A, Lorenzo N, Garrido G, Mazola Y, Falcón V, Torres AM et al (2013) APL-1, an altered peptide ligand derived from human heat-shock protein 60, selectively induces apoptosis in activated CD4 + CD25 + T cells from peripheral blood of rheumatoid arthritis patients. Int Immunopharmacol 17(4):1075–1083

Berns M, Hommes DW (2016) Anti-TNF- therapies for the treatment of Crohns disease: the past, present and future. Expert Opin Investig Drugs 25:129–143

Bielekova B, Goodwin B, Richert N, Cortese I, Kondo T, Afshar G et al (2000) Encephalitogenic potential of the myelin basic protein peptide (amino acids 83-99) in multiple sclerosis: results of a phase II clinical trial with an altered peptide ligand. Nat Med 6(10):1167–1175

Blair HA, Duggan ST (2018) Belimumab: a review in systemic lupus erythematosus. Drugs 78 (3):355–366

Bluestone JA, Bour-Jordan H (2012) Current and future immunomodulation strategies to restore tolerance in autoimmune diseases. Cold Spring Harb Perspect Biol 4(11):1–23

Bluestone JA, Bour-Jordan H (2019) Current and future immunomodulation strategies to restore tolerance in autoimmune diseases. Cold Spring Harb Perspect Biol 4(11):a007542

Böhm M, Luger TA, Schneider M, Schwarz T, Kuhn A (2006) New insight into immunosuppression and treatment of autoimmune diseases. Clin Exp Rheumatol 24(1):S67

Brezinski EA, Dhillon JS, Armstrong AW (2015) Economic burden of psoriasis in the United States a systematic review. JAMA Dermatol 151:651–658

Bucktrout SL, Bluestone JA, Ramsdell F (2018) Recent advances in immunotherapies: from infection and autoimmunity, to cancer, and back again. Genome Med 10(1):79

Burgos-Pol R, Martínez-Sesmero JM, Ventura-Cerdá JM, Elías I, Caloto MT, Casado MÁ (2016) The cost of psoriasis and psoriatic arthritis in 5 European countries: a systematic review. Actas DermoSifiliogr 107(7):577–590. (English Ed)

Carswell EA, Old LJ, Kassel RL, Green S, Fiore N, Williamson B (1975) An endotoxin-induced serum factor that causes necrosis of tumors. Proc Natl Acad Sci 72(9):3666–3670

Carter EE, Barr SG, Clarke AE (2016) The global burden of SLE: prevalence, health disparities and socioeconomic impact. Nat Rev Rheumatol 12:605–620

Chaillous L, Lefèvre H, Thivolet C, Boitard C, Lahlou N, Atlan-Gepner C et al (2000) Oral insulin administration and residual β-cell function in recent-onset type 1 diabetes: a multicentre randomised controlled trial. Lancet 356(9229):545–549

Chandrashekara S (2012) The treatment strategies of autoimmune disease may need a different approach from conventional protocol: a review. Indian J Pharmacol 44(6):665

Chatenoud L (2010) Immune therapy for type 1 diabetes mellituswhat is unique about anti-CD3 antibodies? Nat Rev Endocrinol 6(3):149–157

Chen L, Flies DB (2013) Molecular mechanisms of T cell co-stimulation and co-inhibition. Nat Rev Immunol 13(4):227–242

Chen DS, Mellman I (2013) Oncology meets immunology: the cancer-immunity cycle. Immunity 39(1):1–10

Chen AY, Chonghasawat AO, Leadholm KL (2017) Multiple sclerosis: frequency, cost, and economic burden in the United States. J Clin Neurosci 45:180–186

Cohen JA, Barkhof F, Comi G, Hartung HP, Khatri BO, Montalban X et al (2010) Oral fingolimod or intramuscular interferon for relapsing multiple sclerosis. N Engl J Med 362(5):402–415

Cross M, Smith E, Hoy D, Carmona L, Wolfe F, Vos T et al (2014) The global burden of rheumatoid arthritis: estimates from the Global Burden of Disease 2010 study. Ann Rheum Dis 73(7):1316–1322

De Miguel-Luken MJ, Mansinho A, Boni V, Calvo E (2017) Immunotherapy-based combinations: current status and perspectives. Curr Opin Oncol 29(5):382–394

Derfuss T, Kuhle J, Lindberg R, Kappos L (2013) Natalizumab therapy for multiple sclerosis. Semin Neurol 33(1):26–36

Diabetes Prevention Trial--Type 1 Diabetes Study Group (2002) Effects of insulin in relatives of patients with type 1 diabetes mellitus. N Engl J Med 346(22):1685–1691

Dinarello CA (1996) Biologic basis for interleukin-1 in disease. Blood 87(6):2095–2147

Dominguez MDC, Lorenzo N, Barbera A, Darrasse-Jeze G, Hernández MV, Torres A et al (2011) An altered peptide ligand corresponding to a novel epitope from heat-shock protein 60 induces regulatory T cells and suppresses pathogenic response in an animal model of adjuvant-induced arthritis. Autoimmunity 44(6):471–482

Dubertret L, Sterry W, Bos JD, Chimenti S, Shumack S, Larsen CG et al (2006) CLinical experience acquired with the efalizumab (Raptiva®) (CLEAR) trial in patients with moderate-to-severe plaque psoriasis: results from a phase III international randomized, placebo-controlled trial. Br J Dermatol 155(1):170–181

Ermann J, Fathman CG (2001) Autoimmune diseases: genes, bugs and failed regulation. Nat Immunol 2(9):759–761

Feldmann M, Steinman L (2005) Design of effective immunotherapy for human autoimmunity. Nature 435(7042):612

Fernández O, Calleja-Hernández MA, Meca-Lallana J, Oreja-Guevara C, Polanco A, Pérez-Alcántara F (2017) Estimate of the cost of multiple sclerosis in Spain by literature review. Expert Rev Pharmacoecon Outcomes Res 17:321–333

Fierabracci A (2011) Peptide immunotherapies in type 1 diabetes: lessons from animal models. Curr Med Chem 18(4):577–586

Fife BT, Pauken KE (2011) The role of the PD-1 pathway in autoimmunity and peripheral tolerance. Ann N Y Acad Sci 1217(1):45–59

Francisco LM, Salinas VH, Brown KE, Vanguri VK, Freeman GJ, Kuchroo VK et al (2009) PD-L1 regulates the development, maintenance, and function of induced regulatory T cells. J Exp Med 206(13):3015–3029

Francisco LM, Sage PT, Sharpe AH (2010) The PD-1 pathway in tolerance and autoimmunity. Immunol Rev 236:219–242

Ganapathy S, Vaishnavi Vedam, Vini Rajeev RA. Autoimmune disorders–immunopathogenesis and potential therapies. J Young Pharm 2017;9(1):14–22

Genovese MC, Becker JC, Schiff M, Luggen M, Sherrer Y, Kremer J et al (2005) Abatacept for rheumatoid arthritis refractory to tumor necrosis factor α inhibition. N Engl J Med 353 (11):1114–1123

Gong Z, Pan L, Le Y, Liu Q, Zhou M, Xing W et al (2010) Glutamic acid decarboxylase epitope protects against autoimmune diabetes through activation of Th2 immune response and induction of possible regulatory mechanism. Vaccine 28(24):4052–4058

Grigore A, Inform A (2017) Plant phenolic compounds as immunomodulatory agents. Phenolic Compd Act London, UK IntechOpen. (8):75–98

Guillevin L, Pagnoux C, Karras A, Khouatra C, Aumaître O, Cohen P et al (2014) Rituximab versus azathioprine for maintenance in ANCA-associated vasculitis. N Engl J Med 371(19):1771–1780

Gulácsi L, Brodszky V, Baji P, Kim HU, Kim SY, Cho YY et al (2015) Biosimilars for the management of rheumatoid arthritis: economic considerations. Expert Rev Clin Immunol 11: S43–S52

Harrison LC, Honeyman MC, Steele CE, Stone NL, Sarugeri E, Bonifacio E et al (2004) Pancreatic β-cell function and immune responses to insulin after administration of intranasal insulin to humans at risk for type 1 diabetes. Diabetes Care 27(10):2348–2355

Hartung HP (2009) New cases of progressive multifocal leukoencephalopathy after treatment with natalizumab. Lancet Neurol 8:28–31

Hartung DM (2017) Economics and cost-effectiveness of multiple sclerosis therapies in the USA. Neurotherapeutics 14:1018–1026

Hartung DM, Bourdette DN, Whitham RH (2015) The cost of multiple sclerosis drugs in the US and the pharmaceutical industry: too big to fail? Author response. Neurology 84(21):2185–2192

Hauselmann H (1998) Can collagen type II sustain a methotrexate-induced therapeutic effect in patients with long-standing rheumatoid arthritis? A double-blind, randomized trial. Rheumatology 37(10):1110–1117

Hauser SL, Bar-Or A, Comi G, Giovannoni G, Hartung HP, Hemmer B et al (2017) Ocrelizumab versus interferon beta-1a in relapsing multiple sclerosis. N Engl J Med 376(3):221–234

He XS, Gershwin ME, Ansari AA (2017) Checkpoint-based immunotherapy for autoimmune diseases – opportunities and challenges. J Autoimmun 79:1–3

Heinrich PC, Behrmann I, Müller-Newen G, Schaper F, Graeve L (1998) Interleukin-6-type cytokine signalling through the gp130/Jak/STAT pathway. Biochem J 334:297–314

Hirsch DL, Ponda P (2014) Antigen-based immunotherapy for autoimmune disease: current status. Immunotargets Ther 4:1–11

Hofmann K, Clauder AK, Manz RA (2018) Targeting B cells and plasma cells in autoimmune diseases. Front Immunol 23(9):835

Hogquist KA, Baldwin TA, Jameson SC (2005) Central tolerance: learning self-control in the thymus. Nat Rev Immunol 5:772–782

Huang Z, Yang B, Shi Y, Cai B, Li Y, Feng W et al (2012) Anti-TNF-α therapy improves Treg and suppresses Teff in patients with rheumatoid arthritis. Cell Immunol 279(1):25–29

Imbach P, D'Apuzzo V, Hirt A, Rossi E, Vest M, Barandun S et al (1981) High-dose intravenous gammaglobulin for idiopathic thrombocytopenic purpura in childhood. Lancet 317 (8232):1228–1231

Janikashvili N, Samson M, Magen E, Chikovani T (2016) Immunotherapeutic targeting in autoimmune diseases. Mediat Inflamm 2016:2–4

Joly P, Maho-Vaillant M, Prost-Squarcioni C, Hebert V, Houivet E, Calbo S et al (2017) First-line rituximab combined with short-term prednisone versus prednisone alone for the treatment of pemphigus (Ritux 3): a prospective, multicentre, parallel-group, open-label randomised trial. Lancet 389(10083):2031–2040

Juryńczyk M, Walczak A, Jurewicz A, Jesionek-Kupnicka D, Szczepanik M, Selmaj K (2010) Immune regulation of multiple sclerosis by transdermally applied myelin peptides. Ann Neurol 68(5):593–601

Kaliyaperumal A, Michaels MA, Datta SK (1999) Antigen-specific therapy of murine lupus nephritis using nucleosomal peptides: tolerance spreading impairs pathogenic function of autoimmune T and B cells. J Immunol 162(10):5775–5783

Kappos L, Comi G, Panitch H, Oger J, Antel J, Conlon P et al (2000) Induction of a non-encephalitogenic type 2 T helper-cell autoimmune response in multiple sclerosis after administration of an altered peptide ligand in a placebo-controlled, randomized phase II trial. Nat Med 6(10):1176–1182

Kappos L, Radue EW, O'Connor P, Polman C, Hohlfeld R, Calabresi P et al (2010) A placebo-controlled trial of oral fingolimod in relapsing multiple sclerosis. N Engl J Med 362(5):387–401

Kil LP, De Bruijn MJW, Van Nimwegen M, Corneth OBJ, Van Hamburg JP, Dingjan GM et al (2012) Btk levels set the threshold for B-cell activation and negative selection of autoreactive B cells in mice. Blood 119(16):3744–3756

Kim WU, Lee WK, Ryoo JW, Kim SH, Kim J, Youn J et al (2002) Suppression of collagen-induced arthritis by single administration of poly(lactic-co-glycolic acid) nanoparticles entrapping type II collagen: a novel treatment strategy for induction of oral tolerance. Arthritis Rheum 46 (4):1109–1120

Kimura A, Kishimoto T (2010) IL-6: Regulator of Treg/Th17 balance. Eur J Immunol 40:1830–1835

Kobelt G, Thompson A, Berg J, Gannedahl M, Eriksson J (2017) New insights into the burden and costs of multiple sclerosis in Europe. Mult Scler J 23(8):1123–1136

Koffeman EC, Genovese M, Amox D, Keogh E, Santana E, Matteson EL et al (2009) Epitope-specific immunotherapy of rheumatoid arthritis: clinical responsiveness occurs with immune deviation and relies on the expression of a cluster of molecules associated with T cell tolerance in a double-blind, placebo-controlled, pilot phase II trial. Arthritis Rheum 60(11):3207–3216

Lerner A, Jeremias P, Matthias T (2015) The world incidence and prevalence of autoimmune diseases is increasing. Int J Celiac Dis 3(4):151–155

Linsley PS, Greene JL, Brady W, Bajorath J, Ledbetter JA, Peach R (1994) Human B7-1 (CD80) and B7-2 (CD86) bind with similar avidities but distinct kinetics to CD28 and CTLA-4 receptors. Immunity 1(9):793–801

Ludvigsson J, Faresjö M, Hjorth M, Axelsson S, Chéramy M, Pihl M et al (2008) GAD treatment and insulin secretion in recent-onset type 1 diabetes. N Engl J Med 359(18):1909–1920
Ludvigsson J, Krisky D, Casas R, Battelino T, Castaño L, Greening J et al (2012) GAD65 antigen therapy in recently diagnosed type 1 diabetes mellitus. N Engl J Med 366(5):433–442
Macleod MK, Anderton SM (2015) Antigen-based immunotherapy (AIT) for autoimmune and allergic disease. Curr Opin Pharmacol 23:11–16
Mayer L, Kaser A, Blumberg RS (2012) Dead on arrival: understanding the failure of CTLA4-immunoglobulin therapy in inflammatory bowel disease. Gastroenterology 143:13–17
McKinney EF, Lee JC, Jayne DRW, Lyons PA, Smith KGC (2015) T-cell exhaustion, co-stimulation and clinical outcome in autoimmunity and infection. Nature 523(7562):612–616
Meacock R, Dale N, Harrison MJ (2013) The humanistic and economic burden of systemic lupus erythematosus: a systematic review. PharmacoEconomics 31:49–61
Mennini FS, Marcellusi A, Gitto L, Iannone F (2017) Economic burden of rheumatoid arthritis in Italy: possible consequences on anti-citrullinated protein antibody-positive patients. Clin Drug Investig 37(4):375–386
Merrill JT, Burgos-Vargas R, Westhovens R, Chalmers A, D'Cruz D, Wallace DJ et al (2010) The efficacy and safety of abatacept in patients with non-life-threatening manifestations of systemic lupus erythematosus: results of a twelve-month, multicenter, exploratory, phase IIb, randomized, double-blind, placebo-controlled trial. Arthritis Rheum 62(10):3077–3087
Metzler B, Wraith DC (1993) Inhibition of experimental autoimmune encephalomyelitis by inhalation but not oral administration of the encephalitogenic peptide: influence of MHC binding affinity. Int Immunol 5(9):1159–1165
Miklos D, Cutler CS, Arora M, Waller EK, Jagasia M, Pusic I et al (2017) Ibrutinib for chronic graft-versus-host disease after failure of prior therapy. Blood 130(21):2243–2250
Miller SD, Turley DM, Podojil JR (2007) Antigen-specific tolerance strategies for the prevention and treatment of autoimmune disease. Nat Rev Immunol 7(9):665–677
Monneaux F, Lozano JM, Patarroyo ME, Briand JP, Muller S (2003) T cell recognition and therapeutic effect of a phosphorylated synthetic peptide of the 70K snRNP protein administered in MRL/lpr mice. Eur J Immunol 33:287–296
Muller S, Monneaux F, Schal N, Rashkov RK, Oparanov BA, Wiesel P et al (2008) Spliceosomal peptide P140 for immunotherapy of systemic lupus erythematosus: results of an early phase II clinical trial. Arthritis Rheum 58(12):3873–3883
Musette P, Bouaziz JD (2018) B cell modulation strategies in autoimmune diseases: new concepts. Front Immunol 9:1–5
Nagler-Anderson C, Bober LA, Robinson ME, Siskind GWTG (1986) Suppression of type II collagen-induced arthritis by intragastric administration of soluble type II collagen. Proc Natl Acad Sci U S A 83(19):7443–7446
Nagy G, Huszthy PC, Fossum E, Konttinen Y, Nakken BSP (2015) Selected aspects in the pathogenesis of autoimmune diseases. Mediators Inflamm 2015
Näntö-Salonen K, Kupila A, Simell S, Siljander H, Salonsaari T, Hekkala A et al (2008) Nasal insulin to prevent type 1 diabetes in children with HLA genotypes and autoantibodies conferring increased risk of disease: a double-blind, randomised controlled trial. Lancet 372 (9651):1746–1755
Nipp RD, Sonet EM, Guy GP (2018) Communicating the financial burden of treatment with patients. Am Soc Clin Oncol Educ B 38:524–531
Ochando JC, Yopp AC, Yang Y, Garin A, Li Y, Boros P et al (2005) Lymph node occupancy is required for the peripheral development of alloantigen-specific Foxp3 + regulatory T cells. J Immunol 174(11):6993–7005
Orban T, Bundy B, Becker DJ, DiMeglio LA, Gitelman SE, Goland R et al (2011) Co-stimulation modulation with abatacept in patients with recent-onset type 1 diabetes: a randomised, double-blind, placebo-controlled trial. Lancet 378(9789):412–419
Ostrov BE (2015) Immunotherapeutic biologic agents in autoimmune and autoinflammatory diseases. Immunol Investig 44(8):777–802

Pers YM, Jorgensen C (2016) Perspectives of ofatumumab as CD20 targeted therapy in rheumatoid arthritis and other autoimmune diseases. Immunotherapy 8:1091–1096

Pozzilli P, Pitocco D, Visalli N, Cavallo MG, Buzzetti R, Crinò A et al (2000) No effect of oral insulin on residual beta-cell function in recent-onset type I diabetes (the IMDIAB VII). Diabetologia 43(8):1000–1004

Prakken BJ, Samodal R, Le TD, Giannoni F, Yung GP, Scavulli J et al (2004) Epitope-specific immunotherapy induces immune deviation of proinflammatory T cells in rheumatoid arthritis. Proc Natl Acad Sci 101(12):4228–4233

Puentes F, Dickhaut K, Hofstätter M, Falk K, Rötzschke O (2013) Active suppression induced by repetitive self-epitopes protects against EAE development. PLoS One 8(5):e64888

Raz I, Ziegler AG, Linn T, Schernthaner G, Bonnici F, Distiller LA et al (2014) Treatment of recent-onset type 1 diabetic patients with DiaPep277: results of a double-blind, placebo-controlled, randomized phase 3 trial. Diabetes Care 37:1392–1400

Regueiro M, Kip KE, Baidoo L, Swoger JM, Schraut W (2014) Postoperative therapy with infliximab prevents long-term Crohn's disease recurrence. Clin Gastroenterol Hepatol 12 (9):1494–1502

Rosenblum MD, Remedios KA, Abbas AK (2015) Mechanisms of human autoimmunity. J Clin Invest 125(6):2228–2233

Rosman Z, Shoenfeld Y, Zandman-Goddard G (2013) Biologic therapy for autoimmune diseases: an update. BMC Med 11:88–100

Ruperto N, Lovell DJ, Quartier P, Paz E, Rubio-Pérez N, Silva CA et al (2008) Abatacept in children with juvenile idiopathic arthritis: a randomised, double-blind, placebo-controlled withdrawal trial. Lancet 372(9636):383–391

Salinas GF, Braza F, Brouard S, Tak PP, Baeten D (2013) The role of B lymphocytes in the progression from autoimmunity to autoimmune disease. Clin Immunol 146:34–45

Scalapino KJ, Daikh DI (2008) CTLA-4: A key regulatory point in the control of autoimmune disease. Immunol Rev 223:143–155

Sebastiani GD, Prevete I, Iuliano A, Minisola G (2016) The importance of an early diagnosis in systemic lupus erythematosus. Isr Med Assoc J 18(3–4):212–215

Severs M, Oldenburg B, Van Bodegraven AA, Siersema PD, Mangen MJ, Initiative of Crohn's and Colitis (2016) The economic impact of the introduction of biosimilars in inflammatory bowel disease. J Crohns Colitis 11(3):289–296

Smilek DE, Ehlers MR, Nepom GT (2014) Restoring the balance: immunotherapeutic combinations for autoimmune disease. Dis Model Mech 7(5):503–513

Smyth MJ, Teng MW (2018) 2018 Nobel Prize in physiology or medicine. Clin Transl Immunol 7 (10):e1041

Stone KD, Prussin C, Metcalfe DD (2010) IgE, mast cells, basophils, and eosinophils. J Allergy Clin Immunol 125(Suppl 2):S73–S80

Sugiyama H, McCormick TS, Cooper KD, Korman NJ (2008) Alefacept in the treatment of psoriasis. Clin Dermatol 26(5):503–508

Tanaka Y, Mola EM (2014) IL-6 targeting compared to TNF targeting in rheumatoid arthritis: studies of olokizumab, sarilumaband sirukumab. Ann Rheum Dis 73:1595–1597

Thompson HSG, Harper N, Bevan DJ, Staines NA (1993) Suppression of collagen induced arthritis by oral administration of type II collagen: changes in immune and arthritic responses mediated by active peripheral suppression. Autoimmunity 16(3):189–199

Tian J, Clare-Salzler M, Herschenfeld A, Middleton B, Newman D, Mueller R et al (1996) Modulating autoimmune responses to GAD inhibits disease progression and prolongs islet graft survival in diabetes-prone mice. Nat Med 2(12):1348–1353

Tisch R, Liblau RS, Yang XD, Liblau P, McDevitt HO (1998) Induction of GAD65-specific regulatory T-cells inhibits ongoing autoimmune diabetes in nonobese diabetic mice. Diabetes 47(6):894–899

Vanderpuye-Orgle J, Zhao Y, Lu J, Shrestha A, Sexton A, Seabury S, et al (2015) Evaluating the economic burden of psoriasis in the United States. J Am Acad Dermatol 72(6):961–967.e5

Velasquez MP, Bonifant CL, Gottschalk S (2018) Redirecting T cells to hematological malignancies with bispecific antibodies. Blood 131:30–38

Verazza S, Negro G, Marafon D, Consolaro A, Martini A, Ravelli A (2013) Possible discontinuation of therapies after clinical remission in juvenile idiopathic arthritis. Clin Exp Rheumatol 31 (Suppl 78):S98–S101

Viswanath D (2013) Understanding autoimmune diseases- a review. IOSR J Dent Med Sci 6 (6):08–15

Walczak A, Siger M, Ciach A, Szczepanik M, Selmaj K (2013) Transdermal application of myelin peptides in multiple sclerosis treatment. JAMA Neurol 70(9):1105–1109

Wallace DJ, Gordon C, Strand V, Hobbs K, Petri M, Kalunian K et al (2013) Efficacy and safety of epratuzumab in patients with moderate/severe flaring systemic lupus erythematosus: results from two randomized, double-blind, placebo-controlled, multicentre studies (ALLEVIATE) and follow-up. Rheumatology (United Kingdom) 52(7):1313–1322

Walter M, Philotheou A, Bonnici F, Ziegler AG, Jimenez R (2009) No effect of the altered peptide ligand NBI-6024 on β-cell residual function and insulin needs in new-onset type 1 diabetes. Diabetes Care 32(11):2036–2040

Wang H, Yang J, Jin L, Feng J, Lu Y, Sun Y et al (2009) Immunotherapy of autoimmune diabetes by nasal administration of tandem glutamic acid decarboxylase 65 peptides. Immunol Investig 38(8):690–703

Wang L, Wang FS, Chang C, Gershwin ME (2014) Breach of tolerance: primary biliary cirrhosis. Semin Liver Dis 34(3):297–317

Wang L, Wang FS, Gershwin ME (2015) Human autoimmune diseases: a comprehensive update. J Intern Med 278(4):369–395

Warren KG, Catz I, Wucherpfennig KW (1997) Tolerance induction to myelin basic protein by intravenous synthetic peptides containing epitope P85 VVHFFKNIVTP96 in chronic progressive multiple sclerosis. J Neurol Sci 152(1):31–38

Warrington R, Watson W, Kim HL, Antonetti FR (2011) An introduction to immunology and immunopathology. Allergy, Asthma Clin Immunol 7(S1):1–8

Weiner HL, Mackin GA, Matsui M, Orav EJ, Khoury SJ, Dawson DM et al (1993) Double-blind pilot trial of oral tolerization with myelin antigens in multiple sclerosis. Science 259 (5099):1321–1324

Wherrett DK, Bundy B, Becker DJ, Dimeglio LA, Gitelman SE, Goland R et al (2011) Antigen-based therapy with glutamic acid decarboxylase (GAD) vaccine in patients with recent-onset type 1 diabetes: a randomised double-blind trial. Lancet 378(9788):319–327

Wraith DC (2017) The future of immunotherapy: a 20-year perspective. Front Immunol 8:1668

Wucherpfennig KW (2001) Mechanisms for the induction of autoimmunity by infectious agents. J Clin Invest 108:1097–1104

Ylisaukko-Oja T, Torvinen S, Ventola H, Schmidt S, Herrala S, Kononoff J et al (2019) Healthcare resource utilization and treatment costs of Finnish chronic inflammatory bowel disease patients treated with infliximab. Scand J Gastroenterol 14:1–7

Yoshino S (1995) Antigen-induced arthritis in rats is suppressed by the inducing antigen administered orally before, but not after immunization. Cell Immunol 163(1):55–58

Yu C, Gershwin ME, Chang C (2014) Diagnostic criteria for systemic lupus erythematosus: a critical review. J Autoimmun 48–49:10–13

Zaprutko T, Kopciuch D, Kus K, Merks P, Nowicka M, Augustyniak I et al (2017) Affordability of medicines in the European Union. PLoS One 12(2):e0172753

Zhang Q, Vignali DAA (2016) Co-stimulatory and co-inhibitory pathways in autoimmunity. Immunity 44(5):1034–1051

Zheng MK, Shih DQ, Chen GC (2017) Insights on the use of biosimilars in the treatment of inflammatory bowel disease. World J Gastroenterol 23:1932–1943

Immunotherapy in Neurodegenerative Disorders

5

Dipanjali Kamthe, Netra Gosavi, and Vandana S. Nikam

Abstract

Neurodegenerative disorders are characterized by progressive accumulation of misfolded proteins that eventually advance to the death of a selected population of neurons impairing cognitive function. Neurodegenerative diseases like Alzheimer's disease, Parkinson's disease, and multiple sclerosis have distinct pathophysiology. Initially, therapeutic strategies aimed at resolving the protein aggregates accumulated in the brain or resolving brain inflammation. The immune cells in the brain were considered a threat to neurodegenerative diseases due to the presence of inflammation. Therapies targeted at misfolded proteins or immune cells were failed at large. With the advancement in the profound understanding of neurodegenerative diseases and the discovery of the brain lymphatic system, the cross-talk between the brain and immune cells considered essential, and the inflammation in the brain is considered a normal cellular process. It was proposed that the brain immune system needs assistance from peripheral immune cells during the progression of neurodegenerative disease, and thus immunotherapy emerged as a new therapeutic approach. The number of preclinical and clinical studies harnessing the power of immunotherapy showed promising results. However, these strategies require further refinement with a combinatorial therapeutic approach to curb the progression of neurodegenerative diseases.

Keywords

Alzheimer's disease · Parkinson's disease · Tau neurofibrillary tangles · α-Synuclein · Spinal muscular atrophy · Amyloid precursor protein

D. Kamthe · N. Gosavi · V. S. Nikam (✉)
Department of Pharmacology, STES's Smt. Kashibai Navale College of Pharmacy, S. P. Pune University, Pune, Maharashtra, India

S. P. Sawarkar et al. (eds.), *Immunotherapy – A Novel Facet of Modern Therapeutics*,
https://doi.org/10.1007/978-981-15-9038-2_5

5.1 Introduction

Neurons are a functional unit of the nervous system, which consists of a central nervous system and peripheral nervous system. The central nervous system (CNS) comprises of the brain, spinal cord, whereas the peripheral system consists of the somatic, autonomic, and enteric nervous system. Neurodegenerative diseases are progressive degeneration of neurons, eventually resulting in their death that produces incurable and debilitating conditions characterized by the presence of misfolded protein causing inflammation, oxidative stress in the CNS and ultimately leading to ataxias and dementias. Out of all the neurodegenerative disorders, Parkinson's and Alzheimer's are the most common neurodegenerative disorders affecting most of the people worldwide. These disorders occur due to the loss of neurons in the brain and peripheral nervous system. The damage to the motor neuron causes problems with movement, the process of acquiring thoughts, knowledge, thinking ability, behavior, balance, and talking (Braak and Braak 1991; Selkoe 2004; Dobson 2003; Taylor et al. 2002).

The neurodegenerative diseases include Alzheimer's disease, Parkinson's disease, Prion disease, and many other acute and chronic neurodegenerative diseases having distinct pathophysiology. These diseases are characterized by the accumulation of misfolded protein aggregates that begins with synaptic damages, leading to loss of neuronal population (Selkoe 2004; Taylor et al. 2002). There are more than 30 proteins that cause protein aggregation advancing to neurodegenerative diseases. Alzheimer's disease was first identified in 1906 with two distinct features of beta-amyloid plaque and Tau neurofibrillary tangles (NFT) resulting in impairment in memory with dementia, which also affects mental functioning, a form of dementia (Bondi et al. 2017). Alzheimer's disease (AD) is caused by the progressive generation of senile plaque with neurofibrillary tangles (NFT) in the cerebral cortex, neurons, and synapse. The extracellular beta-amyloid and intracellular hyperphosphorylated tau proteins deposition cause neuronal dysfunction in AD along with reduced acetylcholine in the presynapse (Ravi and Hemachandra 2016).

After Alzheimer's disease, Parkinson's disease (PD) is the second most common neurodegenerative disease that leads to impairment in the body movements due to the degeneration of motor and non-motor neurons. It is characterized by two major events, i.e., loss of dopamine neurons, and accumulation of protein aggregates. The protein aggregates are the synaptic protein, α-synuclein, which misfold, accumulate, and spread trans-cellularly throughout the brains (Mhyre et al. 2012).

Prion disease is featured by various fatal and transmissible neurodegenerative diseases. Prion disease also is known as transmissible spongiform encephalopathy or TSEs which is caused by an infectious agent called a prion. Prion is derived from a misfolded normal host protein known as prion protein. Basically, prion protein, itself is not pathogenic. It resides on the surface of the cell type and protects the brain from damage. When normal prion protein molecules change its shape, form clump together in brain tissue, convert to the infectious prion and cause prion disease. Thus, an infectious, abnormally shaped and aggregated prion proteins are responsible for prion disease (Poggiolini et al. 2013).

Motor neuron diseases (MND) are also called Lou Gehrig's disease, amyotrophic lateral sclerosis, or ALS. These are diseases where in the neurons controlling muscles tend to degenerate, leading to loss of muscle control and finally paralysis. The weakness in the hands, legs, shoulder and grip, slurred speech, a tendency to trip, cramps, and muscles twitching are the main symptoms of MND (Statland et al. 2015).

Huntington's disease (HD) is caused by a genetic disorder (inherited) having problems with both movement and mental functioning. It also causes uncontrolled movements, emotional problems, and loss of thinking ability (cognition). This disease is caused by a genetic defect on chromosome 4 affecting neurons of the caudate nucleus and putamen. The first symptoms of Huntington's are behavioral changes which include antisocial behavior, irritability, moodiness, restlessness or impatience, depression, involuntary movements, poor coordination, trouble in learning new information or making decision, trouble in walking, speaking, and swallowing (Nopoulos 2016).

Spinal muscular atrophy (SMA) is a genetic disorder caused and characterized by a loss of motor neurons that control muscle movement. The primary symptoms of SMA are weakness and wasting (atrophy) of muscles used for movement (skeletal muscles). The weakness in the muscles of the center of the body (proximal) is prominent compared to muscles away from the center (distal). The severity of muscle weakness usually worsens with age. The SMA is classified into different types, based on the age of onset and severity of muscle weakness (Arnold et al. 2015).

Multiple sclerosis is a chronic inflammatory, the demyelinating disease that damages the white matter covering the axons and leads to motor weakness, visual impairment, diplopia, dysarthria, ataxia, and cognitive deficits. Multiple sclerosis can affect different body systems; each person may experience the disease in a specific way (Goldenberg 2012).

Autism spectrum disorder (ASD) is a brain disorder with learning and social disabilities. Biochemical and molecular characteristics of ASD include oxidative stress, activated astrocyte and microglia, neuronal loss, elevated levels of 8-oxo-guanosine, and development of pro-inflammatory cytokines. Children with ASD show skin allergies and different behavioral patterns under stress. Neurotoxicity and focal brain inflammation in ASD are caused by the activation of neurotensin, which stimulates the release of corticotrophin-releasing hormone and activates microglia and mast cells (Faras et al. 2010).

5.2 Neurodegenerative Diseases: Pathophysiology and Causes

Some proteins change their conformations lead to toxicity or loss of their physiological functions. The misfolded proteins (like amyloid protein, tau protein, prion protein) generate small oligomeric or large fibrillary aggregates, which get accumulated in the brain resulting in neurodegeneration (Mroczko et al. 2019). Other factors leading to neurodegeneration include deficient neurotransmitter synthesis, the presence of reactive oxygen species, and defective protein degradation machinery (ubiquitination, proteasome, etc.). The blood-brain barrier (BBB) is a protective, selective barrier shielding the brain from the external milieu. The damage to BBB also leads to various CNS-related disorders. Structural and functional changes or dysfunction in the BBB tends to cause inflammation due to the entry of immune cells, their mediators, and altering the homeostasis of the brain environment. Under normal conditions, autophagy and ubiquitin-proteasome system take care of any misfolded proteins resulting from the complex process of translation and posttranslational modification. The injury or an insult to this machinery impairs the abnormal protein degradation process, and they get accumulated as aggregates in the brain. The formation of abnormal proteins is the hallmark of neurodegenerative diseases. For example, hyperphosphorylated tau proteins (microtubule-associated proteins) are found in AD. Similarly, an abnormal aggregation of α-synuclein protein is the hallmark of PD, whereas abnormal long polyglutamine (Poly Q) leads to Huntington's disease.

Elevated anti-brain protein autoantibodies, high anxiety, increased oxidative stress, and food intolerance, altered levels of reduced glutathione, sulfation, and methylation are the pathological basis for ASD. Acetylcholinesterase (AchE) and butyrylcholinesterase (BchE) both lead to the breakdown of acetylcholine in the synaptic region. It lowers the acetylcholine level and causes age-related disorders with loss of cognitive ability.

The formation of reactive oxygen species (ROS) is another factor for neurodegeneration. An enzyme, COX-1, surrounds amyloids plaque in microglia, and the accumulation of COX-1 enzyme in microglia increases prostaglandin synthesis and local reactive oxidative species (Mahmood et al. 2014).

Most of the neurodegenerative disorders are genetic (hereditary), and other causes include alcoholism, a tumor, a stroke, toxins, chemicals, and viruses.

Some of the neuropeptides, like FMRFamide-related peptides and dynorphins, increases the activity of acid-sensing ion channels (ASICs). Dynorphins are highly expressed in CNS. ASICs are proton-gated cation channels involved in excitatory synaptic transmission and get activated during acidic pH fluctuations. Due to ischemia, autoimmune inflammation, or injury, ASICs become active, leading to pathological acidosis that induces neuronal death or degeneration (Vick and Askwith 2015).

The high content of iron in the brain causes injury to the brain. The release of iron leads to oxidative stress through the formation of oxygen free radical, which is involved in lipid peroxidation and results in membrane fluidity and neuronal cell death (Aaron et al. 2017).

5.3 Milestones in the Treatment of Neurodegenerative Diseases

With the discovery of Alzheimer's disease in 1906, several attempts were made to lower the levels of accumulated proteins in neurodegenerative diseases; however, these approaches failed to curb the progression of the disease (Table 5.1) (Mullard 2016; Masters et al. 2015; Selkoe and Hardy 2016; Panza et al. 2019).

Table 5.1 Different class of drugs used to treat Alzheimer's disease

Class of the drugs with example	Mechanism of action
Anticholinesterase inhibitors Tacrine, donepezil, rivastigmine, galantamine Huperzine	These drugs inhibit cholinesterase enzyme eventually inhibiting the degradation of acetylcholine in the brain and help to improve the memory. They have reversible. Selective cholinesterase inhibition action.
β-secretase inhibitors Verubecestat, solanezumab, lanabecestat	β-secretase enzyme converts amyloid precursor protein into sAPPβ which further leads to aggregation of Aβ plaque. These drugs inhibit the formation of sAPPβ and further Aβ40 and Aβ42 production.
A-secretase activators/modulators Deprenyl	As α-secretase does not produce sAPPβ and Aβ40, Aβ42. Thus increase in α-secretase compete with β-secretase for binding and inhibits binding of β-secretase with sAPPβ
M1 muscarinic agonist Talsaclidine is selective muscarinin M1 agonist	It increases α-secretase activity which inhibits the formation of amyloid plaque. M1 agonist acts through decrease γ-secretase activity and increases α-secretase activity. It decreases Aβ secretion and tau phosphorylation
Aβ-aggregation inhibitor Tramiprosate	It inhibits the aggregation of amyloid plaque by binding with the amyloid plaque.
Aβ-degrading enzymes Neprilysin (NEP), insulin-degrading enzyme (IDE), plasmin, endothelin converting enzyme (ECE) 1 and 2, and angiotensin-converting enzyme (ACE)	These enzyme degrade the amyloid plaque and inhibit its aggregation with amyloid plaque. They also prevent the transport of amyloid protein in and out of the brain
Apolipoprotein E (ApoE) promotes Aβ clearance—bexarotene	It acts indirectly by activating astrocyte which leads to the degradation of amyloid protein and thus improves memory Also activates the nuclear receptor.
Drugs development based on the metals hypothesis—clioquinol, a metal-protein-attenuating compound. Other metal chelators including XH1, DP-109, PBT2	Metal chelators are used as a treatment to AD as its chelate many metals like copper, zinc, iron which are involved in the pathophysiological events of AD. Thus they decrease the concentration of plasma amyloid protein and improve memory in AD patients.
HMG-CoA reductase inhibitor-lovastatin	Decreases level of plasma Aβ protein
Monoamine oxidase inhibitor Deprenyl MAO-B inhibitor rasagiline Ladostigil	It increases the level of dopamine in the brain by inhibiting MAO enzyme. It acts by inhibiting acetylcholinesterase as well as MAO enzymes thus it shows improvement in memory in AD patients. It shows dual activity like rasagiline, i.e., inhibiting cholinesterase as well as MAO enzymes
Prevention of phosphorylation of tau Tau kinase inhibitors could be used as an anti-	When tau protein gets hyperphosphorylated, it forms paired helical filaments and leads to

(continued)

Table 5.1 (continued)

Class of the drugs with example	Mechanism of action
AD treatment. Phosphorylation of tau is controlled by different kinases and phosphatases. Examples—protein phosphatase (PP)-2A, Cyclin-dependent kinase-5 (CDK5), glycogen synthase kinase (GSK)-3β	prevention or impairment in axonal transport. These enzymes prevent either phosphorylation of tau protein or increases the dephosphorylation of tau protein
Prevention of the aggregation of tau Phenothiazines, anthraquinones, polyphenols, thiacarbocyanine dyes, N-phenylamines, thiazolyl-hydrazides, rhodanines, quinoxalines	Neurofibrillary tangles form when the soluble tau protein converted into insoluble aggregates in filamentous form. These drugs prevent the conversion of soluble tau to insoluble form and inhibit aggregation.
Heat shock proteins	During the process of hyperphosphorylation of tau protein if it gets misfolded, it leads to aggregation or deposition. This protein inhibits depositions by activation of chaperon protein and helps to bind the microtubules with tau protein.
Non-steroidal anti-inflammatory drugs (NSAIDs)—ibuprofen, sulindac, flurbiprofen, indomethacin, and diclofenac N-methyl-D-aspartate receptor (NMDA) antagonist	Prolonged intake of NSAIDs decreases the incidence of AD. It improves memory by modifying NMDA receptor activity.
Other pharmacological therapies in clinical AD-estrogen Nicotine Docosahexaenoic acid (DHA) Melatonin Clioquinol Resveratrol, a red wine polyphenol	It increases beta-amyloid degradation by activating metalloproteinase-2 and 9. It decreases neuronal loss and also releases Ach. It acts as anti-amyloid, antioxidant and neuroprotective. It is given with donepezil to improve memory. It acts as a metal chelator. It acts as antioxidant and anti-inflammatory. It inhibits aggregation of amyloid protein by its scavenging property.

The extensive studies of neurodegenerative diseases led to the conclusion that acute and chronic neurodegenerative diseases are associated with inflammation, and hence anti-inflammatory drugs like steroids, non-steroidal anti-inflammatory drugs (NSAIDs) were used to treat these diseases (Table 5.1) (Heneka et al. 2015; Meyer et al. 2019). The strategies used to treat neurodegenerative diseases at large failed to attenuate the progression of the disease that led the scientific community to revise the perspective about brain pathophysiology, especially neurodegenerative diseases (Wang et al. 2017; Stower 2018).

Traditionally CNS is considered as an immune-privileged organ due to lack of primary immune response and limited expression of major histocompatibility complex (Medawar 1948).

There was a prevalent notion that immune cells are detrimental to CNS due to its anatomical structure guarded by barriers and the presence of inflammation in neuropathology. However, the number of evidence through preclinical and clinical findings revisited this view and proposed that it is unlikely that CNS kept away from the power of immune cells. The new discussions and suggestions about the mechanisms that maintain and repair brain injury might be insufficient and need assistance from the immune system during the disease state. This view emerged with a focus on cross-talk between the immune system and the brain. The study by Song E. J. group showed that the cytokine Th2 released by CD4+ T cells is crucial in maintaining the hippocampal neurogenesis and cognitive function (Song et al. 2017). Another study reported that evoking adaptive immunity with Bacillus Calmette-Guérin (BCG) vaccine and enriched environment resulted in the activation and recruitment of T cells in meninges in response to systemic interferon-gamma (IFN-γ). The activated T cells stimulated macrophage M2 polarization and neurotrophic factor expression, which in turn promoted neurogenesis, synaptic plasticity, and cognitive function (Qi et al. 2017). Several reports suggested that adaptive and innate immune cells play an important role in maintaining, repairing, protecting, and healing function in CNS, provided their recruitment, activity, and functions are well controlled across its gateway. There was a paradigm shift for the treatment of neurodegenerative disease by harnessing the power of immune cells, i.e., immunotherapy.

5.4 Immunotherapy in Neurodegenerative Diseases

The neurodegenerative diseases were not considered to be treated with immunotherapy for many years because of the traditional views about CNS pathophysiology. First, it was considered an immune-privileged site, and the second, vaccination is thought for infectious diseases. With the advancement and deeper insight into CNS systems such as cross-talk between the immune system and CNS, the recent discovery of the brain lymphatic drainage system (Louveau et al. 2015; Aspelund et al. 2015) proposed to utilize the benefits of immunotherapy in neurodegenerative diseases.

The striking findings from Schenk and the group (Schenk et al. 1999) demonstrating reduced amyloid aggregates load in transgenic Alzheimer mice (TgPDAPP mice) with active immunization was a turning point for immunotherapy in neurodegenerative disease (Fig. 5.1).

The studies by Janus et al., and Morgan et al. showed improvement in cognition and memory in transgenic mice immunized with Ab peptides and reduced protein aggregate load in CNS (Janus et al. 2000; Nicoll et al. 2003). Thirty AD patients undergone Aβ immunization over period of 1 year showed attenuated cognition decline (Hock et al. 2003). The neuropathological examination of patients undergone immunization with Ab peptide showed the neocortex region devoid of Ab aggregates (Wilcock and Colton 2008). However, the phase III clinical trial with Ab peptide (AN1792) was indefinitely suspended due to detrimental brain inflammation linked with immunization. The clinical trials with Ab peptide failed to reach the clinical endpoints, and hence immunotherapy with tau protein started.

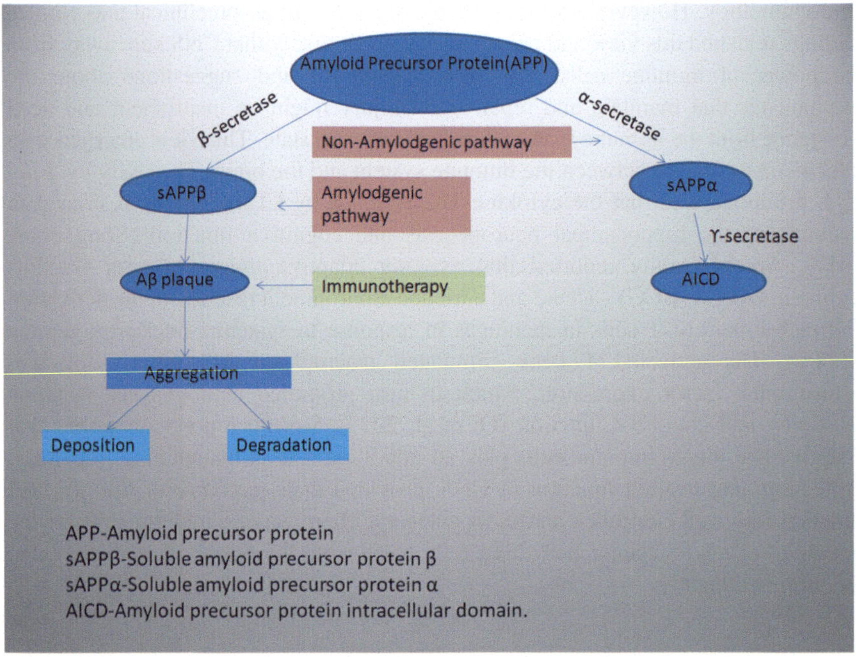

Fig. 5.1 Amyloid beta (Aβ) cascade in Alzheimer disease and immunotherapy

5.5 Mechanisms of Immunotherapy

The first proposed mechanism of immunotherapy for the treatment of AD using iodinated anti-Aβ antibodies demonstrated the role of activated macrophage, and thus anti-Aβ immunotherapy leads to a reduction in Aβ deposition. When anti-Aβ antibodies enter the blood circulation, they bind to the circulating free amyloid and activate macrophages for phagocytosis. This binding between antigen and antibody triggers the infiltration of monocyte and macrophages.

Anti- Aβ antibodies also bind to the Aβ plaque in the CNS to form a complex. This process activates phagocytosis by Fc-gamma receptor-bearing macrophages. The activation of macrophages leads to the digestion of Aβ plaque or the export of complex from CNS (Fig. 5.2).

Other mechanism includes the generation of peripheral sink condition. As shown in Fig. 5.3, circulating antibodies bind to free Aβ aggregates present in the blood, which leads to a difference in the concentration of free Aβ across the blood-brain barrier. This process leads to the efflux of free Aβ from the brain to the circulation due to reduced free concentration of Aβ in blood circulation.

In the third approach, the antibodies modify the secondary structure of fibrillar Aβ plaque. Aβ monomers form aggregates associated with oligomeric or fibrillary

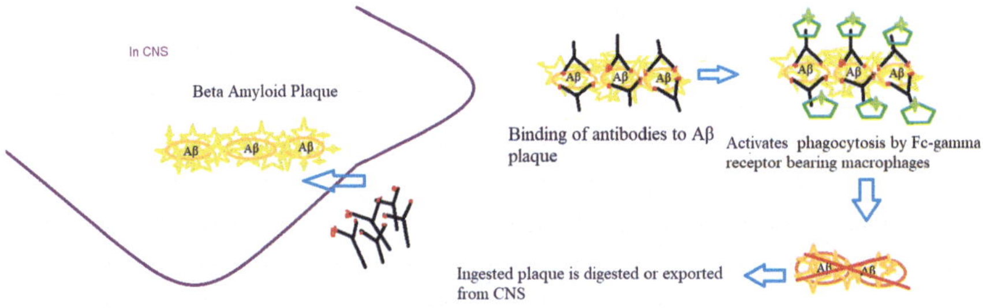

Fig. 5.2 Phagocytosis of beta-amyloid plaque in CNS by activated macrophages

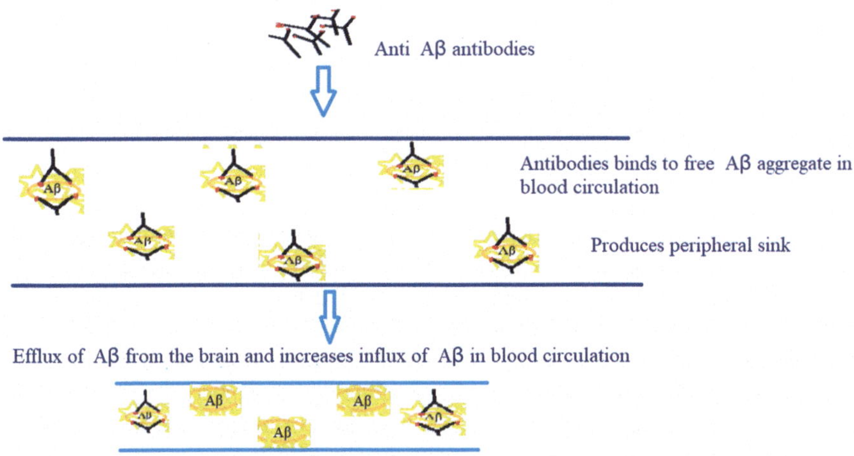

Fig. 5.3 Efflux of amyloid plaque from CNS due to peripheral sink creation

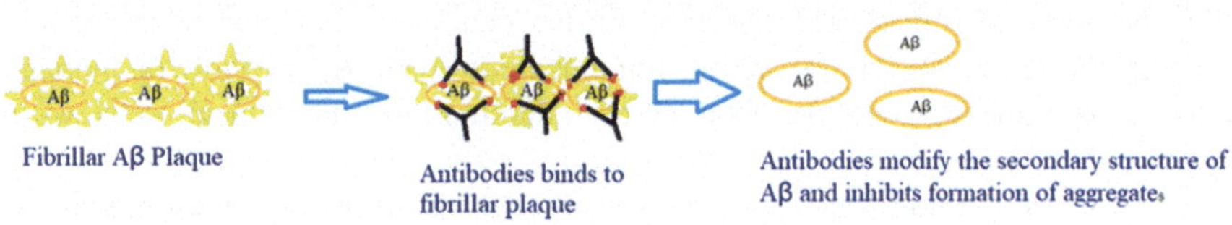

Fig. 5.4 Reduction in aggregates by modification of fibrillar Aβ plaque

forms, and antibodies prevent Aβ fibril formation, thus inhibit their aggregation (Fig. 5.4).

5.6 Immunotherapy for AD: Active and Passive Immunizations

The type of therapy given in AD depends on the cellular or extracellular locations of toxic protein aggregates.

Table 5.2 Antibodies used to treat AD

Name of antibody	Epitopes	Nature of antibody
BAN2401	At AB protofibrillar	Humanized
Crenezumab (mAb)	At conformation epitopes including oligomeric forms and AB protofibrillar	Humanized
GSK 933776	At N-terminal	Humanized IgG1
SAR228810	At protofibrillar and Abs with low molecular weight	Humanized
Bapineuzumab (mAb)	At N-terminal	Humanized
Gantenerumab	At AB central part and n-terminal	Humanized
Solanezumab (mAb)	At central part (amino acids 16 to 24)	
AAB-003	At N-terminal	Fc-engineered Bapineuzumab
BIIB037/ BART	At insoluble fibrillary human AB	Humanized

Active immunization in AD depends upon the complex formation of antigen with strong immune-activating adjuvant, and it leads to long-lasting antibodies production. Active immunization stimulates B cell and T cell to produce antigen-specific antibodies by activation of both humoral and cellular immune systems (Table 5.2). Active immunization shows a large polyclonal antibody response in which many antibodies target multiple epitopes on single antigens (Vandenberghe et al. 2016). In the case of passive immunotherapy, direct injections of monoclonal antibodies are given into the host without raising an immune response.

Active immunizations with aggregated Aβ peptide in disease before or after plaque deposition showed a beneficial effect on plaque pathology and an even more pronounced impact was observed on the behavioral outcome in preclinical settings. The success rate of active immunization with Aβ peptide in humans demonstrated the episodes of fatal brain inflammation and clinical trials were suspended.

The use of peptide vaccines activates T-cells and prevents autoimmune or inflammatory reactions. Examples of these peptide vaccines are- CAD106, ACC001, and Affitope.

The phase 2b study with CAD106 peptide vaccine with 121 AD patients showed well tolerability and antibody response. It also stimulates the production of the antibodies that target the tau protein in the brain (Serrano-Pozo et al. 2011).

Passive immunizations with monoclonal antibodies targeting Aβ reduced plaque pathology and improved the behavioral outcome in animal models (Table 5.2). However, the beneficial effect in animal models did not reflect on humans. The presence of Aβ plaque in the human brain with no dementia denies the causal relationship between Aβ plaque and neurodegeneration. Even though the presence of Aβ plaque in the human brain is a cause of neurotoxicity, and cannot be ignored.

Another pathological hallmark of AD is tau protein, initially identified as NFT in 1998 by Barak and his group (Braak and Braak 1991). The extensive investigation about tau proteins revealed a causal relationship between the spatial distribution of tau in the brain and the extent of cognitive dysfunction and memory loss, unlike Aβ aggregates (Arriagada et al. 1992; Giannakopoulos et al. 2003). Furthermore, with recent studies, the amyloid cascade hypothesis was revised with the addition of tau protein in the cascade, and the convergence of these two cascades results in neurodegeneration. It has been suggested that a reduction in tau protein complements the reduction in Aβ aggregates (Bloom 2014).

5.7 Multiple Sclerosis: Pathology and Immunotherapy

Multiple sclerosis is an autoimmune disease (the body creates antibodies against itself, causing damage). Multiple sclerosis can affect different body systems, with each person may experience the disease in a different way. Depending upon the progression of disease, there are four main types of multiple sclerosis.

1. *Preclinical Stage*: it shows inflammation and damage to the myelin sheath, detected only by using magnetic resonance imaging (MRI).
2. *Relapsing-Remitting MS*: produces attacks due to neuronal dysfunction, mostly follows a predictable pattern with periods in which the condition become worsening and then improve.
3. *Secondary Progressive MS*: among halt of the people with relapsing–remitting stage develop secondary progressive MS. In that, patients may still experience relapses, followed by partial recoveries or stage of remission, but the disorder does not cure between cycles. Instead, it gradually worsens.
4. *Primary Progressive MS*: characterized by slow and constantly progression of the disease with no remission stage. About 10–15% of people have this type of MS. As delineated in Table 5.3, there are different mechanisms of MS ranging from autoimmunity to infection.

The breakdown of the blood-brain barrier, demyelination, multifocal inflammation, oligodendrocytes loss, and axonal degeneration are major pathological processes observed in MS. Based on the presence or absence of complement and immunoglobulins, apoptotic nuclei, and preferential loss of myelin protein, different patterns of MS have been described. This heterogeneity in the lesion pattern is observed between patients, but not within patients. Active lesions are common in relapsing MS patients but become rare during progressive MS (Popescu et al. 2013).

Treatment of Multiple Sclerosis (MS) A widely used approach for treating MS is preventing the entry of immune cells (T cell, lymphocytes, monocytes) in the brain using antibodies.

Natalizumab (Tysabri) and Alemtuzumab (Campath-1H) are the US Food and drug administration (US FDA) approved monoclonal antibodies that are used for the treatment of multiple sclerosis.

Table 5.3 Mechanisms of multiple sclerosis (Baecher-Allan et al. 2018)

Feature	Underlying mechanism
Autoimmunity	Autoreactive T cells are present in normal and pathological conditions and they protect the brain. However, pathogenic Th17- and Th1-type and CD8 myelin autoreactive T cells cause MS
Genetics	Defective MHC genes are responsible for MS. MHC genes control immune cells to supply and responsible for regulatory and tolerance mechanisms
Infection	Various infectious agents produce myelin-reactive pathogenic T cells. The cross-reactivity occurs with CNS myelin antigens which initiate autoreactive immune system or a self-limited infection of the brain, release myelin antigens
B cells	B cells are of two types, mainly pro- and anti-inflammatory. B cells play a central role in MS. Like T cells, there are pro- and anti-inflammatory B cell subsets. In relapsing MS, B cells, drive pathogenic T cells and in progressive MS, B cells increase CNS activity mainly through lymphoid follicles and secreted factors
Microbiome	Microbiome regulates T cell function in the whole body. It contains protective and pathogenic microbial components, which secrete various metabolites and initiate immune set points
Environment	Environmental factors can also responsible for the development of MS and it occurs due to low vitamin D level, smoking, obesity, and lack of UV radiation exposure
Autoantigen	In multiple sclerosis there is spreading of reaction to other organ-specific antigen, hence there was no any single autoantigen to target, so that antigen-specific therapy is not effective to suppress MS or be used as a prophylactic in risk subjects

Natalizumab (Tysabri) (Hutchinson 2007) is a humanized monoclonal antibody that binds with focal adhesion molecule α-4. The focal adhesion molecule α-4 present on T lymphocytes and other immune cells and integrin α 4 facilitate entry of T lymphocytes and other immune cells across BBB at the site of CNS lesion. Blocking the integrin α 4 with antibody prevents the entry of immune cells in the brain and thereby reduces the progression of the disease.

Alemtuzumab (Campath-1H) (Li et al. 2018) acts on an antigen (CD52) present on monocytes and lymphocytes. These antibodies suppress the peripheral immune system and its effect lasts for a year. The phase III clinical trials with RRMS patients showed attenuated relapse rate compared to placebo groups, albeit no improvement on the degree of disability (Lavery et al. 2014). It has drawbacks that it may develop autoimmune thyroid problems and idiopathic thrombocytopenic purpura.

Number of antibodies that are not approved by the US FDA, are used for the treatment of MS. Rituximab (Rituxan) (Felli et al. 2014) was primarily used for the diseases associated with the overproduction of B cells. The B cells express phosphorylated glycoprotein, CD20, and Rituximab binds to it and destroys B cells. It is an alternative treatment for relapsing and progressive MS.

Daclizumab (Zenepax) (Martin 2012) is primarily used as an immunosuppressant to prevent organ rejection. It binds to the interleukin 2 receptor (IL-2), which is expressed by lymphocytes, platelets, neuroblastomas, lymphocytes present in tumors. It eliminates CD4+ and CD8+ T cells population by increasing CD56+

natural killer cells. The patients with multiple sclerosis treated with Daclizumab in combination with B-interferon or as monotherapy showed improved outcome and clinical scores without any fatal side effects (George and Brundin 2015).

Ocrelizumab (Sorensen and Blinkenberg 2016) is next-generation anti CD20 antibody used to deplete circulating B cells rather than plasma cells. It induced apoptosis and eradicate B lymphocytes through complement and antibody-dependent cytotoxicity. It was discontinued in phase III trial for systemic lupus eryhthematous patients due to severe infections after methotrexate exposure.

Ofatumumab (Arzerra) (Barth and Czuczman 2013) is another anti-CD20 human monoclonal antibody used to treat MS. Earlier Ofatumumab in combination with chlorambucil was approved by US FDA for the treatment of chronic lymphocytic leukemia (CLL). It causes $\geq$90% reduction in new T1 gadolinium-enhancing lesions for all doses of Ofatumumab $\geq$30 mg.

5.8 Off Label Use of Polyclonal Intravenous Immunoglobulin's (Irving) for the Treatment of Remitting Multiple Sclerosis (RRMS) Patients

Intravenous Immunoglobulin's (IVIgG) (Afonso and João 2016) is a plasma preparation obtained from two fractions, i.e., 20% from blood donors and 80% from plasma donors. IVIgG is composed of intact IgG molecules along with the varying percentage of subclasses of IgG. It consists of antibodies against a wide spectrum of antigens such as bacterial, viral, and a small quantity of neutralizing antibodies. The number of donors used for pooling the IVIgG is the manufacture's proprietary information. However, large numbers of donors indicate a wide range of individual antibodies to particular antigens, and thus IVIgG constitutes a gamut of reactive IgG of human sera that can target all sorts of foreign antigens as well as a limited set of self-antigens. It is very important to have limited self-reactivity of normal serum IgG with dominant self-antigen to establish self-tolerance.

Infused IVIgG has a half-life of 21 days. IVIgG modulates body's innate and adaptive immunity by affecting the various biological process and thus, infused IVIgG act as immunomodulatory therapy. The numbers of mechanisms by which IVIgG function had been proposed. It causes blockage of Fc receptors present on phagocytes and thereby preventing inflammatory pathways mediated by the Fc receptor present on phagocytes. IVIgG contains antibodies against variable regions of autoantibodies, called an anti-idiopathic antibody. These antibodies bind to the variable region of autoantibodies and this interaction further inhibits production of autoantibodies. IVIgG also attenuates the production of inflammatory cytokines, activation of pathogenic T cells, block the complement system involved in inflammation and membrane attack complex formation. IVIgG therapy at therapeutic dose mediates the balance between Th1 and Th2 cells, thereby maintains the balance between T cells and B cells expansion, activation (Jacob and Rajabally 2009).

IVIgG was licensed for use in immune thrombolytic purpura (ITP). It is also used as replacement therapy for immunodeficiencies disease-like X linked agammaglobulinemia, acquired hypogammaglobulinemia, X–linked hyperimmunoglobulin, severe combined immunodeficiency, HIV infection. Several other disorders such as dermatomyositis, graft-versus-host disease in recipients of allogenic bone marrow transplants and treatments of cellular rejection after organ transplantation, are treated with IVIgGs (Jolles et al. 2005).

5.9 Immunotherapy in PD: Alpha-Synuclein and Parkinson's Disease

Parkinson's disease (PD), mainly caused due to misfolded α-synuclein (α-syn), which is the main component of Lewy bodies and neuritis. Several studies demonstrated that misfolded α-syn is the causative factor for the pathogenesis of PD, in both rare genetic and common idiopathic forms of the disease. The presence of mutations (A53T, A30P, E46K, and H50Q) in the α-synuclein gene increases the risk of PD. The molecular form of α-syn responsible for neuronal toxicity is still unknown in PD (Stefanis 2012).

Recent research found that oligomeric small aggregates of α-syn initiate neuronal death; however, the fibrillar form of α-syn proteins, observed microscopically as Lewy pathology, might be responsible for the pathogenesis of PD. Braak et al. group proposed six pathological steps involved in the spreading of disease from one brain region to another. The research findings reported in 2008 could help to explain it in detail. The study revealed that α-syn proteins released from neurons leak into the extracellular space and interact with neighboring cells inducing aggregation of endogenous proteins (Fields et al. 2019; Rietdijk et al. 2017).

5.10 Neuroinflammation and Glia Activation in Parkinson's Disease

The glia cells constitute a large proportion of brain cells (approximately 10% of all cells) and differentiate to microglia, astrocytes, and oligocytes. Microglia and astrocytes are the resident immune cells that are involved in the repair and maintenance of the brain under normal physiology. In neurodegenerative diseases like in PD, these cells get activated as a primary defensive response, implicating their role in the pathogenesis of PD (Chitnis and Weiner 2017). The microglia cells are an integral part of the innate immune system of CNS. The CNS milieu disturbances caused by different stimuli such as mechanical stress, a neurotoxin, misfolded proteins, transform microglia into pro-inflammatory neurotoxic phenotype cells or anti-inflammatory neuroprotective phenotype cells referred to M1 or M2 type. The presence of activated microglia in a region like basal ganglia in the substantia nigra was detected surrounding the degenerating dopaminergic neurons (Zella and Metzdorf 2019). PD is also associated with certain HLA-variants, and treatment with non-steroidal anti-inflammatory drugs reduces the risk of developing PD (Bartels and Leenders 2010). The study with transgenic mice overexpressing α-syn protein showed that α-syn proteins activate microglia as assessed by the levels of pro-inflammatory cytokines, nitric oxide (NO), and reactive oxygen species (ROS) level. In another study using mice overexpressing human wild type α-syn in the nigrostriatal dopamine neurons, it was observed that microglia takes up released alpha syn protein from nigrostriatal terminals (Ferreira and Romero-Ramos 2018). Activated microglia can engulf different molecular forms of α-syn proteins, as observed in cell-based assays. If the α -syn protein is present in monomeric form, the phagocytosis process is faster. The activated microglia are detrimental to dopamine neurons as α-syn protein aggregates damage dopamine neurons in the presence of activated microglia (Zhang et al. 2018). The activation of microglia and its ability to phagocyte accumulated α-syn proteins needs to be investigated. Another in vitro study showed impaired ability to engulf and degrade α-syn proteins by lipopolysaccharide-induced activated microglia. Hence, it is of paramount importance to delineate the underlying mechanism for microglia activation, their impact in pathology and clearance of protein aggregates in the context of immunotherapy (Ferreira and Romero-Ramos 2018).

Immunotherapy Targeting α-Synuclein in Parkinson's Disease The accumulation of α-syn has been found in the cerebrospinal fluid, plasma of PD patients. The preclinical studies with immunotherapies targeted at α-syn aggregate showed a progressive reduction of α-syn proteins from extracellular space (Menéndez-Gonzále and Padilla-Zambrano 2018). Active and passive immunotherapy had been explored for the treatment of PD.

Active Immunization Therapy In the first study, transgenic mice overexpressing human wild type α-syn under control of platelet-derived growth factor-β (PDFG- β) promoter used to investigate active immunization with recombinant human α-syn. They were immunized with recombinant human α-syn. The immunized mice exhibited the presence of high-affinity α-syn antibodies. Along with significant attenuation of accumulated α-syn in neurons, an increased number of synaptophysin-positive nerve terminals, as well as decreased neuronal death were observed (Amschl et al. 2013). In another study, animals were immunized with recombinant adeno-associated virus α-syn into the substantia nigra. The immunization induced activation of microglia, T cells with a decrease in α-sync inclusions in the substantia nigra (Sanchez-Guajardo et al. 2013). Biotechnology Company AFFIRis AG developed a therapeutic peptide vaccine for active immunization against PD. Peptide vaccine immunization ameliorated long term memory due to the decreased level of alpha sync aggregates in the synaptic terminal. It improved motor function, activated microglia, and elevated the anti-inflammatory cytokines. (Mandler et al. 2014).

Passive Immunization involves direct administration of antibodies against a different epitope of an antigen, and the advantage of this approach is that there is a possibility of titration of dose or termination of treatment if any adverse event appears. The C-terminus of α-syn facilitates the oligomerization of alpha sync proteins and their propagation. Thus, c- terminus is the crucial factor for the progression of the disease. The transgenic mice expressing human α-syn under the PDGF-β promoter were administered a 9E4 antibody, and it resulted in a decrease level of α-syn fragments in axons and synaptic terminals. Bae and group used a different antibody raised against a different region of the C-terminal of α-syn protein using transgenic mice overexpressing α-syn (Ghiglieri et al. 2018). Immunization with 9E4 antibody-induced activation of microglia cleared extracellular α-syn from CNS with improvement in motor function. Transgenic mice expressing human α-syn under the Thy-1 promoter immunized with different antibody (1H7, 9E4, 5D12) against C-terminus of α-syn was studied by another group showed similar results (Games et al. 2014). The study revealed that passive immunization reduced motor deficits, α-syn accumulation in glia, and neurons, as well as slowed down neurodegeneration. Recently, Prothena Biosciences Inc., in collaboration with Roche, completed a Phase 1 safety trial in normal subjects using a humanized form of the antibody 9E4 (PRX002). The phase I clinical trial data showed that the antibody is well tolerated by the subjects with the possibility of identifying the target engagement in peripheral tissue. Followed by this, in 2014, Prothena initiated a multicenter, multiple-ascending-dose trial comparing a six-month course of PRX002 in patients with idiopathic PD to assess the safety and pharmacokinetic parameters. Antibodies raised against the N-terminal and central region of α-syn protein were also evaluated (Dhillon et al. 2017). The antibody raised against the N-terminal of α-syn protein was found to be more effective among two, in an animal model. BioArctic Neuroscience AB has generated monoclonal antibodies selective for oligomeric and protofibril forms of α-syn. The chronic administration of this antibody in transgenic mice expressing human A30P α-syn under the Thy-1 promoter demonstrated a decrease in α-syn protofibrils in the spinal cord with an increase in activated microglia.

Another research group used a recombinant fusion protein consisting of single-chain fragment variable targeted at α-syn oligomers and heavy chains of immunoglobulin, called single-chain fragments variable (Shahaduzzaman et al. 2015). The recombinant single-chain fragment variable targeting α-syn oligomers was linked to the low-density lipoprotein receptor binding domain of apolipoprotein B, to facilitate the cellular clearance of α-syn and passage through the BBB. The treatment with this antibody exhibited reduced accumulation of unphosphorylated and phosphorylated α-syn in the neocortex and hippocampus with an increase in neuron numbers and synapses, and reduced levels of astrocytes.

5.11 Future Outlook: Different Immunotherpay Strategies Used to Treat Neurodegenerative Diseases

Being a complicated process of neurodegeneration and involvement of different cell types, it was proposed to target multiple mechanisms. The distinct cells display distinct functions at the brain border and within the brain. The regulatory T cells and monocytes act within the brain and effector T cells activate and traffic through brain gateway. The strategies to boost peripheral immune cells like induction of attenuating levels of systemic Fox3 + T cells (Baruch et al. 2015), blocking the

PD-1/PD-L1 immune checkpoint pathway (Rosenzweig et al. 2019; Villeda et al. 2014) or vaccination by CNS antigens were proposed to combat neurodegenerative diseases (Kunis et al. 2015).

The neurodegenerative diseases progress with age and diminished immune system. The rejuvenation of the immune system using immunotherapy in combination with other strategies of reducing the aggregate load in the brain would be a promising therapeutic strategy to combat neurodegenerative diseases.

References

Aaron DG, Paraminder D, James S (2017) Neurodegenerative disease: models, mechanisms, and a new hope. Dis Model Mech 10:499–502. https://doi.org/10.1242/dmm.030205

Afonso AFB, João CMP (2016) The production processes and biological effects of intravenous immunoglobulin. Biomol Ther 6:15. https://doi.org/10.3390/biom6010015

Amschl D, Neddens J, Havas D (2013) Time course and progression of wild typeα-Synuclein accumulation in a transgenic mouse model. BMC Neurosci 14:6

Arnold WD, Kassar D, Kissel JT (2015) Spinal muscular atrophy: diagnosis and management in a new therapeutic era. Muscle Nerve 51(2):157–167. https://doi.org/10.1002/mus.24497

Arriagada PV, Growdon JH, Hedley-Whyte ET, Hyman BT (1992) Neurofibrillary tangles but not senile plaques parallel duration and severity of Alzheimer's disease. Neurology 42:631–639

Aspelund A, Antila S, Proulx ST, Karlsen TV, Karaman S, Detmar M, Wiig H, Alitalo K (2015) A dural lymphatic vascular system that drains brain interstitial fluid and macromolecules. J Exp Med 212(7):991–999. https://doi.org/10.1084/jem.20142290

Baecher-Allan C, Kaskow BJ, Weiner HL (2018) Multiple sclerosis: mechanisms and immunotherapy. Neuron 97:742–768. https://doi.org/10.1016/j.neuron.2018.01.021

Bartels AL, Leenders KL (2010) Cyclooxygenase and neuroinflammation in Parkinson's disease neurodegeneration. Curr Neuropharmacol 8:62–68

Barth MJ, Czuczman MS (2013) Ofatumumab: a novel, fully human anti-CD20monoclonal antibody for the treatment of chronic lymphocytic leukemia. Future Oncol 9(12):1829–1839

Baruch K, Rosenzweig N, Kertser A, Deczkowska A, Sharif AM, Spinrad A, Tsitsou-Kampeli A, Sarel A, Cahalon L, Schwartz M (2015) Breaking immune tolerance by targeting Foxp3 (1) regulatory T cells mitigates Alzheimer's disease pathology. Nat Commun 6:7967

Bloom GS (2014) Amyloid-beta and tau: the trigger and bullet in Alzheimer disease pathogenesis. JAMA Neurol 71:505–508

Bondi M et al (2017) Alzheimer's disease: past, present, and future. J Int Neuropsychol Soc 23 (9–10):818–831. https://doi.org/10.1017/S135561771700100X

Braak H, Braak E (1991) Neuropathological stageing of Alzheimer-related changes. Acta Neuropathol 82:239–259

Chitnis T, Weiner HL (2017) CNS inflammation and neurodegeneration. J Clin Invest 127 (10):3577–3587. https://doi.org/10.1172/JCI90609

Dhillon J-KS, Riffe C, Moore BD (2017) A novel panel of α-synuclein antibodies reveal distinctive staining profiles in synucleinopathies. PLoS One 12(9):e0184731. https://doi.org/10.1371/journal.pone.0184731

Dobson CM (2003) Protein folding and misfolding. Nature 426:884–890

Faras H, Al Ateeqi N, Tidmarsh L (2010) Autism spectrum disorders. Ann Saudi Med 30 (4):295–300. https://doi.org/10.4103/0256-4947.65261

Felli V, Di Sibio A, Anselmi M, Gennarelli A, Sucapane P (2014) Progressive multifocal leukoencephalopathy following treatment with rituximab in an HIV-negative patient with non-Hodgkin lymphoma. Neuroradiol J 27:657–664. https://doi.org/10.15274/NRJ-2014-10087

Ferreira SA, Romero-Ramos M (2018) Microglia response during Parkinson's disease: alpha-synuclein intervention. Front Cell Neurosci 12:247. https://doi.org/10.3389/fncel.2018.00247
Fields CR, Bengoa-Vergniory N, Wade-Martins R (2019) Targeting alpha-Synuclein as a therapy for Parkinson's disease. Front Mol Neurosci 12:299. https://doi.org/10.3389/fnmol.2019.00299
Games D, Valera E, Spencer B (2014) Reducing C-terminal-truncated alpha-synuclein by immunotherapy attenuates Neurodegeneration and propagation in Parkinson's disease-like models. J Neurosci 34(28):9441–9454
George S, Brundin P (2015) Immunotherapy in Parkinson's disease: micromanaging alpha-Synuclein aggregation. J Parkinsons Dis 5:413–424. https://doi.org/10.3233/JPD-150630
Ghiglieri V, Calabrese V, Calabresi P (2018) Alpha-synuclein: from early synaptic dysfunction to neurodegeneration. Front Neurol 9:295. https://doi.org/10.3389/fneur.2018.00295
Giannakopoulos P et al (2003) Tangle and neuron numbers, but not amyloid load, predict cognitive status in Alzheimer's disease. Neurology 60:1495–1500
Goldenberg MM (2012) Multiple sclerosis review. Pharm Therapeut 37(3):175–184
Heneka MT, Carson MJ, El Khoury J, Landreth GE, Brosseron F, Feinstein DL, Jacobs AH, Wyss-Coray T, Vitorica J, Ransohoff RM et al (2015) Neuroinflammation in Alzheimer's disease. Lancet Neurol 14:388–405
Hock C, Konietzko U, Streffer JR et al (2003) Antibodies against b-amyloid slow cognitive decline in Alzheimer's disease. Neuron 38:547–554
Hutchinson M (2007) Natalizumab: a new treatment for relapsing remitting multiple sclerosis. Ther Clin Risk Manag 3(2):259–268
Jacob S, Rajabally YA (2009) Current proposed mechanisms of action of intravenous Immunoglobulins in inflammatory neuropathies. Curr Neuropharmacol 7:337–342
Janus C, Pearson J, McLaurin J et al (2000) Ab peptide immunization reduces behavioural impairment and plaques in a model of Alzheimer's disease. Nature 408:979–982
Jolles S, Sewell WAC, Misbah SA (2005) Clinical uses of intravenous immunoglobulin. Clin Exp Immunol 142:1–11
Kunis G, Baruch K, Miller O, Schwartz M (2015) Immunization with a myelin-derived antigen activates the brain's choroid plexus for recruitment of immunoregulatory cells to the CNS and attenuates disease progression in a mouse model of ALS. J Neurosci 35:6381–6393
Lavery AM, Verhey LH, Waldman AT (2014) Outcome measures in relapsing-remitting multiple sclerosis:capturing disability and disease progression in clinical trials. Mult Scler Int 2014:262350. https://doi.org/10.1155/2014/262350
Li Z, Richards S, Surks HK, Jacobs A, Panzara MA (2018) Clinical pharmacology of alemtuzumab, an anti-cd52 immunomodulator, in multiple sclerosis. Clin Exp Immunol 194(3):295–314
Louveau A, Smirnov I, Keyes TJ, Eccles JD, Rouhani SJ, David Peske J, Derecki NC, Castle D, Mandell JW, Lee KS, Harris TH, Kipnis J (2015) Structural and functional features of central nervous system lymphatic vessels. Nature 523(7560):337–341. https://doi.org/10.1038/nature14432
Mahmood R, Arif M, Muhammad SQ, Abdul M, Peter NP, Muhammad A (2014) Recent updates in the treatment of neurodegenerative disorders using natural compounds. Evid Based Complement Alternat Med 2014:979730. https://doi.org/10.1155/2014/979730
Mandler M, Valera E, Rockenstein E (2014) Next-generation active immunization approach for synucleinopathies: implications for Parkinson's disease clinical trials. Acta Neuropathol 127(6):861–879. https://doi.org/10.1007/s00401-014-1256-4
Martin R (2012) Anti-CD25 (daclizumab) monoclonal antibody therapy in relapsing–remitting multiple sclerosis. Clin Immunol 142:9–14
Masters CL, Bateman R, Blennow K, Rowe CC, Sperling RA, Cummings JL (2015) Alzheimer's disease. Nat Rev Dis Primers 1:15056
Medawar PB (1948) Immunity to homologous grafted skin; the fate of skin homografts transplanted to the brain, to subcutaneous tissue, and to the anterior chamber of the eye. Br J Exp Pathol 29:58–69

Menéndez-Gonzále M, Padilla-Zambrano HS (2018) Clearing extracellular alpha-Synuclein from cerebrospinal fluid: a new therapeutic strategy in Parkinson's disease. Brain Sci 8:52. https://doi.org/10.3390/brainsci8040052

Meyer PF, Tremblay-Mercier J, Leoutsakos J, Madjar C, Lafaille-Maignan ME, Savard P, Rosa-Neto J, Poirier PE, Breitner J, Prevent-AD Research Group (2019) INTREPAD: a randomized trial of naproxen to slow progress of presymptomatic Alzheimer disease. [Published erratum appears in 2019 Neurology 93: 371]. Neurology 92:e2070–e2080

Mhyre TR, Boyd JT, Hamill RW, Maguire-Zeiss KA (2012) Parkinson's disease. Subcell Biochem 65:389–455. https://doi.org/10.1007/978-94-007-5416-4_16

Mroczko B, Groblewska M, Litman-Zawadzka A (2019) The role of protein misfolding and tau oligomers (TauOs) in Alzheimer's disease (AD). Int J Mol Sci 20(19):4661. https://doi.org/10.3390/ijms20194661

Mullard A (2016) Pharma pumps up anti-tau Alzheimer pipeline despite first phase III failure. Nat Rev Drug Discov 15:591–592

Nicoll JAR, Wilkinson D, Holmes C, Steart P, Markham H, Weller R (2003) Neuropathology of human Alzheimer disease after immunization with amyloid-b peptide: a case report. Nat Med 9:448–452

Nopoulos PC (2016) Huntington disease: a single-gene degenerative disorder of the striatum. Dialogues Clin Neurosci 18(1):91–98

Panza F, Lozupone M, Dibello V, Greco A, Daniele A, Seripa D, Logroscino G, Imbimbo BP (2019) Are antibodies directed against amyloid-b (Ab) oligomers the last call for the Ab hypothesis of Alzheimer's disease? Immunotherapy 11:3–6

Poggiolini I, Saverioni D, Parchi P (2013) Prion protein misfolding, strains, and neurotoxicity: an update from studies on mammalian prions. Int J Cell Biol 2013:910314. https://doi.org/10.1155/2013/910314

Popescu BFG, Pirko I, Lucchinetti CF (2013) Pathology of multiple sclerosis: where do we stand. Continuum (Minneap Minn) 19(4):901–921

Qi F, Zuo Z, Yang J, Hu S, Yang Y, Yuan Q, Zou J, Guo K, Yao Z (2017) Combined effect of BCG vaccination and enriched environment promote neurogenesis and spatial cognition via a shift in meningeal macrophage M2 polarization. J Neuroinflammation 14:32

Ravi R, Hemachandra PR (2016) Amyloid-beta and phosphorylated tau accumulations cause abnormalities at synapses of Alzheimer's disease neurons. J Alzheimers Dis 57(4):1–25. https://doi.org/10.3233/JAD-160612

Rietdijk CD, Perez-Pardo P, Garssen J, van Wezel RJA (2017) Exploring Braak's hypothesis of Parkinson's disease. Front Neurol 8:1–9. https://doi.org/10.3389/fneur.2017.00037

Rosenzweig N, Dvir-Szternfeld R, Tsitsou-Kampeli A, Keren-Shaul H, Ben-Yehuda H, Weill-Raynal P, Cahalon L, Kertser A, Baruch K, Amit I et al (2019) PD-1/PD-L1 checkpoint blockade harnesses monocyte-derived macrophages to combat cognitive impairment in a tauopathy mouse model. Nat Commun 10:465

Sanchez-Guajardo V, Annibali A, Jensen PH (2013) Alpha-Synuclein vaccination prevents the accumulation of Parkinson disease Y like pathologic inclusions in striatum in association with regulatory T cell recruitment in a rat model. J Neuropathol Exp Neurol 72:7

Schenk D, Barbour R, Dunn W et al (1999) Immunization with amyloid-b attenuates Alzheimer-disease like pathology in the PDAPP mouse. Nature 400:173–177

Selkoe DJ (2004) Cell biology of protein misfolding: the examples of Alzheimer's and Parkinson's diseases. Nat Cell Biol 6:1054–1061

Selkoe DJ, Hardy J (2016) The amyloid hypothesis of Alzheimer's disease. EMBO Mol Med 8:595–608

Serrano-Pozo A, Frosch MP, Masliah E, Hyman BT (2011) Neuropathological alterations in Alzheimer disease. Cold Spring Harb Perspect Med 1:a006189

Shahaduzzaman M, Nash K, Hudson C, Sharif M, Grimmig B, Lin X et al (2015) Anti-human α-Synuclein N-terminal peptide antibody protects against dopaminergic cell death and ameliorates behavioral deficits in an AAV-α-Synuclein rat model of Parkinson'sDisease. PLoS One 10(2):e0116841. https://doi.org/10.1371/journal.pone.0116841

Song EJ, Jeon SG, Kim KA, Kim JI, Moon M (2017) Restricted CD41 T cell receptor repertoire impairs cognitive function via alteration of Th2cytokine levels. Neurogenesis (Austin) 4: e1256856

Sorensen PS, Blinkenberg M (2016) The potential role for ocrelizumab in the treatment of multiple sclerosis: current evidence and future prospects. Ther Adv Neurol Disord 9(1):44–52. https://doi.org/10.1177/1756285615601933

Statland JM, Barohn RJ, McVey AL, Katz JS, Dimachkie MM (2015) Patterns of weakness, classification of motor neuron disease, and clinical diagnosis of sporadic amyotrophic lateral sclerosis. Neurol Clin 33(4):735–748. https://doi.org/10.1016/j.ncl.2015.07.006

Stefanis L (2012) A-Synuclein in Parkinson's disease. Cold Spring Harb Perspect Med 4:a009399

Stower H (2018) Searching for Alzheimer's disease therapies. Nat Med 24:894–897

Taylor JP, Hardy J, Fischbeck KH (2002) Toxic proteins in neurodegenerative disease. Science 296:1991–1995

Vandenberghe R, Riviere ME, Caputo A, Sovago J, Maguire RP, Farlow M et al (2016) Active Aβ immunotherapy CAD106 in Alzheimer's disease: a phase 2b study. Alzheimers Dement 3 (1):10–22

Vick JS, Askwith CC (2015) ASICs and neuropeptides. Neuropharmacology 94:36–41. https://doi.org/10.1016/j.neuropharm.2014.12.012

Villeda SA, Plambeck KE, Middeldorp J, Castellano JM, Mosher KI, Luo J, Smith LK, Bieri G, Lin K, Berdnik D et al (2014) Young blood reverses age-related impairments in cognitive function and synaptic plasticity in mice. Nat Med 20:659–663

Wang J, Gu BJ, Masters CL, Wang YJ (2017) A systemic view of Alzheimer disease—insights from amyloid-b metabolism beyond the brain. Nat Rev Neurol 13:612–623. [published erratum appears in 2017 Nat. Rev. Neurol. 13: 703]

Wilcock DM, Colton CA (2008) Anti-amyloid-beta immunotherapy in Alzheimer's disease: relevance of transgenic mouse studies to clinical trials. J Alzheimers Dis 15(4):555–569. https://doi.org/10.3233/jad-2008-15404

Zella MAS, Metzdorf J (2019) Novel immunotherapeutic approaches to target alpha-Synuclein and related Neuroinflammation in Parkinson's disease. Cell 8:105. https://doi.org/10.3390/cells8020105

Zhang G, Wang T, Xia Y (2018) New perspectives on roles of alpha-synuclein in Parkinson's disease. Front Aging Neurosci 10:370. https://doi.org/10.3389/fnagi.2018.00370

Companion Diagnostics and Clinical Biomarkers for Immunotherapy

6

Vandana S. Nikam

Abstract

The advancement of understanding the complexity of disease biology unraveled many targets, and accordingly, their ligands are investigated. However, patients respond differently to these drugs, which strongly suggests differential expression of biomarkers, albeit with the same disease background. Thus, the concept of companion diagnostic-enable (CDx-enable) therapy evolved and was clinically used to improve the benefits over risk. The non-responders patients are benefitted by non-exposure to the unwanted risk of therapy and cost. The co-development of diagnostic tools to discriminate between non-responders and responders for targeted therapies' had multiple advantages like improved therapeutically benefits, facilitates decision making for physicians, and has a significant economic impact. CDx-enable therapy was first developed and implemented for breast cancer patients. The commercial success of oncology drugs like Herceptin and Imatinib, along with their companion diagnostic assays had accelerated the research in the field. Even though oncology is the major segment for co-development of tests with corresponding targeted therapy, it is gradually expanding in other therapeutic areas too. The biomarkers used as predictive/prognostic markers for co-development of diagnostic tools with therapy require to be validated analytically, and clinically ensuring its clinical utility. The "omic" based tests form the basis for many developed companion diagnostic tools like genomics, proteomics, and metabolomics. Immunotherapy in recent times has witnessed a paradigm shift in the way cancer is treated. The identification of cancer antigen changed the approach for cancer treatment from chemotherapy to anti-tumor response with improved surveillance of immune cells over cancer cells. With an initial handful of CDx-enable oncology drugs, the companion

V. S. Nikam (✉)
Department of Pharmacology, STES's Smt. Kashibai Navale College of Pharmacy, S. P. Pune University, Pune, Maharashtra, India

S. P. Sawarkar et al. (eds.), *Immunotherapy – A Novel Facet of Modern Therapeutics*,
https://doi.org/10.1007/978-981-15-9038-2_6

diagnostic based therapies are emerging in other therapeutic areas like cystic fibrosis, human immune deficiency virus (HIV), severe growth failure, and many more.

Keywords

Companion diagnostics · Biomarkers · Prognostic markers · Predictive markers · Positive predictive value (PPV) · Non-predictive value (NPV)

6.1 Introduction

Companion diagnostics and clinical biomarkers are an integral part of the safe and effective delivery of pharmacotherapy, especially immunotherapy. Considering the heterogeneity of the disease and understanding the underlying molecular, cellular mechanism, the medicines are designed and developed to achieve high benefits over the risk associated with it. The co-developed or companion in vitro diagnostic (in vitro diagnostics (IVD) or companion diagnostics—CDx) is a co-development of the drug and diagnostic assay. Simultaneous process of identification, development, testing of a therapeutic product, along with its corresponding diagnostic tool based on the biomarkers associated with the therapeutic target is the recent trend in immunotherapy for cancer patients to improve benefits to risk ratio (US Food and Drug Administration 2005). Though oncology remains the largest segment for companion diagnostic development, the concept is emerging for other therapeutic areas too.

The strategy of parallel development of therapy and the diagnostic tool was used when selective estrogen receptor modulator, tamoxifen, developed for breast cancer treatment in 1970. The phase II clinical trial with 74 patients with advanced breast cancer showed that a high degree of correlation between response and positive estrogen receptor test suggests the value of the diagnostic test as a means to select patients for tamoxifen treatment (Lerner et al. 1976). Even though the concept of companion diagnostic with definite conclusions emerged four decades back, the adaptation and implementation of it were relatively slow. In 1998, the drug, Trastuzumab (Herceptin), co-developed with an in vitro companion diagnostic test. It was an immunohistochemical test (HercepTest™) to measure the expression of human epidermal growth factor (EGF) in breast cancer tissue and the identification of patients' therapeutic responses. The interest in the prognostic marker, surrogate biomarkers, and establishing the correlation of companion diagnostics and clinical biomarkers with therapeutic response has grown and been studied extensively. However, in the last decade, the CDx implementation gained momentum with widespread acceptance due to the advancement of understanding the underlying pathophysiological mechanism of disease or disorders at the genomic level.

6.2 Consensus on the Terminology and Definition of Term: Companion Diagnostics

Numerous terminologies like pharmacodiagnostics, theranostics, pharmacogenomic biomarkers, and companion in vitro diagnostics are used to describe companion diagnostics. For example, Dako developed the HercepTest™ to treat metastatic breast cancer along with Herceptin. HercepTest™ defines a pharmacodiagnostic, i.e., " a diagnostic test used to measure human EGF expression level in breast cancer patients, and accordingly, decisions were made for the treatment using targeted therapy—Herceptin." The earlier adopted term "theranostics" describes a system combining diagnosis, therapy, and monitoring (Jørgensen 2013). The US FDA and European Union (EU) have adopted the term companion diagnostics (CDx) that is more frequently used in the literature. In 2011, the US FDA issued a draft in which CDx is defined as an in vitro diagnostic device that provides information, and it is essential for the safe and effective use of a corresponding therapeutic product. The US FDA has specified areas where CDx is important, namely (1) to identify the patients who are the most likely to benefit from the therapeutic product (2) to identify the patients likely to be at increased risk of serious adverse drug reaction due to therapeutic product, (3) to monitor the therapeutic response for adjusting the treatment (e.g., schedule, dose, dose regime), and (4) to identify the patients in the population for whom the therapeutic product has been adequately studied and proven to be safe and efficacious, so as to improve efficacy and safety.

6.3 Parallel Road of Drug Development and CDx Development

As shown in Fig. 6.1, the drug/therapeutic product development and companion diagnostic development go hand in hand to ensure effective and safe clinical outcomes with an effective healthcare economy. Drug discovery and development is a complex, costly, and time-intensive process, and if it is coupled with the companion diagnostic, it adds additional benefits concerning efficacy and safety. The CDx development begins with the identification of target molecules based on a thorough understanding of the underlying molecular mechanism of disease or disorder and testing various molecules as a surrogate biomarker. The selection of biomarkers and testing of these markers through prototype assays are performed to build a concrete hypothesis for clinical trial studies.

The biomarkers would be of different types like diagnostic, early detection of disease, monitoring response, risk assessment, predictive (safety and efficacy). The clinical Phase I and Phase II clinical data provide the cut-off value of selected biomarkers to be evaluated in analytical validation. The CDx assay must be evaluated during the analytical validation stage for accuracy, reliability, precision, reproducibility by performing multi-site studies. A golden rule for the application of CDx assay to clinical phase III patients for diagnosis and stratification of therapy is to use analytically validated assay and not the prototype assay (US Food and Drug Administration 2005, 2013). The CDx assay is useful if it differentiates between

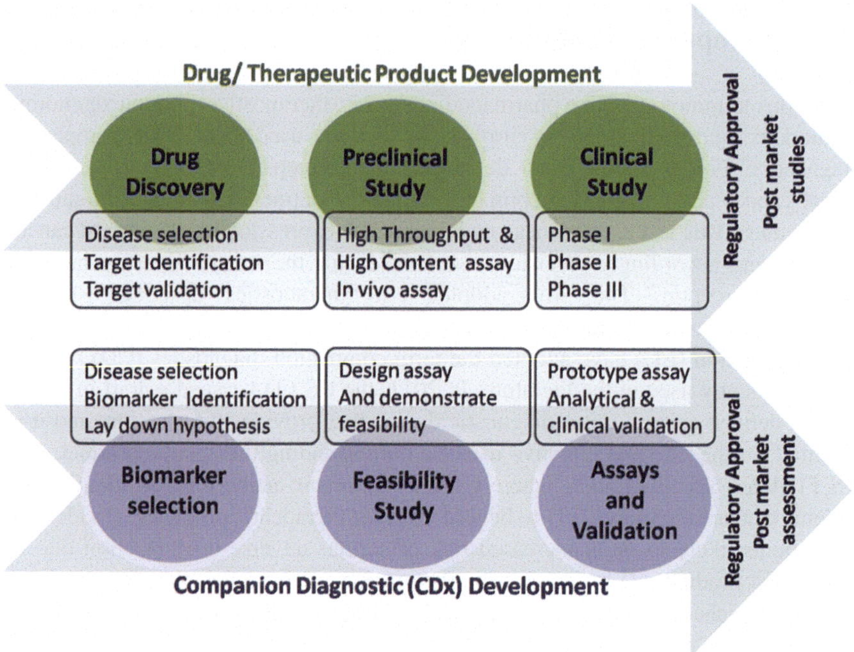

Fig. 6.1 Simultaneous developments of drug/therapeutic product and companion diagnostic

responders and non-responders to the treatment. Thus, the clinical diagnostic matrices comprise specificity, sensitivity, positive predictive value (PPV), non-predictive value (NPV) parameters to confirm the CDx assay accuracy.

6.4 Clinical Trial Designs: Factors Considered in Designing the Protocol for Co-Development of CDx Assay and Drug

To implement CDx-enable therapy routinely, it is crucial to ensure the analytical, clinical validity, and clinical utility of the assay. CDx assay measures a specific biomarker or set of biomarkers having co-relationship with the disease state.

The term biomarker specifies the measurement that tracks and correlates either directly or inversely the progression of the disease before and after the treatment. The surrogate marker is the measurement that indicates the effect of treatment, and, hence, it is a clinically used endpoint that directly correlates with therapeutic outcome (Baker and Kramer 2015; Katz 2004). A prognostic marker is a clinical and biological feature that gives information about the patient's long-term outcome, irrespective of the treatment given or not. Whereas predictive marker is the baseline measurement that cites the likeliness benefits of the treatment to the patients (Baker and Kramer 2015; Simon 2010). The medical use and validation methods of surrogate endpoint marker and prognostic/predictive markers are different. The validation of prognostic and predictive markers is relatively simple in comparison to surrogate markers. The prognostic and predictive markers are validated by three different types of validation methods, viz. analytical validity, clinical validity, and clinical utility (Baker and Kramer 2015; Simon 2010; FDA-NIH 2016).

Analytical validity denotes the robustness and reproducibility of the test for assay performance and tissue handling process. Clinical validity is mostly derived from the data of Phase II clinical studies retrospectively and establishes a correlation with clinical endpoint or characteristics. It gives information about the sensitivity (true positives) and specificity (true negatives) of the test. Clinical validity is done by calculating sensitivity and specificity for the test by identifying responders and non-responders to the treatment, respectively. The receiver operating characteristic (ROC) analysis was originally developed during World War II to analyze and differentiate accurately signal from noise in radar detection, and recently it is adopted in clinical areas for laboratory testing, epidemiology, bioinformatics (Lusted 1971; Lasko et al. 2005).˙ The ROC plot is a useful tool for evaluating the performance of the diagnostic test and for evaluating the accuracy of a statistical model that classifies the outcome into two categories (for example, responders and non-responders in case of the new therapeutic regime). The accuracy of a diagnostic test is indicated by the sensitivity and specificity of the test. Thus, the positive predictive values (PPV) and negative prediction values (NPV) depend on the sensitivity, specificity, and the proportion of responders in the population. The clinical utility defines the improved outcome of the test. The information for therapeutic decision-making depends on the test outcome, and thus, it minimizes the risk and maximizes benefits for responders with new treatment over standard or conventional therapy.

The CDx assays serve as a "security guard" in the decision-making process for the therapy. However, the CDx assay needs to be validated before routine practice. The various designs of clinical trials for co-development of drug and companion diagnostic have been enlisted in the literature; however, three clinical trial designs are discussed here (Fig. 6.2) (Mandrekar et al. 2005; Hoering et al. 2008; Mandrekar and Sargent 2009).

The marker strategy design (Fig. 6.2a) includes the randomization of patients into two groups: one group of patients is tested for test markers and treated based on the test marker and standard prognostic factors, whereas another group of patients is treated based on standard prognostic factors. A large number of patients are required for this strategy design to have statistical significant power for differentiating the outcome of two groups as many patients may receive the same treatment irrespective of which group they belong to and get randomized. This design is very flexible, though, with the use of other designs, the same objectives would be achieved (Table 6.1).

The marker strategy design is modified, as shown in Fig. 6.2b, to overcome the defect of it. As per this modified version, patients are tested for markers, and if the treatment is different from the standard prognostic marker, the patients are randomized into two groups. One group receives marker-based treatment, and the

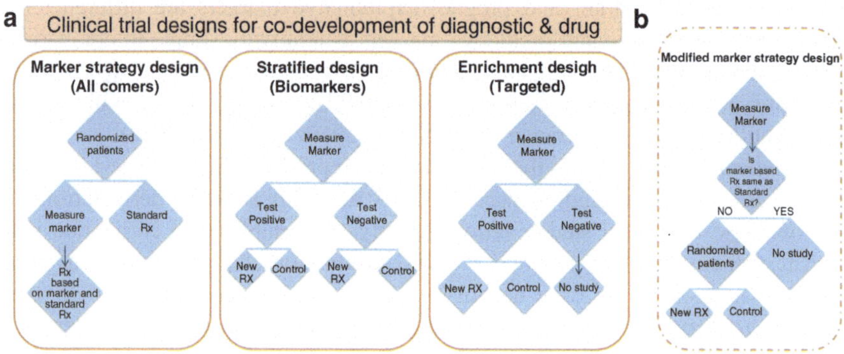

Fig. 6.2 (**a**) Clinical trial designs (Phase III) for co-development of CDx and new Rx (Rx-treatment) and (**b**) modified marker strategy design

other group gets standard treatment. This strategy had been adopted for breast cancer patients with node-negative, and 1–3 positive lymph node disease using a microarray-based 70-gene prognostic signature assay (Bogaerts et al. 2006).

Enrichment or targeted clinical trial design (Fig. 6.2a) is used when there is compelling evidence that test-negative patients are unlikely to be benefitted from a new treatment (Simon and Maitournam 2005, 2006; Maitournam and Simon 2005). For example, the targeted trial design was used for the trastuzumab development for metastatic breast cancer patients based on the expression of HER2 in an immunohistochemistry assay. In this design, the test markers are measured, and patients are segregated into two groups, such as test positive and test negative. The control group patients get standard treatment, whereas the new treatment group patients receive the target-specific treatment. The efficiency of enrichment strategy depends on the number of test-positive patients and the effectiveness of treatment for test-negative patients. The limitation of the enrichment strategy is that it does not provide information regarding the effectiveness of treatment for test-negative patients (Table 6.1).

The stratifying strategy (balancing strategy) includes both test-positive and test-negative patients (Fig. 6.2a) (Simon 2008a, b). The main purpose of this strategy is stratifying randomization so that only patients with test results are eligible and included in the trial. When the predictive classifier has been developed and has not been tested for effectiveness in test-negative patients during the Phase II trial, it is good enough to include both test-negative and test-positive patients. This method is applicable to establish the medical utility of test and therapy but required a well-defined single biomarker with a cut-off point (Table 6.1).

Due to the complexity of disease biology, it is not feasible to have an analytically validated test by the time Phase III clinical trial begins. Various adaptive designs are carefully described to utilize trial data and refine the biomarkers. These designs can be adopted for Phase III clinical trial for co-development to evaluate the effectiveness of the new treatment (Jiang et al. 2007; Freidlin and Simon 2005; Song and Chi 2007).

Table 6.1 Description of various clinical trial designs for co-development of CDx and Rx

Clinical trial design	Description	CDx dependency	Diagnostic matrices	Strengths and limitations
All comers (marker strategy design)	A very large trial if the prevalence of the test is small The diagnostic test could be done retrospectively	Independent of CDx	Sensitivity, specificity, NPV, PPV	Flexible design Same objective would be achieved through other efficient design
Stratified (biomarkers)	Differentially oversample the rarer group (usually test positive) Allows for assessment of treatment by marker interaction	CDx positive and CDx negative	Sensitivity, specificity, NPV, PPV	To establish the effectiveness of therapy and the utility of test Requires well-defined single biomarker and cut-off point Requires large sample size to evaluate the effect in test-negative and test-positive subjects separately
Enrichment (targeted)	Study only the test-positive patients	CDx positive	PPV	Small number of randomized patients required Does not give information about the effectiveness of therapy for test-negative subjects

6.5 Technologies Used for CDx Assay Development

The technologies used for developing companion diagnostic tools encompass multiple molecular disciplines such as genomics, transcriptomics, proteomics, metabolomics to generate a characteristic set of biological molecules (Micheel et al. 2012). For example, genomics deals with gene sequencing, transcriptomics investigates quantification and sequencing of transcripts (mRNA), proteomics refers to quantification and identification of proteins, and metabolomics with metabolites. The disease causality, onset, and progression are a multicomponent, complex process, and keeps on adding a new dimension to it with the advancement in technologies. For example, breast cancer now categorizes into five subtypes and may further get classified with unraveling disease complexity and the advent of advanced technology (Culbertson et al. 2007). Omics translation research produces a plethora of high dimensional data acquired through mining multiple variables and subjected to analytical, clinical validity, and clinical utility.

Genomics based technologies include variant detection, whole-genome sequencing, polymorphism, single or multiple gene panels analysis. Next-generation sequencing (NGS) is a recent breakthrough in the field since NGS based assays are fast, accurate, and cost-effective. The application, strengths, and limitations of NGS based assays need to be considered thoroughly from a regulatory perspective since these assays involved multistep workflow, with each step being the source of variability (Food and Drug Administration 2016, 2017a, b; Genome Web 2017). The first NGS-based test is the Oncomine Dx Target Test that is designed to screen in parallel 23 genes associated with non-small cell lung (NSCL) cancer. Among 23 selected genes, data from three genes can now be used to manage disease therapy options.

Proteomics and metabolomics got revolutionized with a mass spectroscopic analytical tool that provides to identify, characterize, and quantify the proteins in disease and healthy tissue samples (Rivers et al. 2014; Duarte and Spencer 2016; Bonislawski 2017; Baylor Genetics 2018; Beger et al. 2016). The rapid advancement and protocol optimization in proteomics-based approaches serve as a critical tool in oncology biomarker identification and may spread its reach in other therapeutic areas too (Rivers et al. 2014; Duarte and Spencer 2016; Bonislawski 2017). The first companion diagnostic test was the immunohistochemistry (IHC) based detection method of HER2 protein in breast tissue samples. Tables 6.2 and 6.3 enlist various platform technologies like real-time polymerase chain reaction (RT-PCR), IHC, fluorescence in situ hybridization (FISH), chromogenic in situ hybridization (CISH), magnetic resonance imaging (MRI) as companion diagnostic assay along with targeted therapy.

6.6 Impact of Companion Diagnostics

The companion diagnostics-enable drugs mostly belong to the oncology segment, and it is expanding every year. Despite the selected small group of patients based on diagnostic results, the sale of these drugs achieves more than $1 billion annually. For example, in 2012, the worldwide sale of Herceptin and Gleevac was $6.5 billion and $4.7 billion, respectively (Roche 2012; Novartis 2012).

The therapies based on companion diagnostic not only select the patient population, which is likely to be benefited from therapy but also exclude the non-responder from unwanted risk and the cost of therapy. The co-development of CDx-enabled drugs has a significant impact on R&D spending and drug approval rate. The increase in precision of CDx assay reduces the trial size by tenfold, and in turn, cut down the cost of R&D investment up to 60%. It had been reported that CDx-enable therapy cost savings would be $50 billion across all therapeutic areas, the major contributors being oncology and CNS disorders (ARK Investment Management, LLC n.d.).

Additionally, the companion diagnostic-enable therapy provides a new tool to the physician and facilitates the decision-making and assessing the benefit to risk ratio. The therapies with companion diagnostics had been shown to improve patients' safety and tolerability. These observations were significant for cutaneous, gastrointestinal, and neuronal toxicity (Ocana et al. 2015).

Table 6.2 Companion diagnostic assay based on real-time polymerase chain reaction (RT-PCR) platform

Drug trade name (generic name)	Device trade name	Disease	Company	Platform technology
Erbitux (cetuximab); Vectibix (panitumumab)	The cobas KRAS Mutation Test	Colorectal cancer	Amgen	RT-PCR
Erbitux (cetuximab); Vectibix (panitumumab)	Therascreen KRAS RGQ PCR kit	Colorectal cancer	Qiagen	RT-PCR
Gilotrif (afatinib)	Therascreen EGFR RGQ PCR kit	Non-small cell lung cancer	Qiagen	RT-PCR
Gleevec (imatinib mesylate)	KIT D816V mutation detection by PCR for Gleevec Eligibility in aggressive systemic mastocytosis (ASM)	Aggressive systemic mastocytosis	Aruplab	RT-PCR
Ressa (gefitinib)	Therascreen EGFR RGQ PCR Kit	Non-small cell lung cancer	Qiagen	RT-PCR
Lynparza (olaparib)	BRAC Analysis CDx	Ovarian cancer	Myriad Inc.	RT-PCR
Mekinist (trametinib); Tafinlar (dabrafenib)	THxID BRAF kit	Melanoma	BioMerieux	RT-PCR
Tagrisso (osimertinib)	Cobas EGFR Mutation Test v2	Non-small cell lung cancer	Roche	RT-PCR
Tarceva (erlotinib)	Cobas EGFR Mutation Test	Non-small cell lung cancer	Roche	RT-PCR
Zelboraf (vemurafenib)	Cobas 4800 BRAF V600 Mutation Test	Melanoma	Roche	RT-PCR

The drugs of other therapeutic areas, including cystic fibrosis, human immu-no-deficiency virus (HIV), and severe growth failure, have been approved along with their CDx assays. For example, cystic fibrosis Kalydeco® (ivacaftor) is paired with COBAS 4800 BRAF V600® by Roche diagnostics. The company, LabCorp, developed Trofile®, an HIV co-receptor tropism assay with the drug Selzentry® (maraviroc) for HIV.

6.7 Companion Diagnostic and Immunotherapy

Immunotherapy has evolved and revolutionized in the past few years for the treatment of solid tumors, including lung, neck, head, and melanoma. The role of cancer antigen, interferon-gamma adaptive immunity in cancer immunosurveillance, dendritic cells, and their receptors for sensitizing microorganisms led to having a new dimension for immunotherapy (Steinman and Cohn 1973; van der Bruggen et al. 1991; Schreiber et al. 2011). The identification of cancer antigen accelerated

Table 6.3 Companion diagnostic assay based on other technologies

Drug trade name (generic name)	Device trade name	Disease	Company	Platform technology
Herceptin (trastuzumab)	Bond Oracle Her2 IHC System	Breast cancer	Dako	IHC
Herceptin (trastuzumab); Perjeta (pertuzumab); KADCYLA (ado-trastuzumab emtansine)	HercepTest	Breast cancer	Dako	IHC
Erbitux (cetuximab); Vectibix (panitumumab)	Dako Egfr PharmDx kit	Colorectal cancer	Dako	IHC
GLEEVEC (imatinib mesylate)	DAKO C-KIT PharmDx	Gastrointestinal stromal tumor	Dako	IHC
Herceptin (trastuzumab)	Bond Oracle Her2 IHC system	Breast cancer	Leica	IHC
Exjade (deferasirox)	FerriScan	Thalassemia	FerriScan	MRI
Gleevec (imatinib mesylate)	PDGFRB FISH for Gleevec Eligibility in myelodysplastic syndrome/ myeloproliferative disease (MDS/MPD)	Myelodysplastic syndrome/ myeloproliferative disease	Aruplab	FISH
Herceptin (trastuzumab)	Pathvysion HER-2 DNA Probe Kit	Breast cancer	Abott	FISH
Venclexta (venetoclax)	VYSIS CLL FISH Probe Kit	Chronic lymphocytic leukemia	Abott	FISH
Rubraca (rucaparib)	Foundation focus CDxBRCA test	Ovarian cancer	Foundation Medicine, Inc.	NGS

research in the direction of anti-tumor response strategies such as peptide vaccine, dendritic cell vaccine, cytokine therapy, adoptive T cell therapy. However, these therapies had their limitations due to a lack of understanding of the immune checkpoints (Pardoll 2012). The patients benefited from improved and novel immunotherapy drugs aimed at a wide range of malignancies. The monoclonal antibodies targeted at program cell death protein (PD-1), programmed cell death protein ligand 1 (PD-L1), and cytotoxic T-lymphocyte antigen 4 (CTLA-4) are clinically used immunotherapy currently available for cancer patients. (Ribas and Wolchok 2018; Bansal et al. 2016)

The balance between immune response and immune tolerance is mediated by regulating the activation and inhibition of T cell receptors. The T cell receptors activation requires two signals, viz., T cell receptors ligation with an antigen-presenting signal and co-stimulatory receptor CD28 activation with co-stimulatory molecules CD80 or CD86. The ligation of CD28 receptors present on T cell with CD80/86 sends a positive stimulatory signal, whereas CTLA-4 receptors send an inhibitory signal. Like CTLA-4, PD-1 belongs to the CD28 family, binds to PDL-1 ligand, and delivers an inhibitory signal (Okazaki et al. 2013; Freeman 2000; Latchman 2001). Both CTLA-4 and PD-1 are expressed upon activation of T cells; however, the CTLA-4 activation occurs in the early phase of the immune response, and PD-1 activation happens at the effector phase. PD-1 and CTLA-4 are immune checkpoints having inhibitory activity at different stages of T cell activation (Parry 2005). The gene, PD-1, was discovered in 1992 by isolating it from apoptosis-induced immune T cells. (Ishida et al. 1992) PD-1 receptors selectively found on CD4 and CD8 positive T cells, B cell, natural killer cells, dendritic cells (Agata 1996; Iwai et al. 2002). Its ligand, PDL-1, is expressed by tumor cells, and thus tumor cells escape the immunosurveillance by negatively regulating the immune response (Iwai et al. 2009). The blockades of PD-1/PDL-1 or CTLA-4 receptor with respective antibodies elicit the activation of T cell, and in turn, augment anti-tumor response (Fig. 6.3).

CDx-enable immunotherapy with checkpoint inhibitors such as pembrolizumab (Merck & Co., Inc., Kenilworth, NJ), nivolumab (Bristol-Myers Squibb, Lawrenceville, NJ), and atezolizumab (Genentech/Roche, South San Francisco, CA) approved by the US Food and Drug Administration (FDA) for advanced-stage non-small lung cancer, had resulted in promising outcome showing a high

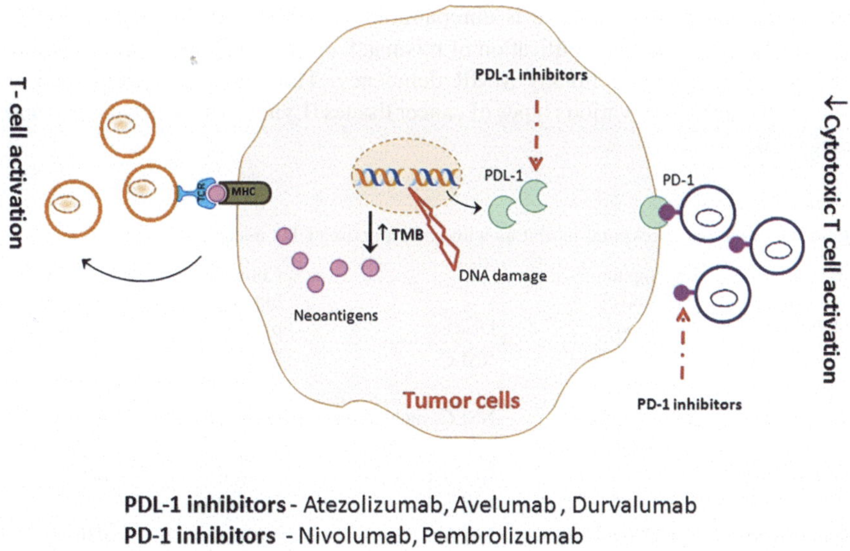

Fig. 6.3 Cancer cells express PDL-1 ligand and escape immunosurveillance. PD-1/PDL-1 inhibitors activate anti-tumor response via cytotoxic T cell activation

degree of correlation between the level of PDL-1 expression on tumors and magnitude of the efficacy of these drugs (Merck and Company, Inc. n.d.; U.S. Food and Drug Administration n.d.-a, b; Roche n.d.; Dako n.d.-a, b). CDx assay developed for estimation of program cell death ligand uses immunohistochemistry based platform technology.

The initial trial checkmate 057 was carried out as a second-line therapy when first-line platinum-based chemotherapy failed to halt the progression of cancer. The patients who exhibited higher expression PDL-1 on tumor cells, and received nivolumab, showed improved response rate, increase overall survival compared to docetaxel treatment (Borghaei et al. 2015; Brahmer et al. 2015).

Another trial, Keynote 024, with advanced-stage NSLC patients, previously untreated and having >50% PDL-1 expression, exhibited significant overall progression-free survival compared to patients who received platinum-based chemotherapy (Reck et al. 2016).

In March 2019, FDA had approved Ventana PD-L1 (SP142) assay as a companion diagnostic assay along with atezolizumab + nab-paclitaxel combination for the treatment of PDL-1 positive metastatically advanced triple-negative breast cancer (TNBC) patients in a clinical trial (Impassion130) (U.S. Food and Drug Administration n.d.-c). Table 6.4 enlists the immunohistochemistry based CDx-enabled immunotherapy for various types of cancer.

Incorporation of mismatch nucleotide during DNA replication, recombination, or repairing DNA damage caused due to chemicals, drugs, or UV exposure is edited by mismatch repair (MMR) machinery that generates a cascade of signal from cell cycle arrest to apoptosis (Arora et al. 2017; Iyer et al. 2006; Neguteanu and Salsbury Jr 2012). When the DNA damage is unrepairable by MMR machinery, such MMR defects (dMMR) lead to amplification of mismatch errors, resulting in microsatellite instability (MSI) and eventually MMR deficiency. The MSI is clinically used to measure the dMMR in various types of cancer tissues (Lynch et al. 2010; Boland and Goel 2010).

Table 6.4 Immune checkpoint inhibitors with corresponding CDx assay

Immune checkpoint inhibitors	Companion diagnostic antibody	Disease	Company biomarker platform	Platform technology
Pembrolizumab	22C3 mouse antibody	NSLC	Dako Link48	IHC
Nivolumab	28–8 rabbit antibody	NSLC, melanoma	Dako Link48	IHC
Atezolizumab	SP142 rabbit antibody	Bladder cancer, NSLC	Ventana BenchMark	IHC
Durvalumab	SP263 rabbit antibody	NSLC	Ventana BenchMark	IHC
Avelumab	73–10 RAb	Urothelial and Merkel cell carcinoma	Dako	IHC

The tumor tissue generally lacks MMR proteins indicating the insertion of mutation in DNA, in turn synthesizing abnormal neoantigen proteins, causing infiltration of immune cells and high PD-1/PDL-1 expression. The tumor exhibits a high tumor mutation burden (TMB) and responds well to immune checkpoint blockade drugs. Gandara et al. (2018) used blood tumor mutation burden (bTMB) as a marker for treatment of NSCLC patients with atezolizumab. There was a high correlation between bTMB and progression-free survival as well as TMB.

Tumor mutation burden (TMB), tumor infiltration lymphocytes are emerging biomarkers along with predictors may receive FDA approval as a companion diagnostic for selected tumor types (Galon et al. 2006; Hamada et al. 2018; Pages et al. 2018; Sato et al. 2017). Similarly, other biomarkers like mutated epidermal growth factor receptors (EGFR), oncogenes K-RAS, tumor infiltration lymphocytes, chemokines expression profile (CXCL9, 10, 11), interferon-gamma are under vigorous investigation for their clinical usefulness as biomarker for developing CDx-enable immunotherapy for various types of cancer (Ayers et al. 2017; Ji et al. 2012).

References

Agata Y (1996) Expression of the PD-1 antigen on the surface of stimulated mouse T and B lymphocytes. Int Immunol 8:765–772

ARK Investment Management, LLC (n.d.) https://ark-invest.com/analyst-research/companion-diagnostics-cdx/

Arora S, Huwe PJ, Sikder R, Shah M, Browne AJ, Lesh R et al (2017) Functional analysis of rare variants in mismatch repair proteins augments results from computation-based predictive methods. Cancer Biol Ther 201718:519–533

Ayers M, Lunceford J, Nebozhyn M, Murphy E, Loboda A, Kaufman DR et al (2017) IFN-gamma-related mRNA profile predicts clinical response to PD-1blockade. J Clin Investig 127:2930–2940

Baker SG, Kramer BS (2015) Evaluating surrogate endpoints, prognostic markers, and predictive markers—some simple themes. Clin Trials 12(4):299–308. https://doi.org/10.1177/1740774514557725

Bansal P, Osman D, Gan GN, Simon GR, Boumber Y (2016) Recent advances in immunotherapy in metastatic NSCLC. Front Oncol 6:239

Baylor Genetics (2018) Medical genetics metabolic test. https://www.bcm.edu/research/medical-genetics-labs/test_detail.cfm?testcode=4400

Beger RD, Dunn W, Schmidt MA, Gross SS, Kirwan JA, Cascante M et al (2016) Metabolomics enables precision medicine: a white paper, community perspective. Metabolomics 12(10):149

Bogaerts J, Cardoso F, Buyse M, Braga S, Loi S, Harrison JA et al (2006) Gene signature evaluation as a prognostic tool: challenges in the design of the MINDACT trial. Nat Clin Pract Oncol 3 (10):540–551

Boland CR, Goel A (2010) Microsatellite instability in colorectal cancer. Gastroenterology 138:e3

Bonislawski A (2017) FDA, NCI memorandum indicates growing interest in proteogenomics as clinical approach. https://www.genomeweb.com/proteomics-protein-research/fdanci-memorandum-indicates-growing-interest-proteogenomics-clinical#.WwWS1q_rvct

Borghaei H, Paz-Ares L, Horn L, Spigel DR, Steins M, Ready NE et al (2015) Nivolumab versus docetaxel in advanced nonsquamous non-small small-cell lung cancer. N Engl J Med 373:1627–1639

Brahmer J, Reckamp KL, Baas P, Crino L, Eberhardt WE, Poddubskaya E et al (2015) Nivolumab versus docetaxel in advanced squamous-cell nonsmall-cell lung cancer. N Engl J Med 373:123–135

Culbertson AW, Valentine SJ, Naylor S (2007) Personalized medicine: technological innovation and patient empowerment or exuberant hyperbole? In: Drug discovery, pp 16–31

Dako (n.d.-a) PD-L1 IHC 28-8 pharmDX IFU. http://www.agilent.com/en-us/products/pharmdx/pd-l1-ihc-28-8-overview. Accessed 10 Nov 2016

Dako (n.d.-b) PD-L1 IHC Pharm testing. http://www.agilent.com/en-us/products/pharmdx/pd-l1-ihc-22c3-pharmdxtesting. Accessed 10 Nov 2016

Duarte TT, Spencer CT (2016) Personalized proteomics: the future of precision medicine. Proteomes 4:29

FDA-NIH (2016) Biomarker Working Group. BEST (Biomarkers, EndpointS, and other tools). Resource [Internet]

Food and Drug Administration (2016) Foundation Focus CDxBRC: premarket approval (PMA) next generation sequencing oncology panel, somatic or germline variant detection system. Food and Drug Administration, Washington

Food and Drug Administration (2017a) FDA grants regular approval to dabrafenib and trametinib combination for metastatic NSCLC with BRAF V600E mutation. https://www.fda.gov/drugs/informationondrugs/approveddrugs/ucm564331.htm

Food and Drug Administration (2017b) FDA granted marketing approval to the Praxis Extended RAS Panel. https://www.fda.gov/drugs/informationondrugs/approveddrugs/ucm565785.htm

Freeman GJ (2000) Engagement of the PD-1 immunoinhibitory receptor by a novel B7 family member leads to negative regulation of lymphocyte reactivation. J Exp Med 192:1027–1034

Freidlin B, Simon R (2005) Adaptive signature design: an adaptive clinical trial design for generating and prospectively testing a gene expression signature for sensitive patients. Clin Cancer Res 11:7872–7878

Galon J, Costes A, Sanchez-Cabo F, Kirilovsky A, Mlecnik B, Lagorce-Pages C et al (2006) Type, density, and location of immune cells within human colorectal tumors predict clinical outcome. Science 313:1960–1964

Gandara DR, Paul SM, Kowanetz M, Schleifman E, Zou W, Li Y et al (2018) Blood based tumor mutational burden as a predictor of clinical benefit in non-small-cell lung cancer patients treated with atezolizumab. Nat Med 24(9):1441–1448

Genome Web (2017) Thermo Fisher next-gen sequencing panel wins FDA approval as companion test. https://www.genomeweb.com/molecular-diagnostics/thermo-fisher-nextgen-sequencing-panel-wins-fda-approval-companion-test

Hamada T, Soong TR, Masugi Y, Kosumi K, Nowak JA, da Silva A et al (2018) TIME (tumor immunity in the micro environment) classification based on tumor CD274 (PD-L1) expression status and tumor infiltrating lymphocytes in colorectal carcinomas. Onco Targets Ther 7: e1442999

Hoering A, Leblanc M, Crowley JJ (2008) Randomized phase III clinical trial designs for targeted agents. Clin Cancer Res 14(14):4358–4367

Ishida Y, Agata Y, Shibahara K, Honjo T (1992) Induced expression of PD-1, a novel member of the immunoglobulin gene superfamily, upon programmed cell death. EMBO J 11:3887–3895

Iwai Y, Ishida M, Tanaka Y, Okazaki T, Honjo T, Minato N (2009) Involvement of PD-L1 on tumor cells in the escape from host immune system and tumor immunotherapy by PD-L1 blockade. Proc Natl Acad Sci USA 122:93–97

Iwai Y, Okazaki T, Nishimura H, Kawasaki A, Yagita H, Honjo T (2002) Microanatomical localization of PD-1 in human tonsils. Immunol Lett 83:215–220

Iyer RR, Pluciennik A, Burdett V, Modrich PL (2006) DNA mismatch repair: functions and mechanisms. Chem Rev 106:302–323

Ji RR, Chasalow SD, Wang L, Hamid O, Schmidt H, Cogswell J et al (2012) An immune-active tumor microenvironment favors clinical response to ipilimumab. Cancer Immunol Immunother 61:1019–1031

Jiang W, Freidlin B, Simon R (2007) Biomarker adaptive threshold design: a procedure for evaluating treatment with possible biomarker-defined subset effect. J Natl Cancer Inst 99 (13):1036–1043

Jørgensen JT (2013) Companion diagnostics in oncology–current status and future aspects. Oncology 85:59–68. https://doi.org/10.1159/000353454

Katz R (2004) Biomarkers and surrogate markers: an FDA perspective. NeuroRx 1(2):189–195

Lasko TA, Bhagwat JG, Zou KH, Ohno-Machado L (2005) The use of receiver operating characteristic curves in biomedical informatics. J Biomed Inform 38(5):404–415

Latchman Y (2001) PD-L2 is a second ligand for PD-1 and inhibits T cell activation. Nat Immunol 2:261–268

Lerner HJ, Band PR, Israel L, Leung BS (1976) Phase II study of tamoxifen: report of 74 patients with stage IV breast cancer. Cancer Treat Rep 60(10):1431–1435

Lusted LB (1971) Signal detectability and medical decision making. Science 171(3977):1217–1219

Lynch HT, Jascur T, Lanspa S, Boland CR (2010) Making sense of missense in Lynch syndrome: the clinical perspective. Cancer Prev Res 3:1371–1374

Maitournam A, Simon R (2005) On the efficiency of targeted clinical trials. Stat Med 24:329–339

Mandrekar SJ, Sargent DJ (2009) Clinical trial designs for predictive biomarker validation: one size does not fit all. J Biopharm Stat 19(3):530–542

Mandrekar SJ, Grothey A, Goetz MP, Sargent DJ (2005) Clinical trial designs for prospective validation of biomarkers. Am J Pharmacogenomics 5(5):317–325

Merck and Company, Inc. (n.d.) FDA approves Merck's KEYTRUDA (pembrolizumab) in metastatic NSCLC for first-line treatment of patients whose tumors have highPD-L1 expression (tumor proportion score [TPS] of 50 percent or more) with no EGFR or ALK genomic tumor aberrations. http://www.mercknewsroomcom/news-release/prescription-medicine-news/fda-approvesmercks-keytruda-pembrolizumab-metastatic-nsclc. Accessed 10 Nov 2016

Micheel CM, Nass SJ, Omenn GS, Institute of Medicine (2012) Evolution of translational omics: lessons learned and the path forward. The National Academies Press, Washington, DC. https://doi.org/10.17226/13297

Negureanu L, Salsbury FR Jr (2012) The molecular origin of the MMR-dependent apoptosis pathway from dynamics analysis of MutSalpha-DNA complexes. J Biomol Struct Dyn 30:347–361

Novartis (2012) Annual report [webpage on the Internet]. http://www.novartis.com/downloads/investors/reports/novartis-annual-report-2012-en.pdf. Accessed 26 Aug 2014

Ocana A, Ethier JL, Díez-González L, Corrales-Sánchez V, Srikanthan A, Gascón-Escribano MJ et al (2015) Influence of companion diagnostics on efficacy and safety of targeted anti-cancer drugs: systematic review and metaanalyses. Oncotarget 6(37):39538–39549

Okazaki T, Chikuma S, Iwai Y, Fagarasan S, Honjo T (2013) A rheostat for immune responses: the unique properties of PD-1 and their advantages for clinical application. Nat Immunol 14:1212–1218

Pages F, Mlecnik B, Marliot F, Bindea G, Ou FS, Bifulco C et al (2018) International validation of the consensus immunoscore for the classification of colon cancer: a prognostic and accuracy study. Lancet 391:2128–2139

Pardoll DM (2012) The blockade of immune checkpoints in cancer immunotherapy. Nat Rev Cancer 12:252–264

Parry RV (2005) CTLA-4 and PD-1 receptors inhibit T-cell activation by distinct mechanisms. Mol Cell Biol 25:9543–9553

Reck M, Rodriguez-Abreu D, Robinson AG, Hui R, Csoszi T, Fulop A et al (2016) Investigators K-: pembrolizumab versus chemotherapy for PD-L1-positive non-small-cell lung cancer. N Engl J Med 375:1823–1833

Ribas A, Wolchok JD (2018) Cancer immunotherapy using checkpoint blockade. Science 359:1350–1355

Rivers RC et al (2014) Linking cancer genome to proteome: NCI's investment into proteogenomics. Proteomics 14:2633–2636

Roche (2012) Annual reports archive; [webpage on the Internet]. http://www.roche.com/investors/annual_reports/annual_reports_archive.htm. Accessed 26 Aug 2014
Roche (n.d.) Media release: Roche receives FDA approval for novel PD-L1 biomarker assay. http://www.ventana.com/pd-l1-biomarker-assay-news. Accessed 10 Nov 2016
Sato H, Niimi A, Yasuhara T, Permata TBM, Hagiwara Y, Isono M et al (2017) DNA double-strand break repair pathway regulates PD-L1 expression in cancer cells. Nat Commun 8:1751
Schreiber RD, Old LJ, Smyth MJ (2011) Cancer immunoediting: integrating immunity's roles in cancer suppression and promotion. Science 331(6024):1565–1570
Simon R (2008a) Using genomics in clinical trial design. Clin Cancer Res 14:5984–5993
Simon R (2008b) Designs and adaptive analysis plans for pivotal clinical trials of therapeutics and companion diagnostics. Expert Rev Mol Diagn 2(6):721–729
Simon R (2010) Clinical trial designs for evaluating the medical utility of prognostic and predictive biomarkers in oncology. Pers Med 7(1):33–47
Simon R, Maitournam A (2005) Evaluating the efficiency of targeted designs for randomized clinical trials. Clin Cancer Res 10:6759–6763
Simon R, Maitournam A (2006) Evaluating the efficiency of targeted designs for randomized clinical trials: supplement and correction. Clin Cancer Res 12:3229
Song Y, Chi GYH (2007) A method for testing a prespecified subgroup in clinical trials. Stat Med 26:3535–3549
Steinman RM, Cohn ZA (1973) Identification of a novel cell type in peripheral lymphoid organs of mice. I Morphology, quantitation, tissue distribution. J Exp Med 137(5):1142–1162
U.S. Food and Drug Administration (n.d.-a) Nivolumab (Obdivo). http://www.fda.gov/Drugs/InformationOnDrugs/ApprovedDrugs/ucm436566.htm. Accessed 10 Nov 2016
U.S. Food and Drug Administration (n.d.-b) FDA approves Keytruda for advanced non-small cell lung cancer. http://www.fda.gov/NewsEvents/Newsroom/PressAnnouncements/ucm465444.htm. Accessed 10 Nov 2016
U.S. Food and Drug Administration (n.d.-c) FDA approves atezolizumab for PD-L1 positive unresectable locally advanced or metastatic triple negative breast cancer. https://www.fda.gov/drugs/drug-approvals-and-databases/fda-approves-atezolizumab-pd-l1-positive-unresectable-locally-advancedor-metastatic-triple-negative. Accessed 9 Aug 2019
US Food and Drug Administration (2005) Drug—diagnostic co-development concept paper. Food and Drug Administration, Rome. http://www.fda.gov/downloads/Drugs/ScienceResearch/ResearchAreas/Pharmacogenetics/UCM116689.pdf
US Food and Drug Administration (2013) Paving the way for personalized medicine: FDA's role in a new era of medical product development. US Food and Drug Administration, Washington. http://www.fda.gov/downloads/ScienceResearch/SpecialTopics/Personalized Medicine/UCM372421.pdf
van der Bruggen P, Traversari C, Chomez P, Lurquin C, De Plaen E, Van den Eynde B et al (1991) A gene encoding an antigen recognized by cytolytic T lymphocytes on a human melanoma. Science 254(5038):1643–1647

Novel Drug Delivery Systems for Immunotherapeutics

7

Krishna Baxi, Munira Momin, and Sujata Sawarkar

Abstract

Immunotherapy is one of the most upcoming therapeutic approaches explored presently for management of diseases and disorders such as cancer, Type 1 Diabetes mellitus and other autoimmune disorders. However, commercial formulations such as solution containing antigens have shown limited therapeutic efficacy, non-specific targeting and have led to occurrence of adverse reactions. In order to overcome the challenges associated with these traditional formulations, researchers are exploring novel targeted drug delivery systems in order to enhance the safety and efficacy of immunotherapy. These novel drug delivery systems have shown to enhance the delivery of antigen into cytosols of targeted antigen-presenting cells which further elicited strong immune response in both in vitro as well as in vivo studies when compared to conventional formulations. These systems hold potential to re-establish the natural tolerogenic immune response. Several novel drug carriers such as nanoparticles, liposomes, polymeric micelles, hydrogels and nanorods have been discussed extensively in this chapter. These systems have been investigated worldwide to engineer and orchestrate the delivery of immunotherapeutic at the target site.

Keywords

Immunotherapy · Novel drug delivery system · Natural tolerogenic immune response · Antigen-presenting cells

K. Baxi · M. Momin · S. Sawarkar (✉)
SVKM's Dr. Bhanuben Nanavati College of Pharmacy, University of Mumbai, Mumbai, India
e-mail: sujata.sawarkar@bncp.ac.in

S. P. Sawarkar et al. (eds.), *Immunotherapy – A Novel Facet of Modern Therapeutics*,
https://doi.org/10.1007/978-981-15-9038-2_7

7.1 Introduction

Immunotherapy is the upcoming therapeutic domain for the management of various diseases and disorders which mainly includes cancer, diabetes mellitus Type-1, neurodegenerative disorders, and autoimmune diseases. In the previous sections we have discussed the concept of immunotherapy, its need and its application in clinical practice for various ailments. The major clinical applications of immunotherapy have been in the field of cancer therapeutics. The efficacy of commercially available immunotherapy mainly in cancer has been limited due to lack of its specificity towards tumor cells. Conventionally designed formulations like solutions of immunotherapeutic agents have been often related to off-target actions causing various adverse reactions and autoimmune reactions, thereby leading to less number of patient responding to these therapies.

Researchers in order to enhance the efficacy and safety of immunotherapeutic agents have looked upon novel targeted delivery approaches. One of the objectives in developing novel delivery system is to widen the therapeutic window by target specific delivery and thereby decreasing their biodistribution in normal tissues, overcoming the chances of causing the immune related adverse events. In addition to achieving target specific delivery, these novel approaches are designed to exhibit controlled released profile, which render them safe and less toxic for wide range of patient population. These novel carriers also enhance the stability of actives against various degradation processes till they reach target site (Riley et al. 2019). These advantages associated with targeted drug delivery system are attributed to characteristic features like large surface area, reactivity, strength, sensitivity, and stability by protecting the therapeutic agent from undesired degradation pathways. Especially these systems have been looked upon as one of the most promising strategies for delivering immunotherapeutic targeting to central nervous system (CNS). These systems aid to cross the major biological barrier Blood brain barrier (BBB) when the particles are in the range of 10–200 nm. Present section highlights the various delivery systems investigated for improving the safety profile and efficacy of immunotherapeutics which includes polymeric nanoparticles, lipid based drug delivery, implants and scaffold for various ailments at preclinical and clinical stages.

7.2 Novel Drug Delivery Approaches

7.2.1 Nanoparticles

Nanoparticles are ultrafine carrier systems, measured in the range of 10–1000 nm. Therapeutic agents encapsulated in the nanoparticle have shown enhanced safety profile; prolonged systemic circulation and also improved bioavailability, resulting in better therapeutic outcomes. Additionally nanoparticles have been looked upon as the promising tool for delivery of immunotherapeutic due to their ability to provide multiple coordinated signals to shape and tailor the immune response by carrying multiple immune regulatory moieties. Nanoparticles are being extensively

Table 7.1 Physicochemical properties of nanoparticles (Park et al. 2018a; Reddy et al. 2006)

Sr. no	Physicochemical properties	Inferences
1.	Particle size	Particle size in range around 5–100 nm is ideal for delivering tumor antigen at lymph nodes because below 5 nm size nanoparticles are prone to leak out of blood vessels during circulation, whereas particles larger than 100 nm get entrapped by extracellular matrix (ECM) and remain in lymph nodes. It has been stated that in a study conducted by Reddy S et al. that after intradermal administration of nanoparticles, particles of size 25 nm efficiently reached the target lymph nodes through interstitial flow while only 10% of particle measuring 100 nm were successfully administered at target site
2.	Particle shape	Non-spherical particles usually have higher aspect ratio than spherical entities which attributes to their enhanced margination effect, prolonged circulation time and easy penetration into solid tumor
3.	Surface charge	Surface charge has direct implication on cellular uptake of nanoparticles. It has been observed that cationic nanoparticles elicit better immune response as compared to anionic and neutral nanoparticles; however, the permeability of cationic particles is less because they get immobilized in negatively charged ECM. It has been observed that cationic nanoparticles are efficiently taken up by dendritic cells (DCs) present at administration site. However, these cationic particles face challenge in the delivery to lymphatic system as they can tend to cause hemolysis and platelet aggregation leading to impulsive release of antigens and variable cellular uptake
4.	Hydrophobicity	Nanoparticles fabricated from polymers having hydrophobic chain like chitosan and poly(lactic-co-glycolic acid) (PLGA) have additional advantage of initiating immune response by helping to activate immune cells

investigated by the researchers worldwide to engineer and orchestrate the delivery of immunotherapeutic at the target site.

Nanoparticles are fabricated using various range of encapsulating agents such as poly(amino acids), polysaccharides, and poly(alpha-hydroxy acids) and non-degradable compounds such as gold, silver, iron, etc. One of the most common polymers utilized for encapsulating therapeutic agents into NPs is poly(lactic-co-glycolic acid) (PLGA). PLGA is USFDA approved and basically it is biodegradable and biocompatible in nature. Another polymer which is frequently used is poly (propylene sulfide) (PPS), a hydrophobic polymer capable of releasing hydrophobic drug by undergoing series of oxidative changes and getting converted to hydrophilic water soluble polymer.

Park W et al. has collectively enlisted the ideal properties that nanoparticles should have for delivery of antigens at target site. These properties are summarized in Table 7.1:

Apart from characteristics mentioned above, another important aspect that also needs to be highlighted in case of nanoparticulate delivery is its biodistribution in the body which is controlled by size and route of administration. Getts D et al. and Neef T et al. have highlighted that the NPs delivered by systemic route get accumulated in organs such as liver, spleen, lungs, kidney. NPs delivered by intradermal, intramuscular, and intra-lymphatic route get drained into lymph nodes. NPs less than 200 nm get drained easily while larger sized NPs (>200 nm) undergo phagocytosis by APCs and eventually carried to lymph nodes. Nanoparticles of size more than 100 nm diameter are unable to permeate endothelial barriers like the BBB or the endothelial layer of the heart, reproductive organs, and gastrointestinal route (Fig. 7.1) (Getts et al. 2015; Neef and Miller 2017).

Nanoparticles are currently being widely explored in the field of cancer immunotherapy. Tumor-specific and tumor-associated antigens are the most common immunotherapeutic moieties used for cancer management. The major pre-requisite for these tumor-specific and tumor-associated antigens is to get effectively transferred to antigen-presenting cells and trigger the activation of the immature cytotoxic T cells at the lymph nodes to initiate tumor-specific immune response. It has been, however, observed that such tumor-specific and tumor-associated antigens have low efficacy due to their possible degradation by the specific enzymes before reaching the secondary lymphoid organs (Park et al. 2018a). In order to overcome this, nanoparticles are explored for delivery of antigens at antigen-presenting cells (APCs) present in the lymph nodes.

In the study performed by Zhang Z. et al., the researchers successfully fabricated PLGA nanoparticles of size 80 ± 27 nm encapsulating antigenic peptides-hgp10025e33, TRP2180e188obtained from murine melanoma and also amonophosphoryl lipid A (MPLA), toll-like receptor 4 (TLR4) agonist to demonstrate capability of this system to elicit cytotoxic T lymphocyte (CTLs) reaction against tumor-associated self-antigens in in vivo study using C57BL/6 mouse models. The research group prepared three sets of NPs comprising of—(a) NPs incorporating hgp10025e33 and TRP2180e188, (b) NPs incorporating MPLA, and (c) NPs incorporating MPLA and TRP by double emulsion method. The study concluded that osmotic pressure gradient across two aqueous phases is a critical parameter to be considered while formulating the antigens or peptides by double emulsion method. It was observed that there was enhancement in entrapment efficiency due reduction in osmotic pressure on addition of 15% solution of glucose or 5% solution of sodium chloride in the outer phase. In vitro cellular uptake of NPs labeled with the help of fluorescent lipophilic dye DiD1(1-Dioctadecyl-3,3,3,3-tetramethylindodicarbocyanine) by APCs was demonstrated using dendritic cells extracted from bone marrow cells of murine model, respectively. In vitro study showed that after 24 h of incubation 100% of cells were stained. In vivo study involved injecting NPs intradermally in the inguinal area and then dissecting the draining lymph nodes and its analysis, post 48 hours of administration of injection. It was concluded that NPs were efficiently taken up by APCs such as DCs and macrophages. Also another in vivo comparative study showed that these peptide incorporated nanoparticles had higher capability to induce antigen-specific T cell

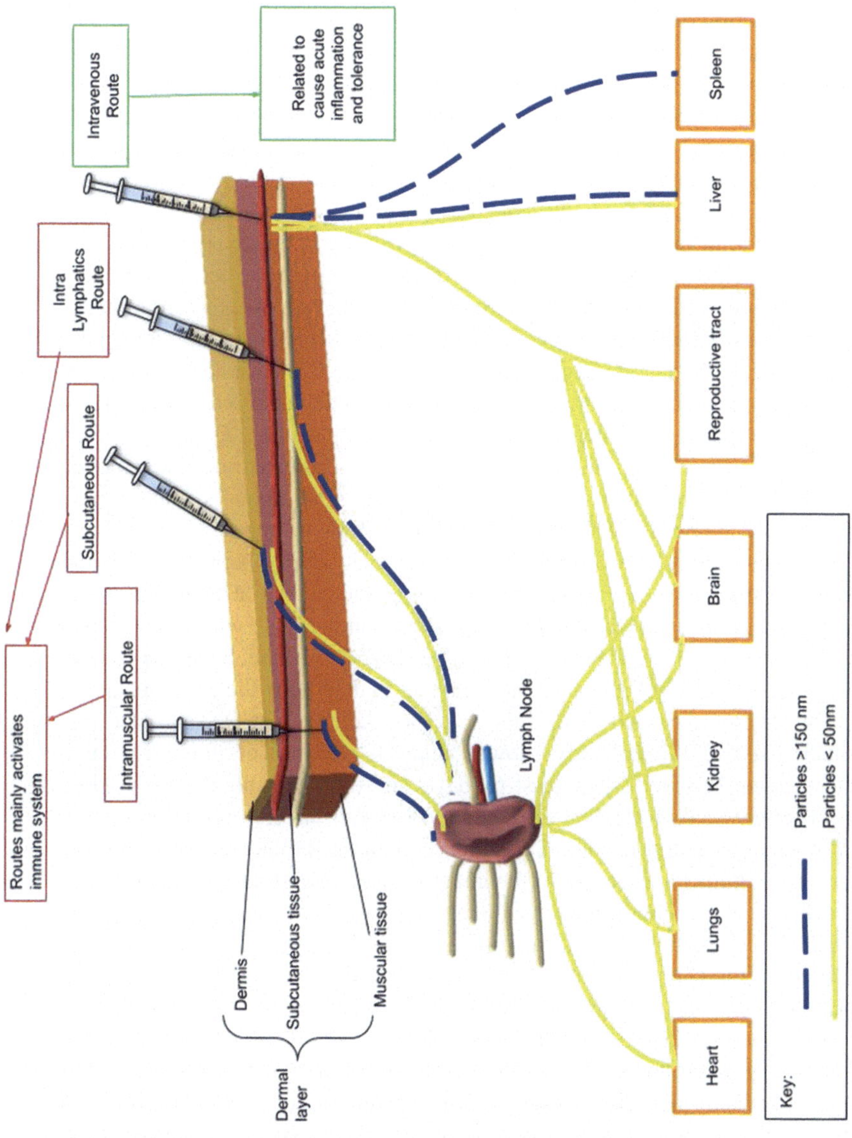

Fig. 7.1 Schematic illustration Fate of nanoparticles in body. Getts D et al. mentioned that both route of administration and size of NP are crucial parameters which determine target site and immune response final outcome. NPs of size less than 50 nm fate are depicted by yellow solid line while for NPs of greater than 150 nm depicted in blue dotted line. NPs administered through subcutaneous, intramuscular or intra-lymphatic route elicit immune activation while NPs delivered via intravenous route causes acute inflammation and tolerance (Getts et al. 2015)

response in murine model compared to peptides simply delivered in combination with Freund's adjuvant (Zhang et al. 2011).

In another study, Chen Q et al. formulated in situ sprayed bioresponsive immunotherapeutic fibrin gel comprising of calcium carbonate ($CaCO_3$) nanoparticles encapsulating anti-CD47 antibody to obstruct cancer recurrence and metastasis effectively post-surgery. Fibrin gel was prepared by interaction between solutions of fibrinogen and thrombin. The gel is biocompatible in nature, easy to administer and it enhances wound healing post-surgery. The product has been approved by US FDA. Calcium carbonate particles are biocompatible in nature. Their main role was to serve as reservoir for delivery of anti-CD47 antibody and acts as proton scavenger to regulate acidic of the tumor environment. In situ gel is formed when fibrinogen solution having anti-CD47 antibody-loaded CaCO3 nanoparticles is sprayed simultaneously along with thrombin solution in equal volume in tumor resection cavity post-surgery. CaCO3 nanoparticles tend to release the immunotherapeutic agent anti-CD47 antibody in tumor micro-environment thus elevate the activation rate of M1-type tumor-associated macrophages. The nanoparticles brought about phagocytosis of cancer cells by macrophages via blockade of the CD47 and SIRPα interaction as well as enhanced antitumor T cell responses. They also reduced the toxic effect when compared to systemic administration of plain anti-CD47 antibody. Therapeutic efficacy of the bio responsive gel was evaluated in mice model. B16F10 cells were implanted into their right flanks. The animal's tumors were resected, leaving approximately 1% residual cancer cells to mimic residual micro tumors post-surgery. Comparative evaluation was done between different formulations comprising of Fibrin gels containing IgG with CaCO3, Fibrin gels containing IgG with aCD47, and Fibrin gels containing IgG with CaCO3 and aCD47. The dose was 1 mg CaCO3 and 50 μg aCD47 per mouse. Bioluminescence signals obtained from B16F10 cancer cells helped to determine the extent of tumor growth. Mice group treated with Fibrin gel containing CaCO3 nanoparticles encapsulating anti-CD47 antibody showed improved control of tumor regrowth with half of them survived for at least 60 days, and there was no change in body weights. The presence of residual tumors was observed on fifth day after surgery using flow cytometry and immunofluorescence staining. Initial findings indicated that there was significant increase in the count of Tumor Infiltrating Lymphocytes (CD3+ cells) in the residual tumor post-treatment with Fibrin gels containing IgG with CaCO3 nanoparticles encapsulating anti-CD47 antibody. Also there was enhancement in the count of CD8+ T cells in the residual tumors in this treatment group as compared to other groups. Additionally secretion of cytokines such as IFN-γ, IL-6, and IL-12p70 concluded that inherent as well as adaptive immune response were brought about by the gel containing all the three components. The fibrin gel comprising of CaCO3 nanoparticles encapsulating anti-CD47 antibody was found to be potentially effective to prevent tumor regrowth and metastasis post-surgery (Chen et al. 2019).

Nanoparticles, apart from cancer immunotherapy, also have been investigated for inducing or restoring antigen-specific immune tolerance in various autoimmune disorders such as diabetes type 1, encephalomyelitis and multiple sclerosis because of their ability to target APCs easily. Such nanoparticles are termed as tolerogenic nanoparticles (tNPs) or synthetic vaccine particles.

Kishimoto T et al. mentioned about three approaches by which tNPs are functionally categorized such as:

- tNPs that carry antigens singly and bring about natural tolerogenic process or environment.
- tNPs providing antigen while targeting protolerogenic receptors.
- tNPs using pharmacological immunomodulators to bring about tolerogenic immune response against target antigen (Kishimoto and Maldonado 2018).

LaMothe R et al. studied tNPs prepared from polylactic acid (PLA) and poly (lactic-co-glycolic acid) (PLGA) polymers incorporating peptides such as OVA323–339 peptide, 2W1S peptide or PLP139–151 peptide, and Rapamycin. The objective of formulating the tNPs was to induce expansion of antigen-specific regulatory T cells (Tregs) to furnish immune tolerance in an in vivo model of relapsing experimental autoimmune encephalomyelitis (rEAE). The study demonstrated that EAE was not diagnosed in mice group pretreated with this tNPs. In addition it was also observed that tolerance was adoptively transferred from tNP treated animal to unaffected animals. These results were revealed from following points-A single injection of tNP comprising OVA323 and Rapamycin administered in Rag−/− mice, 1 day before OTII cell transfer, inhibited total antigen-specific T cell proliferation and induced expansion antigen-specific Tregs. It was also observed that during splenocytes assay conducted to determine 2 W-specific CD4 cells, there was an increase in endogenous 2 W-specific Foxp3 + CD4+ T cells in mice treated with tNP containing 2W1S peptide and Rapamycin compared to control group treated with PBS or NPs containing Rapamycin, when challenged with free 2 W peptide. It was observed that when splenocytes from mice treated with tNP containing PLP139-Rapa, transferred to naïve animals, there was significant attenuation and delay in disease compared to group in which splenocytes were transferred from both donor groups treated with placebo nanoparticles and nanoparticles containing RAPA, respectively (LaMothe et al. 2018).

Pei et al. (2018) prepared tNPs which directly and specifically depleted and modulated myelin-autoreactive T cells in the EAE murine model as an alternative to tolerogenic based strategy. These NPs were prepared using biodegradable Poly (lactic-co-glycolic acid) (PLGA) which carried numerous regulatory actives-co-coupled antigens MOG_{40-54}/H-$2D^b$-Ig dimer, MOG_{35-55}/I-A^b multimer, regulatory molecules (anti-Fas, PD-L1-Fc), self-marker CD47-Fc and also incorporated transforming growth factor-β1. MOG_{35-55} peptide-induced EAE mice were treated with tNPs intravenously. These NPs directly modulated myelin-autoreactive T cells and improved the therapeutic condition of EAE. It was observed that particles of size 217 nm were capable of crossing the blood–brain barrier in EAE induced mice. This can be attributed to the leaky vasculature formed due to disease progression (Pei et al. 2018).

Apart from above mentioned ailments NPs have also been investigated for the treatment of Diabetes type 1, as an alternative to cell based therapies like administration of ex vivo-differentiated FoxP3(+) regulatory T (Treg) cells or tolerogenic dendritic cells (DCs) which are associated with various clinical challenges. Yeste A et al. has studied an alternative approach for cell therapies for management of the Diabetes Type 1. They prepared gold nanoparticles (NPs) comprising of layer of thiol–polyethylene glycol to co-administer a tolerogenic molecule, the aryl hydrocarbon receptor (AhR) ligand 2-(1′H-indole-3′-carbonyl)-thiazole-4-carboxylic acid methyl ester (ITE), and the β cell antigen proinsulin (NPITE+Ins). AhR activation has been linked to generation of tolerogenic DCs, which enhances the naïve CD4+ T cells to get differentiated into Treg cells. Surface modification with thiol–polyethylene glycol enhanced solubility and stability of NPs. In the non-obese diabetic mice of 8 week old were treated with these NPs. It was observed that there was increase in DC of tolerogenic phenotype, FoxP3+ Treg cells differentiation was enhanced simultaneously reducing stimulation of inflammatory effector T cells and inhibited development of diabetes type 1 in mice. Thereby this data summed up the idea that NPs hold potential to re-establish tolerance in Diabetes type 1 (Yeste et al. 2016).

These abovementioned preclinical studies have demonstrated that nanoparticles can be employed as one of the promising nanocarrier platforms for targeted delivery of therapeutic agent at target site enabling enhanced clinical efficacy and safety in the field of immunotherapy.

7.2.2 Liposomes

Liposomes are small vesicular nanocarrier composed of hydrophilic core enclosed by hydrophobic lipid bilayer. Hydrophobic bilayer mainly comprises of phospholipids and cholesterol. Liposomes are the second nanocarriers which have been extensively studied to deliver immunotherapeutic agents. Liposomes are biodegradable, biocompatible, nontoxic flexible, and non-immunogenic in nature. Additionally they encapsulate wide range of compounds such as peptides having extreme solubility profiles—hydrophilic, lipophilic or amphiphilic. Liposomal carriers protect the therapeutic agent against degradation caused by external and physiological pathways. Liposome have been reported to exhibit an effective tolerogenic-signal induction and can also help in achieving active targeting by surface modification strategy using site-specific ligands. Highlight some advantages and unique features.

Phospholipids and cholesterol are building blocks of lipid bilayer. Phospholipids commonly used in preparation of liposomes include phosphatidylcholine, phosphatidylglycerol, phosphatidic acid, phosphatidylethanolamine, and phosphatidylserine. Phosphatidylserine (PS) is commonly used in fabricating liposomes which can mimic apoptotic cell mechanism (apoptotic cells are capable of inducing immunological tolerance). PS is used because it is a phospholipid membrane component that is exposed in apoptotic cells and modulates immune responses. Other than phospholipids, positively charged lipids such as *N*-

[1-(2,3-dioleyloxy)propyl]-*N*,*N*,*N*-triethylammonium (DOTMA) and 1,2-dioleoyl-3-trimethylammoniopropane (DOTAP)) are also used.

Pujo-Autonell et al. formulated PS-rich liposomes encapsulating myelin-oligodendrocyte glycoprotein peptide 40–55 (MOG40–55 peptide). Efficacy of this formulation against multiple sclerosis was evaluated using experimental autoimmune encephalomyelitis (EAE) model by inducing EAE in C57BL/6 female mice. PS used in fabricating liposomes interact with receptor present on APCs thereby enables recognition and phagocytosis by APCs like DCs. Thereby PS acts as "eat me" and "tolerate me" signal. This also suppresses DCs maturation and allows presentation of antigens in a tolerogenic manner inducing specific immune tolerance. It was observed during in vivo study that incidence of EAE was remarkably lesser in group treated with PS-liposomes containing MOG40 (45.45%) as compared to placebo liposomes treated group (100%, $p < 0.05$) and sham group (92.31%, $p < 0.05$), while it was slightly less than the MOG-treated group (75%). Mice treated with PSMOG-liposomes also demonstrated delayed onset in EAE at day 16.00 ± 4.56 post-immunization (p.i.) when compared with empty PS-liposomes treated (14.80 ± 2.70) and sham groups (13.17 ± 4.73). Intracellular FoxP3 staining was carried out on splenocytes of animal after completion of 15 day study period to study its effect on T cell subtypes. It was observed that percentage of CD25+ FoxP3-T cells in PSMOG-liposomes group ($14.83 \pm 5.47\%$) was elevated as compared to sham group ($9.65 \pm 2.84\%$). Thus immunotherapy was reported to be safe and effective when compared to treatment with MOG peptide, since liposomes fabricated from PS imitated apoptotic cells thereby enabled specificity and tolerogenic signals to dendritic cells (Pujol-Autonell et al. 2017).

Thenuis et al. investigated liposome based vaccine as one of the treatment strategies for tauopathies such as Alzheimer disease. Therapeutic efficacy of these formulations was evaluated using wild type mice and Tau.P301L mice. Liposome vaccine incorporated with a synthetic peptide with16 amino acids sequence corresponding to an assumed pathological Tau epitope: (Tau393–409 = 8) elicited an effective antibody response in both wild group and Tau P301L mice groups (Theunis et al. 2013).

Yuba E et al. prepared surface-modified egg yolk phosphatidylcholine/dioleoylphosphatidylethanolamine liposomes using pH-sensitive fusogenic polymer 3-methylglutarylated poly(glycidol) of linear (MGlu-LPG) or hyperbranched structure (MGlu-HPG) for encapsulation of antigenic ovalbumin (OVA). Polymer-modified pH-sensitive liposomes engulfed through endocytic pathways by DCs and were further entrapped into endosomes containing mild acidic environment. Liposomes further get fused with endosomes and destabilize their structure. As a result antigenic molecules are released into cytosol. This further culminates into antigen presentation through MHC class I molecules and sensitization of antigen-specific cytotoxic T lymphocyte (CTLs). In vitro study using dendritic cells extracted from murine bone marrow was conducted to understand the ability of liposomes labeled with fluorescent agents Lissamine rhodamine B-sulfonyl phosphatidylethanolamine (Rh-PE) and OVA molecule labeled with fluorescein isothiocyanate (FITC)-OVA presence inside DCs. It was observed that when cells are subjected to unmodified liposomes fluorescence signals of Rh-PE and FTIC-OVA were observed from the same position in cell revealing that both OVA and liposomes are entrapped in the endosomes as well as lysosomes. However, it was observed that when cells were treated with MGlu-HPG-modified liposomes, there was diffuse fluorescence of FTIC-OVA showing partial delivery of OVA to cell cytosol. In case of cells exposed to MGlu-LPG modified liposomes fluorescence signals were more intense than that of the MGlu-HPG-modified liposomes, reiterating enhance delivery of FITC-OVA with the MGlu-LPG-modified liposomes. Also it was observed that there was sufficient regression of tumor growth in animal group subjected to polymer-modified liposomes containing OVA in comparison to the unmodified liposomes containing OVA or free OVA when administered by nasal route. However, when the drug delivery was administered subcutaneously, it gave more promising results than nasal route in terms of tumor growth inhibition and survival rate of immunized mice. In order to understand the therapeutic efficiency of all types of liposomes the study was carried out in tumor mice model implanted with E.G7-OVA cells. Tumor volume increased exponentially in case of non-immunized mice at the end of fifth day of study. In case of treatment group administered with OVA-loaded liposomes, tumor volume decreased after 12th day confirming the fact that OVA-specific immunity was elicited by treatment with OVA-loaded liposomes. OVA-loaded unmodified liposomes showed modest 30% tumor growth inhibition at the end of 17th day, however, there was reoccurrence of tumor growth thereafter. In case of MGlu-HPG modified OVA-loaded liposomes treatment group, tumor was found to be undetectable on 16th day, but there were relapses after 21st day. However, for MGlu-LPG-modified OVA-loaded liposomes treatment group there was maximum suppression of tumor. This aspect was attributed to the polymer-modified liposomes having high efficiency of inducing antigen-specific immune response, causing significant reduction of tumor burden (Yuba et al. 2013).

Lu J et al. developed liposomal carrier which targeted breast cancer by efficiently triggering immunogenic cell death (ICD) and obstructing locally overproduced immunosuppressive effect of indoleamine 2,3-dioxygenase (IDO-1) at the tumor. Liposomes were designed to deliver chemo-immunotherapeutic drug Doxorubicin (DOX) and immunotherapeutic drug Indoximod(IND) in encapsulated form. Intravenous injection showed improved pharmacokinetics and drug concentrations of both the therapeutic agent at tumor site in an orthotopic 4T1 tumor model in syngeneic mice. On reaching the threshold level of ICD effect dying breast cancer cells were phagocytosized by DCs which resulted in presentation of tumor antigen naive T cells leading to its activation to perforin and IFN-γ releasing CTLs. Eventually these activated CTLs induced anti-neoplastic response at both, local as well as metastatic tumor. It was also demonstrated that this dual carrier when co-delivered with PD-1 blocking antibodies eliminated metastatic lung cancer (Lu et al. 2018).

Koshy S et al. formulated cationic liposome to deliver anionic 2′3′-cyclic GMP-AMP(cGAMP), a stimulator of interferon genes (STING) agonist, because STING agonist being anionic in nature is associated with poor membrane permeability leading to limited interaction with STING in the cytosol and poor efficacy. In order to enhance its permeation they encapsulated the anionic cGAMP into cationic liposome carrier system composed of 1,2-dioleoyl-3-trimethylammonium-propane (DOTAP), cholesterol, and 1,2-distearoyl-sn-glycero-3-phosphoethanolamine-N-[methoxy(polyethyleneglycol)-2000] (DSPE-PEG-2000) by thin film rehydration method which when delivered to the tumor microenvironment stimulated innate and adaptive immunity against tumor. Encapsulation efficiency was observed to be about 20% and the liposomes exhibited controlled release pattern. Particle stability study of the non-PEGylated liposome showed considerable increase in diameter when suspended in complete cell culture medium due to probable interaction between anionic serum protein and cationic liposomes causing aggregation. On other hand it was observed that formulations with 5 mol% and 10 mol% of DSPE-PEG(200) showed minimal enhancement in liposome size suggesting that PEGylation prevented unwanted interaction and segregation. It was also concluded that more positive the zeta potential of the liposomes, higher was the rate of cellular binding and subsequent uptake of liposomes. In vitro study using bone marrow-derived dendritic cells (BMDC) showed that liposomes enhanced cellular association of cGAMP compared to free cGAMP, and -PEGylated liposomes fastened cellular-association internalization of cGAMP. It was also concluded that liposome encapsulated cGAMP was more potent to induce cellular maturation of DCs via STING route compared to not encapsulated cGAMP. Systemic delivery of liposome encapsulated and free cGAMP in B16-F10 melanoma model with metastatic lung tumors showed that liposomal cGAMP induces increased pro-inflammatory gene expression and showed anti-tumor activity whereas free cGAMP did not show significant results. In an orthotopic skin B16-F10 melanoma model cGAMP encapsulated liposomes induced recession of implanted tumors and produced immunological memory that protected previously treated mice from futuristic challenge again with tumor cells (Koshy et al. 2017).

Immunotherapeutic drugs are also now being extensively investigated for management of insulin dependent diabetic cases. The main focus of this approach is to protect remnant beta cells from destruction. Current therapeutic approach of injectable insulin is unable to target the underlying immunological root cause of destruction of beta cells by auto reactive T lymphocytes. The therapy merely provides a temporary balance in insulin level in body. Liposomes have been studied as carriers of immunotherapy for diabetes by Pellegrino et al. They developed cationic liposomes based carrier delivery system for siRNA targeting C1858T mutation in the PTPN22 gene, and deliver the therapeutic moiety efficiently to Jurkat T cells and human peripheral blood mononuclear lymphocytes (PBMC) of healthy subjects. Both these cells were efficiently transfected by lipoplexes. Experimental results revealed that Jurkat T cells were unaffected. Lipoplexes got incorporated in both CD3+ and CD3- peripheral blood immunotypes without showing any indication of toxicity, injury or cell death. Lipoplexes showed effect on Lyp protein expression in both transfected Jurkat T cells as well as PBMC. Also there was no further mutilation of Lyp inhibitory activity. There was significant increase of Interleukin −2 (IL-2) secretion in supernatants of PBMC cultures following anti-CD3/CD28 T cell receptor-driven stimulation The same researcher group also conducted study on PBMC of Type 1 diabetes patients. The experimental study revealed that CD3 + and CD3 − immunotypes were efficiently transferred following lipoplexes treatment with the cellular structure and viability of cells maintained. There was specific target mRNA down-modulation. Increase in level of IL2 in cultures indicated that Lyp function in lipoplexes treated PBMC was restored, following T cell receptor. The researchers concluded that lipoplexes can be potential candidates that can be taken up for further trials involving study of autoimmunity based on the specific inhibitory targeting of C1858TPTPN22 (Pellegrino et al. 2019; Perri et al. 2017).

Rodriguez-Fernandez S et al. developed phosphatidylserine liposomes to promote tolerogenic feature in type 1 diabetes by apoptotic mimicry. Apoptotic cells are responsible for immunological tolerance by process of efferocytosis. PS-Liposomes of size 690 nm and 788 nm separately encapsulating two separated chains of insulin-Chain A (Peptide A) and chain B (peptide B), respectively, were formulated using the thin film hydration method from a lipid mixture of 1,2-dioleoyl-sn-glycero-3-phospho- l-serine, 1,2-didodecanoyl-sn-glycero-3-phosphocholine and cholesterol at 1:1:1.33 molar ratio. These liposomes were analyzed on the DCs isolated from T1D subject s and control age-related subjects group, respectively. PS-liposomes cellular uptake by DCs was comparatively higher than the liposomes without PS in the first 2 h of co-culture, which was constant after 6 h. Thus it can be observed that presence of PS in liposomes promoted phagocytosis of liposomes. Phagocytosis of these liposomes, further led to the expression of multiple pattern of moieties that were involved in process such as efferocytosis, antigen presentation, immunoregulation and activation in DCs having tolerogenic ability, in both the groups. In addition DC had reduced capacity to elicit related T cell proliferation. Thus it was stated that immunoregulatory pro-life can be achieved which is attributed to transcriptional modification in DCs obtained from TID patients post PS-liposomes phagocytosis. This study demonstrated that liposomes mimicking apoptotic β-cells induced to lerogenic dendritic cell (DC) generation and thereby inhibited autoimmune reaction against to β-cells and prevented experimental occurrence of T1D r (Rodriguez-Fernandez et al. 2018).

These studies have overall highlighted that Liposomes serve to be safe and nontoxic carrier system which can be engineered to deliver autoantigens and elicit tolerogenic response in autoimmune disorder and anti-tumor responses in cancer, respectively.

7.2.3 Micelles

Micelles are nanoparticles ranging in size 2–20 nm, usually formed by self-assembly of lipids and surfactants-amphiphilic molecules. Micelles are spherical structure formed in aqueous environment by these amphiphilic molecules at concentration above critical micelle concentration with polar end of molecules facing outwards and non-polar end facing inwards to form core. They are capable to encapsulate both hydrophilic as well as hydrophobic therapeutic agents. This encapsulation of the agent thereby enables to enhance chemical and physical stability, pharmacokinetic of drug, bioavailability and distribution of agent at target site.

Jiang D et al. formulated novel pH/redox dual-sensitive micellar vaccine encapsulating ovalbumin (OLM-D) designed to produce series of lysosomal responsive events which would enhance lysosomal escape and exhibit immunotherapy against cancer. Micelles were prepared by cleavable conjugation of an antigen with in-house synthesized amphiphilic poly (L-histidine)-poly (ethylene glycol) (PLH-PEG). The basis of this study was "Cascade cytosol delivery" which states to design and prepare nanocarrier which is engineered in such a way that allow proton influx into the carrier and aids in accelerating disassembly of micellar structure, this sequential event together referred to as proton sponge effect further leads to faster delivery of antigen in the cytosol. This approach presented advantage of reduction of antigen retention time in lysosome, combining rate and antigen "cross-presentation" at MHC I molecules, overall exhibiting potential immunotherapy activity. In this designed micelle system, peptide PLH performs numerous roles such as exhibiting proton pump effect, modifying solubility profile (presence of imidazole rings in PLH makes it susceptible to undergo protonation). PLH remains as uncharged molecule at pH 7.4 when in circulation and undergoes protonation at pH 5.0 in lysosomes serving as positively charged hydrophilic moiety. It was observed that after uptake of micelles into the lysosomes of APCs, series of steps occurred which are as follow: (1) first step was release of OVA caused by redox reaction mediated cleavage of di-sulfide bond thereby open up the channels for protons influx, (2) this proton influx enables disorganise and mediate additional proton influx, (3) further proton influx enables lysosome to break and deliver the OVA into the cytosol of the cells. It was observed that there was more accumulation, retention, and distribution order in different LNs after inguinal subcutaneous administration of this novel micellar vaccine immunization studies in vivo in C57Bl/6 mice showed greater differentiation of CD3 + CD8+ T cell and CD3 + CD8 + 25D11.6+ T cells ($p < 0.05$), IFN-γ and IL-2 secretion ($p < 0.05$) (Jiang et al. 2018).

In another study, Liu Z et al. prepared polyethylene glycol-phosphatidylethanolamine (PEG-PE) micelles incorporating antigen peptides and Monophosphoryl lipid A (MPLA) for co-delivery at target tumor site. It has been stated that polypeptide with alpha helical structure mediate membrane fusion and thereby can be easily delivered into cytosol. However, here the therapeutic exogenous soluble antigenic peptides were non-alpha-helix structured and cannot be delivered easily in cytosol. In this study therefore polyethylene glycol-phosphatidylethanolamine (PEG-PE) micelles were prepared which facilitated conversion of non-α-helical structure of peptide into α-helix and thus gained cytosolic antigen delivery, the mechanism behind this is not clear but is hypothesized that PEG-PE micelles furnishes an electrostatic surrounding to peptide enabling to undergo conformational changes thus establishing a stable conformation state. In addition to delivery of antigen into APC cytosol, the same APCs also need to be coordinately stimulated by adjuvants. Monophosphoryl lipid A (MPLA), a detoxified derivative of lipid A from lipopolysaccharide (LPS), is commonly used as vaccines adjuvant. Though MLPA is safe and FDA approved but has comparatively less adjuvant activity than LPS. Hence more enhancement adjuvant effect was needed. In this study it was also observed that PEG-PE micelles acts as a companion to MPLA for TLR signaling and DC stimulation. This micelles also elevated MPLA activity approximately 100-times more, which was evident from more production of TNF-α, IL-6 and IL-12. The co-stimulatory moieties production was also more by micelle encapsulated MPLA compared to free MPLA. It was stated that MPLA are usually present as aggregates and therefore cannot bind to target receptor TLR4/MD-2. However it was observed that this PEG-PE micelles enhanced signalling activity of MPLA by keeping MPLA in monomerized state and thereby permitting its efficient binding at the target site. This novel micelle-based vaccine served to efficiently co-deliver tumor-related antigens as well as MPLA into the same APC and elicit enhanced CTL effect (Liu et al. 2017).

Senapati S et al. also designed a novel amphiphilic pentablock copolymer micelle (PBC) adjuvants which are biocompatible and forms reversible pH- and temperature-sensitive micelle to enhance the cytosolic delivery of associated antigens to APCs. Pentablock copolymer (PDEAEM-PEO-PPO-PEO-PDEAEM) comprised of Pluronic F127®, poly(ethylene oxide) (PEO) and poly(propylene oxide) (PPO), and cationic blocks such as poly(diethylaminoethylmethacrylate) (PDEAEM) and was synthesized by atom transfer radical polymerization (ATRP). It has been noted that "aqueous solution of these spherical micelles (about 30 nm in diameter) and at higher concentrations (>20 wt.% polymer) form physical hydrogels in response to both temperature and pH." It has been observed that when PBC are formulated as micelles it waives out probability of any potential inflammatory response occurrence which is usually associated to occur at higher concentration of gel. This study has highlighted additional advantage of the PBC micelles, that it causes less release of nitric oxide and reactive oxygen species by APCs when compared to traditional TLR adjuvants. Thus PBC based micelles are safe, produce low inflammatory response and efficiently deliver the antigen at the cytosol. In order to evaluate ability of micelles to induce humoral immune responses, PBC micelle-Ova formulation was delivered into mice. Additionally different groups of mice were treated either with soluble ovalbumin (sOVA), alum+Ova or a hydrogel system, each containing 50 μg of Ova29. It was observed that there was nine times rise in anti-Ova antibody titers in sera of mice exposed to micelles when compared to sOVA at 2 weeks post-immunization (p.i.); however, there was no remarkable distinction observed among mice group treated with micelle and hydrogel formulation, respectively. This indicates that even at low polymer concentrations PBC can induce significant humoral immune response in the form of micelles compared to hydrogels. Antibody titre in animals treated with alum+Ova was not significantly different from PBC micelles treated group at 2 weeks post-immunization but there was four time higher at 4 weeks post-immunization. It has been concluded that, although these micelles do not activate APCs unlike traditional TLR agonists but these micelles enhanced antibody response by causing less inflammatory effect and improved delivery of antigen into the APCs cytosol by acting as an adjuvant (Senapati et al. 2019).

These aforementioned studies overall highlights the ability of micelles to enhance the transfer and deliver antigen and adjuvants into APCs compared to traditional approaches.

7.2.4 Nanorods

Nanorods are nanoparticle having rod like shape measuring in the range 10–120 nm. They are most commonly fabricated from metals such as gold and silver. These nanorods have been investigated to provide dual therapy like photothermal therapy (PTT) and immunotherapy (Zhou et al. 2018).

Zhou B et al. combined PTT and immunotherapy for management of skin cancer. The group fabricated novel bovine serum albumin (BSA)-bioinspired gold nanorods (GNRs) coated along with cetyltrimethylammonium bromide (CTAB) and PEG along with incorporation of immunoadjuvant imiquimod (R837) These nanocomplexes showed effective anti-tumor activity by destroying cancerous cells and triggered effective immune response in metastatic melanoma mice model when subjected to near-infrared (NIR) radiation. In vitro PTT of these nanocomplexes was studied on B16-F10 cells treated with gold concentration of 58.4 μM for 6 h followed by irradiation with a 1064 nm NIR laser (1.0 W/cm228) for 10 min. It was observed that cell viability was 27.45% post-incubation with complexes containing 11.5 μg Au/mL in combination with exposure to the NIR laser irradiation at the power density of 0.85 W/cm^2, whereas the cell viability was 83.5% when subjected to only laser irradiation under the same test condition. It was also observed that when BMDCs were treated with nanoplexes with or without immunoadjuvant R837 or R837 alone for 24 h percentages of mature DCs estimated using flow cytometry was 65.1% for nanoplexes containing R837 which was close to group treated with free R837 that has 60% of mature DC, while 37.9% for nanople*xes* without R837 was similar to that of PBS control-34.9%. Thus these nanoplexes containing R837 were capable to induce a strong immune stimulatory response. The immunogenic cell death was studied by measuring release of the HSP70/β-Actin protein and ATP after apoptosis from B16-F10 cells. It was observed to be higher in nanocomplexes containing R837 plus laser irradiation than only with laser. In the in vivo B16-F10 cells induced mice model it was observed that group exposed to *m*PEG-GNRs@BSA/R837-PTT under laser radiation showed significant tumor growth inhibition compared to groups treated individually by *m*PEG-GNRs@BSA/R837-PTT and laser radiation and PBS treated control group. This was anticipated because gold nanorods based PTT destroyed partially cells in tumor, these killed cells can subsequently provide tumor-related antigens. Antigens undergo further processing and are presented to APCs to activate and elicit tumor-specific effector T cells proliferation in lymphoid organs with the help of immunoadjuvant R83. Induced T cells could initiate anti-tumor activity against both localized as well as metastatic tumor cells. Mice group exposed to*m*PEG-GNRs@BSA/R837-PTT could notably inhibit the growth of re-inoculated tumor; with all mice survival time reported to be greater than 100 days post second tumor inoculation. On other hand none of the animals remained alive in the both the groups age-matched and group subjected to laser alone. These out-turns illustrated that a powerful long-term immune memory could be induced by this novel complex to prevent tumor recurrence. PTT delivers adequate thermal energy to kill cancerous cells at target site and subsequently releases TSA. Immunotherapy therapy involving h local delivery of R837 in this study, simultaneously allows the exposed tumor antigens to activate, enhance, and direct the host immune system to initiate a tumor-specific immunity. This synergistic action of PTT and immunological stimulation exhibited significant anti-tumor activity (Zhou et al. 2018).

In another study conducted by Li W et al. prepared Amantadine surface-modified PVP-PEG coated Silver nanorods to enhance HIV vaccine immunotherapeutic activity against HIV-infected cells (Li et al. 2018). It has been reported in literature that in in vivo *studies*, immune action of cytotoxic lymphocytes (CTLs), natural killer (NK) cells, and complement membrane attack complex (MAC) have crucial role in death of HIV-infected cells. This is attributed to the fact that HIV-specific CTLs produces TNF-α which induces cell death in HIV-infected cells. Based on this matter of fact of CTLs-derived TNF-α, the researcher group aimed at preparing a nanomaterial which enhanced production of CTLs-derived TNF-α in vivo and thereby promoted the death of HIV-infected cells. Amantadine (Ada) was selected as it has been reported in many literatures as immunoregulatory molecule which has capability to induce enhanced TNF-α molecule both in the in vivo as well as in the in vitro models. This moiety is also approved clinically by FDA. Further amino group present in ADA undergoes direct condensation reaction with carboxyl (-COOH) group thereby enables to surface modify the PVP-PEG coated silver nanorods. These surface engineered nanorods were capable to stimulate CTLs to produce around 8-timesmoretumor necrosis factor alpha (TNF-α) in vivo. This increase in CTLs-derived TNF-α which were specific to HIV could remarkably induced inhibition of HIV-infected cells (from 28.86% to 84.19%) and also decreased HIV production (around 6 times). In vivo study was conducted to ascertain the ability of Ada-PVP-PEG silver nanorods on regulating HIV vaccine-triggered CTLs. There were four groups each having six animals as follows: Group 1: 50 μg HIV DNA vaccine and 10 μg Ada-PVP-PEG silver nanorods per one mouse, Group 2: 50 μg HIV DNA vaccine and 10 μg PVP-PEG silver nanorods per one mouse, Group 3: 50 μg HIV DNA vaccine per one mouse, Group 4: Mice do not receive any vaccination (blank group). Flow cytometry was used to measure the percentage of two important subgroups of T cells which can produce HIV-specific TNF-α (CD3 + CD4+ T cells/T helper cells and CD3 + CD8+ T cells/CTLs). Results demonstrated that modification by Ada significantly improved production of TNF-α –by HIV vaccine-triggered CTLs, instead of T helper cells. Additionally that percentage of IL-2 and IFN-γ released by CD3 + CD4+ T cells/T helper cells and CD3 + CD8+ T cells/CTLs in all four groups was also studied. IL-2 derived from T Helper cells remarkably elevated in Group 1 and Group 2, in comparison to Group 3. CTL-derived IFN-γ in Group 1 and Group 2 showed a more production than that in Group 3. Thus this study indicates that surface modification of nanomaterial allows to provide artificially engineered carrier system with assumed immunoregulatory function (Li et al. 2018).

7.2.5 Hydrogels and Nanogels

Hydrogels are three-dimensional network formed by physical or chemical cross-linking of hydrophilic or amphiphilic polymers. Nanogels are type of hydrogels ranging in nanoscale size (20–200 nm). Polymers commonly used to prepare hydrogels include either natural polymer or synthetic polymers, such as PLGA and poly(ε-caprolactone). Polymers which are biocompatible and biodegradable are usually used to fabricate there gels. Hydrogel and nanogels are attributed to have high loading capacity and also exhibits control released pattern due to presence of microporous structure with tunable porosity and size. Unlike nanoparticles, these system can encapsulate more than one active moiety.

Park C G et al. demonstrated that when innate immunity agonists such as Toll-like receptor 7/8 (TLR7/8) or stimulator of interferon genes (STING) were formulated into biodegradable hydrogel and delivered into the tumor resection site it exhibited extended release profile and cured more percentage of animals compared to animals treated with systemic or local administration. A scaffold was prepared from biodegradable Hyaluronic acid and the ability of this scaffold to exhibit extended release of agonists of innate immunity was studied in vivo. Female BALB/cJ mice were injected orthotopically with 4 T1-Luc2 breast tumor cells in their fourth mammary fat pad. Mice were subjected to bioluminescent IVIS imaging 9 days later, to confirm tumor size was uniform among all animals and can be randomized in the study. Tumor of size approximately $100mm^3$ was resected from animal on tenth day after tumor inoculation and hydrogel containing immunomodulatory compound—anti-PD-1, anti-CTLA-4 (cytotoxic T lymphocyte antigen 4), IL-15sa, lenalidomide, celecoxib, 2′3′-c-di- AM(PS)2 (Rp,Rp) ("STING-RR"), or R848—was placed in the tumor resection site. IVIS imaging used to measure tumor burden showed that recurrence of tumor locally was suppressed successfully when innate immunity agonist such as STING-RR or R848 was delivered through hydrogel formulation. Also perioperative administration of R848 or STING-RR via hydrogel was greater compared with intratumoral injection of either compound. The results were found to be remarkable as cyclic dinucleotide STING agonists could be delivered clinically into the tumor as well as by perioperative route. The perioperative delivery was found to access not only superficial but also deep lesions (Park et al. 2018b).

Li P et al. prepared novel bioreducible cationic alginate-polyethylenimine (PEI) nanogels as a novel vaccine delivery system. Initially alginate-polyethylenimine (AP) nanogel network was prepared by process of electrostatic interaction between negatively charged carboxyl group of alginate sodium with positively charged amine group of branched Polyethylenimine (Mw = 200)(PEI2k) in ratio of 1:3. This AP nanogel network was further stabilized by disulfide cross-linking using 3,3′-dithiobis (sulfosuccinimidylpropionate (DTSSP) to produce bioreducible nanogels (AP-SS).

DTSSP was selected because it increases the stability in non-reducing extracellular matrix and also allows self-disassembly in the reducing surrounding present in cytosol and subcellular organelles, unlike the other cross linker such as suberic acid bis(N-hydroxysuccinimide ester) (DSS) which was used to prepare non-reducible AP nanogels (AP-CC) for comparison. It has been stated that at higher mass ratio of AP nanogel to cross-linking, 10:1 sufficient stability and also entrapment of model antigen OVA was observed compared to lower ratio 1:1 where aggregation was observed. It was observed that for both AP-SS and AP-CC placebo nanogels exhibited the particle size and zeta potential of around 80 nm and + 40 mV. While incorporation of OVA antigen slightly increased the particle size (100 nm) and also dropped the zeta potential(+20 mV). At ratio of nanogels to OVA at 5:1 the encapsulation efficiency and antigen loading capacity of AP-SS gel was over 90% and 38.99%, respectively, indicating that nanogels serves to be potential carrier of antigens. For any vaccine mediated immune response it is very much necessary to evaluate the antigen uptake by APCs. This study was conducted using mouse BMDCs and Raw 264.7 mouse macrophages at 37 °C. It was observed that both AP-SS as well AP-CC markedly enhanced cellular uptake of FITC labeled OVA antigen. Intracellular processing of antigens after the cellular uptake by BMDCs was studied using DQ albumin (DQ-OVA), a self-quenched bright green fluorescent OVA conjugate, containing nanogel formulation. It was observed that after cellular uptake DQ-OVA underwent proteolytic degradation to produce small peptides by enzyme proteases which are further presented through MHC class I or II molecules to stimulateCD8+ and CD4+ T cells, respectively, in order to activate APC-primed T cell when BMDCs were subjected to free DQ-OVA, AP-SS, and AP-CC incorporated DQ-OVA nanogels it was observed that only AP-SS showed significant fluorescence intensity than other free form and AP-CC nanogel treated cells. This high intensity is attributed to inherent capability of DTSSP to undergo cleavage by bioreductants present in the cytosol and later in endosomes, thereby increasing the rate of nanogel disassembly and antigen release and degradation unlike non-reducible DSS containing AP-CC nanogel. In contrast to AP-CC nanogels, the bioreducible AP-SS nanogels greatly increased vaccine-induced antibody production of IgG production and CD8+ T cell-mediated tumor cell lysis, which was attributed to their ability of encouraging intracellular antigen processing and allowing MHC class I/II antigen presentation. The AP-SS nanogel increased antibody IgG responses and Th1 cytokine production in mice model when compared with AP-CC treated group. This bioreducible nanogel acts as potent adjuvant to increase vaccine-induced humoral and cellular immune responses in infections and cancers (Li et al. 2013).

7.3 Conclusion

In the present chapter we have discussed various novel drug delivery systems and their applicability in immunotherapeutic domain. Undoubtedly it can be concluded that novel systems such as nanoparticles, liposomes, micelles, hydrogels, and nanorods have shown to spectacularly enhance the delivery of antigens to APC's in lymph nodes and induce a strong immune response in preclinical studies. However there is an urgent need to have relevant clinical data which would completely allow easy transition of this approach from clinical trial to current clinical application. In near future it can be seen that these nanocarriers will enhance the efficacy of immunotherapeutic and thereby enhance patient's quality of life.

References

Chen Q, Wang C, Zhang X et al (2019) In situ sprayed bioresponsive immunotherapeutic gel for post-surgical cancer treatment. Nat Nanotechnol 14:89–97

Getts DR, Shea LD, Miller SD, King NJC (2015) Harnessing nanoparticles for immune modulation. Trends Immunol 36:419–427

Jiang D, Mu W, Pang X, Liu Y, Zhang N, Song Y, Garg S (2018) Cascade cytosol delivery of dual-sensitive micelle-tailored vaccine for enhancing cancer immunotherapy. ACS Appl Mater Interfaces 10:37797–37811

Kishimoto TK, Maldonado RA (2018) Nanoparticles for the induction of antigen-specific immunological tolerance. Front Immunol. https://doi.org/10.3389/fimmu.2018.00230

Koshy S, Cheung A, Gu L, Graveline A, Mooney D (2017) Liposomal delivery enhances immune activation by STING agonists for cancer immunotherapy. Adv Biosyst 1:1–24

LaMothe RA, Kolte PN, Vo T et al (2018) Tolerogenic nanoparticles induce antigen-specific regulatory T cells and provide therapeutic efficacy and transferrable tolerance against experimental autoimmune encephalomyelitis. Front Immunol 9:1–11

Li P, Luo Z, Liu P, Gao N, Zhang Y, Pan H, Liu L, Wang C, Cai L, Ma Y (2013) Bioreducible alginate-poly(ethylenimine) nanogels as an antigen-delivery system robustly enhance vaccine-elicited humoral and cellular immune responses. J Control Release 168:271–279

Li W, Balachandran YL, Hao Y, Hao X, Li R, Nan Z, Zhang H, Shao Y, Liu Y (2018) Amantadine surface-modified silver nanorods improves immunotherapy of HIV vaccine against HIV-infected cells. ACS Appl Mater Interfaces 10:28494–28501

Liu Z, Zhou C, Qin Y et al (2017) Coordinating antigen cytosolic delivery and danger signaling to program potent cross-priming by micelle-based nanovaccine. Cell Discov 3:1–14

Lu J, Liu X, Liao YP, Wang X, Ahmed A, Jiang W, Ji Y, Meng H, Nel AE (2018) Breast Cancer chemo-immunotherapy through liposomal delivery of an immunogenic cell death stimulus plus interference in the IDO-1 pathway. ACS Nano 12:11041–11061

Neef T, Miller SD (2017) Tolerogenic nanoparticles to treat islet autoimmunity. Curr Diab Rep. https://doi.org/10.1007/s11892-017-0914-z

Park W, Heo Y-J, Han DK (2018a) New opportunities for nanoparticles in cancer immunotherapy. Biomater Res 22:1–10

Park CG, Hartl CA, Schmid D, Carmona EM, Kim HJ, Goldberg MS (2018b) Extended release of perioperative immunotherapy prevents tumor recurrence and eliminates metastases. Sci Transl Med. https://doi.org/10.1126/scitranslmed.aar1916

Pei W, Wan X, Shahzad KA, Zhang L, Song S, Jin X, Wang L, Zhao C, Shen C (2018) Direct modulation of myelin-autoreactive CD4+ and CD8+ T cells in EAE mice by a tolerogenic nanoparticle co-carrying myelin peptide-loaded major histocompatibility complexes, CD47 and multiple regulatory molecules. Int J Nanomedicine 13:3731–3750

Pellegrino M, Ceccacci F, Petrini S, Scipioni A, De Santis S, Cappa M, Mancini G, Fierabracci A (2019) Exploiting novel tailored immunotherapies of type 1 diabetes: short interfering RNA delivered by cationic liposomes enables efficient down-regulation of variant PTPN22 gene in T lymphocytes. Nanomed Nanotechnol Biol Med 18:371–379

Perri V, Pellegrino M, Ceccacci F, Scipioni A, Petrini S, Gianchecchi E, Lo Russo A, De Santis S, Mancini G, Fierabracci A (2017) Use of short interfering RNA delivered by cationic liposomes to enable efficient down-regulation of PTPN22 gene in human T lymphocytes. PLoS One 12: e0175784

Pujol-Autonell I, Mansilla MJ, Rodriguez-Fernandez S et al (2017) Liposome-based immunotherapy against autoimmune diseases: therapeutic effect on multiple sclerosis. Nanomedicine 12:1231–1242

Reddy ST, Rehor A, Schmoekel HG, Hubbell JA, Swartz MA (2006) In vivo targeting of dendritic cells in lymph nodes with poly (propylene sulfide) nanoparticles. J Control Release 112:26–34

Riley RS, June CH, Langer R, Mitchell MJ (2019) Delivery technologies for cancer immunotherapy. Nat Rev Drug Discov 18:175–196

Rodriguez-Fernandez S, Pujol-Autonell I, Brianso F et al (2018) Phosphatidylserine-liposomes promote tolerogenic features on dendritic cells in human type 1 diabetes by apoptotic mimicry. Front Immunol. https://doi.org/10.3389/fimmu.2018.00253

Senapati S, Darling RJ, Loh D, Schneider IC, Wannemuehler MJ, Narasimhan B, Mallapragada SK (2019) Pentablock copolymer micelle nanoadjuvants enhance cytosolic delivery of antigen and improve vaccine efficacy while inducing low inflammation. ACS Biomater Sci Eng 5:1332–1342

Theunis C, Crespo-Biel N, Gafner V et al (2013) Efficacy and safety of a liposome-based vaccine against protein Tau, assessed in tau.P301L mice that model tauopathy. PLoS One. https://doi.org/10.1371/journal.pone.0072301

Yeste A, Takenaka MC, Mascanfroni ID et al (2016) Tolerogenic nanoparticles inhibit T cell-mediated autoimmunity through SOCS2. Sci Signal. https://doi.org/10.1126/scisignal.aad0612

Yuba E, Harada A, Sakanishi Y, Watarai S, Kono K (2013) A liposome-based antigen delivery system using pH-sensitive fusogenic polymers for cancer immunotherapy. Biomaterials 34:3042–3052

Zhang Z, Tongchusak S, Mizukami Y, Kang YJ, Ioji T, Touma M, Reinhold B, Keskin DB, Reinherz EL, Sasada T (2011) Induction of anti-tumor cytotoxic T cell responses through PLGA-nanoparticle mediated antigen delivery. Biomaterials 32:3666–3678

Zhou B, Song J, Wang M, Wang X, Wang J, Howard EW, Zhou F, Qu J, Chen WR (2018) BSA-bioinspired gold nanorods loaded with immunoadjuvant for the treatment of melanoma by combined photothermal therapy and immunotherapy. Nanoscale 10:21640–21647

Discovery, Screening Methods, Design Considerations, and Scale-up Aspects of Immunotherapeutic Drugs

8

Pratiksha Palahe and Vinal Pardhi

Abstract

Over the past decade, immunotherapy has tremendously augmented the traditional treatment strategies in cancer cure. It has provided newer opportunities to research and development of enormous varieties of lifesaving drugs—each representing its strengths and weaknesses. None-the-less, it has improved treatment protocols, provided simple solutions to complex biological problems, and increased the overall disease-free survival within patients. However, the beneficiaries of these technological advancements are very few. This may be due to the heterogeneity within individuals (which is generally not considered for development of Immunotherapeutics) or due to the fact that it is high risk, high investment, and high returns technology, supplemented with side effects. Some of these manageable factors like cost can be worked upon by newer strategies or improvising existing manufacturing processes—to scale up within regulatory compliances. Process scale up is an integral step in large scale manufacturing that demands for critical understanding of process attribute, analytical tool, and product quality. Designing of scalable technology . and manufacturing machineries with emphasis on process invariability, minimized time, and cost of therapy will provide the success of any immunotherapeutic drug. This chapter is an attempt to provide a bird's eye view on the various opportunities available in the drug designing, scale up and analytics of immunotherapeutic drugs.

Keywords

Immunotherapy · Monoclonal antibodies · Car-T cell, Cellular therapies · Bioreactor design · Closed loop Automation

P. Palahe (✉) · V. Pardhi
National Facility for Biopharmaceuticals, G N Khalsa College, Nathalal Parekh Marg, Mumbai, India
e-mail: pratiksha@nfbindia.in

S. P. Sawarkar et al. (eds.), *Immunotherapy – A Novel Facet of Modern Therapeutics*,
https://doi.org/10.1007/978-981-15-9038-2_8

8.1 Introduction

The immune system has played an integral role in defense, in grafts or in presence of tumors (Miller and Sadelain 2015). This has been regulated by the soldiers of the immune system, primarily B and T cells. This led to an intricate linkage between immunotherapy and oncology where the primary evidence on thymopoiesis and lymphocyte subsets came from leukemogenesis within mice. This further led to the understanding of antigen recognition and presentation, activation, and costimulation among T cells. Since then, focus has been on T-lymphocytes from their discovery to genetic engineering in cancer immunotherapy. These T cells were then exploited as cell therapy within melanoma patients further leading to the development of cancer vaccines (Miller and Sadelain 2015). This trajectory has led to the development of a newer therapy, popularly known as Immunotherapy. This chapter mainly deals with the evolution of immunotherapeutic drugs in oncology.

Immunotherapy is using body's own immune system to fight a disease. This can be done either by stimulating the immune system or supplementing it with proteins or antibodies to assist the immune system.

Types of immunotherapy

1. Immunotherapy vaccines (Cancer.net 2020; Cancer.net 2018): utilizes certain proteins which when administered into the body boost the immune system. This can help fight certain cancers. Within therapy, these proteins can improve overall survival.
2. Checkpoint blockade inhibitors.
3. Adaptive cell transfer: WBC's from patient's body are conditioned to recognize cancer cells and then reintroduced as therapy.
4. Antibody therapy (Cancer.net 2018) and tumor agnostic therapy (Cancer.net 2020).
5. Oncolytic viral therapy (Cancer.net 2020).

Immunotherapy works by

1. Stopping or slowing the growth of cancer cells.
2. Stopping the spread of cancer to other parts.
3. Boosting the immune system to destroy cancer (Cancer.net 2020).

8.1.1 Development of Immunotherapeutic Vaccines

Cancer vaccines work by increasing the T cell population and augmenting its function. There is a fine regulation between T cell proliferation for immune reaction and excessive activation causing lymphomas. This same regulation is important for vaccines as well. These vaccines are developed mainly as dendritic cell activators, adjuvants, T cell stimulators and growth factors, checkpoint inhibitors and vaccines which neutralize cytokines and oncogenic enzymes. These vaccines can be of two major types.

1. Prevention vaccines.
2. Treatment vaccines (Schilsky 2018).

Vaccine inducing T cell response has proven to be effective. The efficacy of vaccine depends on the numerical increase in T cell population and also the persistent time of T cell in circulation. These vaccines present antigen to the T cell causing their recognition clonal expansion to target cancer. The major problem facing immunotherapy today is the lack of available agents with established immunologic function, especially vaccine adjuvants. For example, GM-CSF (Granulocyte macrophage colony stimulating factor) is mostly used as adjuvant in academic clinical trials for vaccines, however, it is FDA approved as a growth factor and not as vaccine. The same is true for imiquimod. Some of the lead candidates for cancer vaccine development are

1. IL-15: T cell growth factor which inhibits antigen induced cell death of T cells. IL-15 is made by dendritic cells and macrophages and other stromal cells and their target cells include CD4+ and CD8+ T cells, NK cells, and mast cells. As vaccine candidate, it promotes enhanced life span of CD8+ T cells. It also improves the sensitivity of T cells towards allergens and cause in vitro differentiation of monocyte derived dendritic cells. It can also recruit CD4+ T cytotoxic cells and has shown tumor regression in mice models. Additionally, it can also be used as a cytokine therapy to treat cancers (discussed later in the chapter). Similarly IL-7 is another T cell growth factor required for T cell development and survival of naïve T cells. Clinical trials provide evidence that administration of IL-7 causes substantial increase in CD4+ and CD8+ T cells, NK cells, and δγ T cells. No significant impact was observed on mature B cells or Treg cells. In mice, it has been demonstrated as vaccine adjuvant. Both IL-7 and IL-15 can be used as vaccine or adjuvants as they both work on different population of T cells. The former proliferate naïve T cells whereas the latter focuses on T effector population/T memory cells.
2. The second class of cancer vaccines is represented by T cell checkpoint inhibitors represented by PD-1 (Programmed death −1). PD-1 blockage is known to have long lasting effect on tumor regression. The details of this therapy are discussed further in the chapter.
3. The next category of molecule is the CD40 agonists which work as APC stimulators. Activation of APC has potential therapeutic action as it works directly on tumor inhibition and prevents angiogenesis. It also activates the NK cells and macrophages for tumor targeting. In vitro studies on tumor cells have shown regression of B-cell lymphomas causing heightened apoptosis and necrosis. These CD40 agonists are used as either recombinant trimeric ligand or as monoclonal antibody. It has been used in clinical trials against melanomas, Non-Hodgkin's lymphoma, and some solid tumors. It can be combined with

other therapies like chemotherapy, radiation, TLR therapy or in combination with cytokines.

4. Enzyme inhibitors as vaccine candidates: immunosuppressive enzymes like indole2,3-dioxygenase (IDO) are overexpressed in tumor cells. These enzymes degrade tryptophan thereby making it unavailable for activation of T cells. 1-methyl tryptophan (1MT) is an inhibitor of IDO and thus is a strong candidate as immunotherapeutic drug. 1MT in combination with chemotherapy limits the tumor growth. Clinical trials are underway for this therapy.
5. T cell stimulators like anti-CD137 have shown considerable stimulatory and anti-apoptotic activity. CD 137 is present on T cells, NK cells, and NKT cells but absent on tumor cells. Anti-CD137 could be specifically used to promote T cell proliferation either as a monotherapy or in conjugation. Its anti-tumor activity has been observed in in vivo murine models though with some side effects, while phase I human trials are underway where anti-CD137 (anti-4-1BB) is used in combination with paclitaxel and carboplatin and in another with radiotherapy.

There are still many other leads like DC growth factors (Flt3 ligand) which can act as vaccine adjuvants. Others being IL-12, CpG, monophosphoryl lipid A, resiquimod and 852A(which are imidazoquinolinamines) and polylC and polylCLC (potent TLR agonists) (Mac Cheever 2008).

8.1.2 Development of Checkpoint Inhibitors

Cancer cells have several mechanisms to evade the immune system. Checkpoint molecules constitute of stimulatory molecules which activate the immune system and inhibitory molecules which suppress the immune system. Stimulatory checkpoint molecules promote activation of naive T cells, effector T cells, memory T cells, and regulatory T cell responses. Inhibitory checkpoint molecules limit the threshold of T cell activation, duration of immune response which in turn affects the inflammatory status, disease tolerance, and homeostasis. The multiple ways in which the immune system may regulate the disease status is as follows:

1. They may secrete proinflammatory cytokines to recruit T_{regs} and MDSCs (myeloid derived suppressor cells) to cause an immune compromised condition.
2. They may lose their antigenicity by loss of MHC expression or dysregulating the antigen processing machinery.
3. By decreasing the T cell mediated killing by expressing PD-L1 (programmed cell death ligand 1)—an inhibitory checkpoint molecule.

PD1 is a negative regulator of T cell function. It belongs to the immunoglobulin superfamily and structurally related to CTLA-4 and CD28. Its ligand is expressed on T and B cells, macrophages, and DCs, as well as parenchymal and tumor cells (Mac Cheever 2008). Activation of PD1 causes impaired cytokine production and loss of cytotoxicity of activated T cells (Ni and Dong 2017). PD-1 L is expressed by esophageal, colon, lung, and ovarian cancer and melanoma by immunohistochemistry. Thus, these tumors can be used as appropriate candidates for tumors having high infiltration of T cells for initial testing of anti-PD-1. T cell checkpoint inhibitors like CTLA4 and PD-1 have shown remarkable response as standalone as well as combination therapies (Dougall et al. 2017). Even though the results are positive only on a subset of patients, they have triggered greater research in utilizing checkpoint as immunotherapeutic targets for treatment of cancer. Anti-CLA4 blocking antibodies have shown heightened anti-tumor immune response. This has clinically led to the development of anti-CTLA4 antibody—Ipilimumab which is an US FDA approved immunotherapy for melanoma.

Same is true for PD-1 blocking antibodies like Nivolumab and Pembrolizumab which have shown 30–50% response rates in melanoma, non-small cell lung cancer, kidney cancer, Hodgkin's disease, head and neck and bladder cancer. This has provoked researchers into identifying other potential checkpoints to be used as immunotherapeutic targets (Sharpe 2017).

TIGIT (T cell Immunoglobulin and ITIM domain) and CD96 are coinhibitory receptors. Along with CD226 (costimulatory receptor), they trigger a pathway that is analogous to CD28/CTLA4. These could be explored as potential targets as combination to existing therapy. Regulatory and effector T cells within the tumor have an enriched expression of TIGIT in comparison to normal peripheral tissue. This makes them more as promising candidates as it reduces the risk of autoimmune like toxicity. Preclinical studies support co-targeting of TIGIT with PD-1/Tim3.A similar combination could be used along with anti-CD96 antibodies (Dougall et al. 2017).

Another category of checkpoint inhibitor is the B7 family which consists of cell surface protein ligands which are structurally related. These ligands bind to the receptor on lymphocytes and mediate immune response. These can either initiate costimulatory or coinhibitory response (Collins et al. 2005). The B7 family has at least ten members (Fig. 8.1). Each of these could be explored for their role as novel checkpoint regulators in cancer therapy. B7-H3 in mice has demonstrated coinhibitory functions on T cells and NK cells. Anti-B7-H3 antibodies are currently under phase I/II clinical trials as promising check point immunotherapeutic targets. B7-S1 is a potential biomarker in many solid tumors. However, it is currently not being explored for therapeutic application. This can definitely display potential in cancer treatment through targeted inhibition using anti-B7-S1 antibodies. VISTA (V-domain Ig suppressor of T cell activation) is another such target which can be explored for potentials. It is predominantly expressed on myeloid cells and can suppress T cell activation between APC and T cells.

Each of these proposed targets can either be used singly or in combination therapy for treatment of cancers (Ni and Dong 2017).

Similar to adaptive immunity, cells of the innate immunity also express some checkpoint molecules which can be employed to enhance or suppress an immune response.

For example, APC which are involved in Ag recognition, acquisition, processing, and presentation to T cells express many ligands and costimulatory molecules. Some of the inhibitory molecules like CD47, TAM receptors (Tyro, Axl, and MerTK of

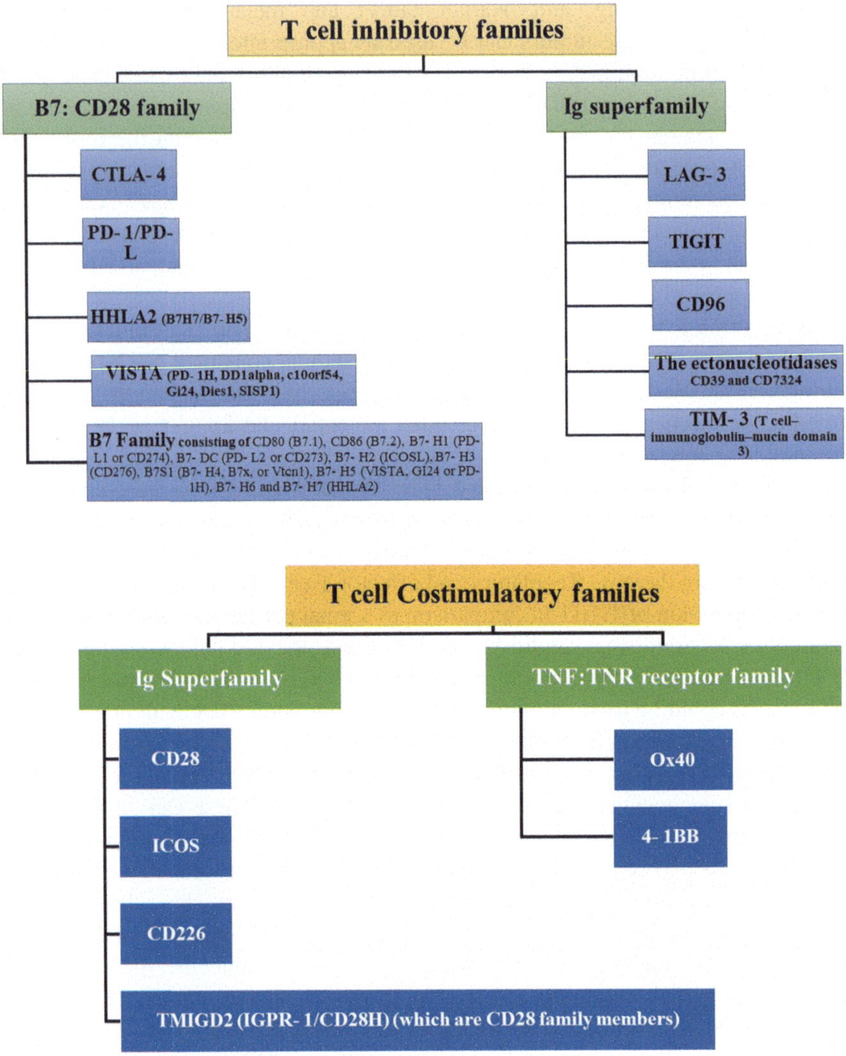

Fig. 8.1 The figure above depicts the major classes of molecules which have the capability to either stimulate or inhibit T cell activation for tumor recognition. Each of these molecules depicts potential to be converted into immunotherapeutic targets. However, they may pose some limitation based on their limitations (Ni and Dong 2017; Sharpe 2017)

tyrosine kinase family), and Siglecs can act as potential targets for immunotherapy and stimulate the innate anti-tumor immunity (Sharpe 2017).

8.1.3 Development of Adaptive Cell Transfer Technology

This technology involves exploiting the indigenous cells of the immune system and enhancing their functionality using various recombinant therapy approaches. It has been successfully used on T cells and NK cells.

T cell therapy is most commonly known as Chimeric Antigen Receptor (CAR)—T cell Therapy (CAR-T) (Schilsky 2018). Understanding costimulatory pathways have led to many therapeutic advances in T cell therapy. One of them being the checkpoint inhibitors which are based on monoclonal antibodies. Patient derived T cells can be genetically engineered to target specific tumors. This is done through specialized receptors—T cell receptors (TCR) present on the T cells for recognizing tumor specific antigens or through synthetic receptors called CAR (Chimeric antigen receptors) (Fig. 8.2).

CARs are complex receptors designed by integrating B and T cell antigen recognition sites. They offer immunogenic advantage over TCR as they can evade HLA recognition thus making them applicable to all types of patients. They not only target the tumor but also enhance T cell function. Clinically both these approaches have shown promising results in treatment of lymphoblastic leukemia (targeting CD19). 2nd generation CARs offer potential leads as immunotherapeutic drugs for cancer. They may be combined with cytokines such as IL-15 or IL-12, or ligands with costimulatory receptors which will enhance their T cell potency, specificity, and safety (Miller and Sadelain 2015). Currently there are three FDA approved treatments using CAR-T cell therapy. They are Tisagenlecleucel (KymriahTM)—approved for the treatment of B-cell precursor acute lymphoblastic leukemia (ALL), Axicabtagene ciloleucel (Yescarta™)—FDA approved for the treatment of adult patients with relapsed or refractory large B-cell lymphoma and Tocilizumab (Actemra®)—approved for the treatment of adults and pediatric patients 2 years of age and older with severe or life-threatening cytokine release syndrome (CRS).

NK cells target tumor cells via germline encoded cell surface receptors. This makes them the part of the innate immunity system (Childs and Carlsten 2015). The γ chain family of cytokines interleukins IL-2 and IL-15 as well as proinflammatory cytokine IL-12 are being characterized for their capability to stimulate NK cells for their anti-tumor immunity. Most of them are in their early preclinical development phase (Childs and Carlsten 2015).

Ex Vivo Manipulation of NK Cells

This immunotherapeutic strategy involves manipulating NK cells ex vivo for pre-activation prior to infusion. This involves

(a) Short-term ex vivo NK cell activation: Haploidentical NK cells stimulated with IL-2 prior to infusion has shown positive clinical response in AML and multiple myeloma patients. Immunosuppression using chemotherapy (fludarabine and cyclophosphamide) is performed to prevent rejection of infused NK cells and also to facilitate homing and expansion of NK cells. IL-15 can also be preferred to suppress activation of T_{reg} cells.

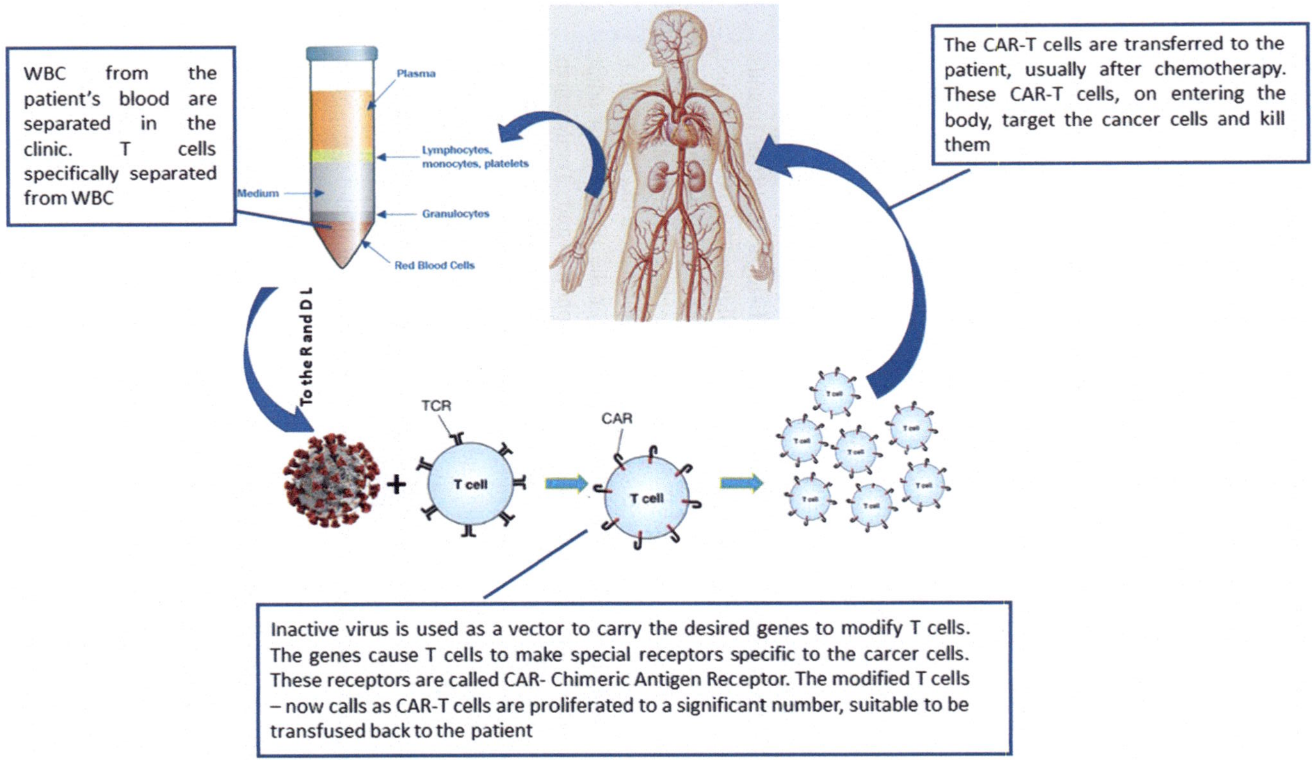

Fig. 8.2 depicts the design of a typical CAR-T cell therapy. Autologous T cells are isolated from patient using leukopheresis and genetically modified ex vivo using viral and non-viral transfection methods. First designed in 1989, the extracellular domain of CAR is composed of an antigen binding moiety and a spacer arm. These antigen binding moieties could be a single chain Fv of an antibody, a human Fab fragment isolated from phage libraries or any natural ligand. The CARs have the capacity to recognize carbohydrate and glycolipid molecules present on the tumor cells. They are capable of bypassing the MHC mediated lysis pathway and predominantly execute cytolysis of tumor cells by exocytosis using perforins and granzyme or sometimes triggering Fas-L ot TNF-R death receptor

signalling pathways. Usually the spacer region is of IgG1. T cells care engineered to produce CAR by multiple gene transfer techniques such as viral transduction, using transposons, mRNA mediated infection using various delivery routes like nanoparticles, liposomes or electroporation and CRISPR-Cas9 technology. These re-engineered CAR-T cells are transfused back to the patient under immune compromised condition to prevent rejection. Once successfully infused, they will eradicate the cancer cells and provide long-term remission (Miliotou and Papadopoulou 2018). Pic credits Leukemia and Lymphoma Society (https://www.lls.org/treatment/types-of-treatment/immunotherapy/chimeric-antigen-receptor-car-t-cell-therapy)

(b) Ex vivo expansion is carried out by enriching cells from peripheral blood, umbilical cord blood, hematopoietic progenitors, embryonic stem cells or iPSC. This is primarily done by promoting ex vivo expansion using IL-2 and IL-15. Modified protocols now use feeder cells like PBMC, T cells, EBV transformed lymphoblastoid cell lines, and K562 cells.
(c) Genetic manipulation in NK cells is a challenging task as the efficiency of viral transduction is poor. Alternatively, mRNA electroporation is preferred. CD19 specific chimeric antigen receptor coded mRNA when transfected into NK cells displayed enhanced activity towards B-cell malignancies both in vitro and in preclinical animal models.

Pharmacological and Cellular Approaches

Cytokine activated ex vivo expanded NK cells when glycosylated using TZ101 (recom. α1–3 flucosyltransferase VI) enhances their ability to bind to recombinant E selectin. This improved their homing capacity to the bone marrow. This could be used as a novel approach to treat hematological malignancies. Another approach involves coculturing NK cells with K562 cells. This caused genetic modification of ex vivo expanded NK cells to express CC chemokine receptor type 7 [CCR7]. These CCR7 expressing NK cells when infused into mice showed improved homing capacity to lymph nodes (Childs and Carlsten 2015).

Improvements such as integrating thymopoiesis with cell engineering, generating T cells from hiPSC or in vitro maturation of T cells will open greater avenues as immunotherapeutic targets and increase disease outcomes with additional functional features (Miller and Sadelain 2015).

8.1.4 Development of Immunotherapeutic Monoclonal Antibody Therapy and Tumor Agnostic Therapy

These are used as targeted therapies to block abnormal proteins in cancer (Cancer.net 2020). Binding of Mabs to cancer cells flags them to the immune system which can then destroy them (Cancer.net 2020). Some Mabs release the brakes on the immune system. These brakes majorly form the checkpoint inhibitors like PD1 and CTLA4. Anti-checkpoint inhibitors antibodies bind to these brakes and release them so that the immune system can act (Cancer.net 2020) (Table 8.1).

Even though checkpoint therapy has been approved for certain type of cancers, they can be used to treat tumors elsewhere in the human body—called as tumor agnostic therapy. For example, Pembrolizumab has been approved for most metastatic tumors having high microsatellite instability (MSI-H) or DNA mismatch repair deficiency (dMMR). Nivolumab (Opdivo) has also been approved for MSI-H and dMMR metastatic colorectal cancer. FDA approved Larotrectinib is also a targeted therapy for solid tumors having gene alterations known as neurotrophic receptor tyrosine kinase (NRTK) gene fusion (Schilsky 2018).

Monoclonal antibodies like Rituximab (CD20 specific mAb) were used in the treatment of lymphoma patients homozygous for CD16-125 V gene in comparison

Table 8.1 The above table shows the potential targets for development of immunotherapeutic drugs and the implications they are being used for with their commercial trade names. It shows the successful development of these drugs from bench to bedside (Cancer.net 2020)

Checkpoint inhibitors	Trade name	Target	Disease
Ipilimumab	Yervoy	CTLA4	Late stage metastatic melanoma (Cancer.net 2020; Ni and Dong 2017)
Nivolumab	Opdivo	PD1	Melanoma, lung and renal cell carcinoma, Hodgkin lymphoma, head and neck, colon, and liver cancer.
Pembrolizumab	Keytruda	PD1	Melanoma, lung cancer, head and neck cancer, Hodgkin lymphoma and stomach cancer.
Atezolizumab	Tecentriq	PD-L1	Urothelial carcinoma, non-small and small cell lung cancer, triple negative breast cancer.
Avelumab	Bavencio	PD-1	Non-small cell lung cancer, Merkel cell carcinoma in adults,
Durvalumab	Infinzi	CD274	Non-small cell lung cancer, cancer of bladder or urinary tract.

to CD16-158F polymorphism. Cetuximab showed similar effects on polymorphism for patients under metastatic colon cancer treatment. This CD16 gene is known to have importance in NK cell mediating anti-tumor activity via ADCC. Another such MAb is GD2 which is used in the treatment of neuroblastomas with the absence of one or more KIR ligands (Childs and Carlsten 2015).

Combining mAb with cytokines such as IL-2 or IL-15 could theoretically improve the activity of NK cells.

Another therapeutically potential molecule is IPH2102 (also known as Lirilumab) which is a fully human monoclonal Ab that blocks the HLA-C binding to KIR2D receptor which are expressed on the surface of NK cells. This binding enhances the cytotoxicity of NK cells towards HLA class I mediated tumor targets. It has shown much positive results in preclinical mouse models. Phase II clinical trials in AML patients with multiple myeloma have rendered the product safe (Childs and Carlsten 2015).

Bispecific mAbs also called as bispecific/trispecific killer engagers (BiKEs or TriKEs) are engineered molecules that cross link epitopes on tumor cell and DC16 receptor on NK cells initiating ADCC. They are advantageous over mAb as they bind to different epitopes, e.g. CD16-CD33 BiKE boost NK cell cytokine production in myeloid malignancies (Childs and Carlsten 2015).

8.1.5 Development of Oncolytic Viral Therapy

Oncolytic viral therapy uses genetically modified virus to kill cancer. Viruses are injected into cancer cells and allowed to undergo lysis, releasing antigens. These antigens trigger the immune system, scavenging for cancer cells having those specific antigens. In 2015, US FDA approved the first oncolytic viral therapy called Talimogene laherparepvec (Imlygic) or T-Vec for the treatment of melanoma

(Kaufman et al. 2015). It is a GM version of Herpes simplex virus (Schilsky 2018). The mechanism of action of oncolytic viruses is not clearly understood. However, they are thought to mediate anti-tumor activity by selectively replicating within the tumor cells and causing lysis, leading to trigger the immune system. The one advantage that the tumor cells offer for selectively hosting the virus is the absence of protein kinase R (PKR). This is present in normal cells and allows clearance of virus upon infection. Also the IFN pathway is abnormal in cancer cells.

Many viruses have been proposed as immunotherapeutic agents like adenovirus, poxvirus, HSV-1, coxsackievirus, poliovirus, measles virus, New Castle disease virus, reovirus—many of which have entered early phase trials. H101, a genetically modified oncolytic adenovirus has been approved for treatment of nasopharangeal carcinoma along with chemotherapy (Kaufman et al. 2015).

8.1.6 Development of Non- Specific Immunotherapeutic Drugs

These therapies, like Mabs, can also boost the immune system. They can be administered as adjuvant therapy along with chemo and radiation or post-treatment to mitigate disease progression. These therapies include:

1. Interferons: which slow the cancer growth. INF α [recom. INFα [2a]—Roferon, 2b—Intron A and N3—Alferon] are mostly used for treatment of cancer with significant side effects.
2. Interleukins—boost the immune system to destroy cancer. Recom. IL-2 [Aldesleukin] is used to treat metastatic kidney cancer/metastatic melanoma. This was one of the first cytokines to be used to boost immunity in cancer patients. Preclinical proof of concept studies revealed therapeutic potential of this drug with relevant toxicity at high doses and recruitment of T_{reg} cells at low doses (Childs and Carlsten 2015).

Modifications are underway to develop variants of IL-2 constructs which selectively bind to IL-2β receptor expressed on NK cells. This may provide better in vivo boost to NK cells anti-tumor activity.

Another lead candidate can be IL-15 which preferentially stimulates $CD8^+$ T cells and non-terminally differentiated NK cells. However, single chain recombinant IL-15 (scIL-15) evaluated on cancer patients showed dose dependent grade 3/4 toxicity. While, AML patients on scIL-15 therapy supplemented with NK cell infusion showed persistence and proliferation of NK cells. Improved therapy involved using heterodimer IL-15 (IL15-scIL-15Rα) was found to more potent. Heterodimer IL-15 (IL-15N72D and IL-15RαSu/Fc) showed improved half-life with reduced dosing (Childs and Carlsten 2015).Similarly, other proinflammatory cytokine (IL-12) can be considered as potential targets.

3. Thalidomide derivatives like lenalidomide (Revlimid) and pomalidomide (Pomalyst/Imnovid) are immunomodulatory drugs used in the treatment of

multiple myeloma and myodysplastic syndromes. They directly/indirectly act by stimulating anti-tumor immunity like boosting NK cell production and cytotoxicity. These drugs can be coupled with MAb to further enhance their response (Childs and Carlsten 2015).

Drugs that sensitize tumors to NK cells

(a) Proteasome inhibitors and antracyclins: drugs like bortezomib (Velcade) and carfilzomib (Kyprolis) upregulate TRAIL receptors (TNF related apoptosis inducing ligands) on tumor cell surface. This causes recruitment of NK cells initiating apoptosis through cleavage of caspase 8. This therapeutic approach has been successful in animal models. This is a targeted therapy as normal cells expressing decoy TRAIL receptors are insensitive to it. These proteasome inhibitors can also sensitize tumor cells by upregulating NKG2D receptor ligands on tumor cells. Bortezomib with NK cells infusion have shown positive results in renal cell carcinoma and chronic lymphocytic leukemia. Doxorubicin (anthracycline antibiotic) works similar to proteasome inhibitors with activating TRAIL receptors.
(b) HDAC inhibitors have proved anti-tumor activity in vitro. Valproic acid and Romidepsin (Istodax) upregulates NKGZD ligand and sensitizes tumor cells to killing via NK cells. Valproic acid has shown positive results in monoblastic leukemia and hepatoma cells.
(c) Selenite—a selenium derivative reduces the tumor cell expression of HLA-E making them susceptible to CD94-NKGZA$^+$ NK cells. This also uses a target based approach as normal cells are rendered harmless (Childs and Carlsten 2015).

4. Nucleic acids have been used as drugs (conceptualized over three decades now) where in vitro transcribed mRNA or plasmid DNA encoded corresponding proteins when injected into mice skeletal muscles. In the 1960s, in vitro transcribed mRNA (IVT mRNA) was being explored for its application in protein substitution and vaccine development in cancer and various infectious diseases. The therapeutic approach does not involve entry into the nucleus. IVT mRNA gets translated into proteins on entering into cytoplasm in contrast to DNA therapeutics which need entry into the nucleus. It also offers other advantages over viral DNA based drugs that they do not integrate into the genome causing insertional mutagenesis, are transiently active, simple to produce and relatively inexpensive. As therapeutically active cancer vaccines, IVT mRNA has been successful in preclinical phase and has reached Phase III testing. For other applications as surrogate protein therapy in oncology, cardiology, endocrinology, hematology, and pulmonary disease, IVT mRNA are still under preclinical evaluation.

Approaches for IVT mRNA

1. Ex vivo transfer into patient's cells where the transfected cells are administered back to the patient. Such approach has been investigated for genome engineering, genetic reprogramming, T cell, and dendritic cell bases immunotherapy for cancer and other infectious diseases.
2. Direct delivery using various routes. This is being exploited for application in oncology, infectious diseases, allergies, and as protein-replacement therapies.

Currently, mRNA based drugs are being used as immunotherapeutic, protein-replacement therapies, and as regenerative medicines. It has been used in cancer treatment since 1995 where it was successfully used to elicit antigen-specific and T cell based immune response. Since then, many clinical trials have been undertaken to prove its efficacy. Argos therapeutics had initiated a phase III clinical trial for advanced renal cell carcinoma. This technology is relatively versatile, robust, and cost effective. Its combination with TCR and CAR is being looked into for personalized treatment and is opening greater avenues for treatment strategy (Sahin et al. 2014).

8.2 Design Considerations and Scale-up Aspects of Immunotherapeutic Drugs

Introduction

Immunotherapeutics has been recognized by differing molecular concepts associated with it. It is generally referred as immunological molecules or cells that have therapeutic effect towards classified diseases or disorders. It includes molecules like Monoclonal antibodies, vaccines, fusion proteins, soluble cytokine receptors, recombinant cytokines, small-molecule mimetics, and cellular therapies like CAR-T therapy. It could be also defined as the molecules that are used in Immunotherapies which is one of the most successful forms of therapies today. The efficiency of these therapeutics impacted its drastic increment in demand globally. These agents could successfully engage in treatment of cardiovascular, inflammatory, autoimmune, and other life-threatening diseases.

Due to higher number of patients suffering from these diseases worldwide, the demand for immunotherapeutics is always on a higher side. With the advent of recombinant DNA technology and synthetic biology the manufacturing of immunotherapeutics in bulk amount has become possible. To meet the market demand, numerous efforts have been taken in order to increase the production and market ready formulation of these therapeutic agents. Hundreds of molecules are currently in development and many are in pipeline stage. But the question arises whether the currently available molecules in the market could fulfill the need of the market? A stable formulation delivers its therapeutic action with similar or equivalent efficacy? All these questions have one answer, i.e. A Robust manufacturing processes. At R & D level, the optimized process needs to be scaled up at pilot level and then production level. The process robustness should remain within the threshold set by regulatory bodies.

8.2.1 Monoclonal Antibody Therapeutics

Monoclonal antibodies (mAbs) are immunological molecules with varying applications in all fields of biological sciences. Its first ever production was reported to be discovered in 1975 by Georges Kohler of West Germany and Cesar Milstein of Argentina using hybridoma technology. Antibodies are antigen-specific towards particular type and it has the ability to provide continuous resistance against it. Due to this special characteristic of antibodies, we use it in the treatment of various diseases.

Development of In vitro techniques for antibody production affected the mAbs production for theragnostic applications (Mahmuda et al. 2017).

Numerous classical methods for monoclonal antibody production have been listed so far, known to us. Hybridoma technique is one of the efficient technique for monoclonal Antibody out of the available classical techniques. The ever increasing need of mAb for theranostic applications in various sectors has led to the development of scalable bio-manufacturing processes with enhanced final yield of target.

8.2.2 Types of mAbs and Problems Associated with its Production

Different types of mAbs with slight or complete variation in design to have similar or elevated efficacy are developed with the advent of protein engineering techniques. Murine mAbs, Chimeric mAbs, Humanized mAbs, and Fully human mAbs, different types of mAbs differ not only in its application but also in the source. A critical attribute of a therapeutic is structural similarity, and Murine, chimeric and humanized mAbs does different slightly to great extent as its development was from non-human source. mAbs obtained from different sources presents greater process invariability, adverse events such as allergic response, mild cytotoxicity, problems related to transplantation rejection. And undoubtedly, these problems have been tackled with some modifications in the approaches. After the first mAb got its licence, lot of effort was put into the mAb research and development. High global demand of recombinant mAbs, viz. rituximab (Rituxan), infliximab (Remicade), trastuzumab (Herceptin), etc. provoked many companies and contract manufacturing organizations (CMOs) to enter into large scale production. Amalgamation of high yielding processes with larger capacity bioreactor designs has given significant output in terms of mAb productivity and meets rising market demands. Manufacturing process of recombinant therapeutic mAbs shares common steps, i.e. the outline of the process would be similar to one another. Despite this the process reports the variation due to many other factors that have listed in this chapter.

Table 8.2 List of approved mAbs by FDA during last 2 years and its approved use for the treatment of various diseased/disorder conditions (U.S. FDA 2017)

Sr. no.	Trade name	Summary of FDA approved use on approval date
Year 2018		
1	Trogarzo	Multidrug resistant HIV-1
2	Crysvita (burosumab-twza)	With x-linked hypophosphatemia (XLH), a rare, inherited form of rickets
3	Aimovig (erenumab-aooe)	Adult patients for the prevention of migraine
4	Ajovy (fremanezumab–vfrm)	
5	Emgality (galcanezumab-gnlm)	
6	Gamifant (emapalumab-lzsg)	Patients with primary hemophagocytic lymphohistiocytosis (HLH)
7	Libtayo (cemiplimab-rwlc)	Cutaneous squamous cell carcinoma
8	Lumoxiti (moxetumomab pasudotox-tdfk)	Hairy cell leukemia
9	Poteligeo (mogamulizumab–kpkc)	Relapsed or refractory mycosis fungoides (MF)
10	Takhzyro (lanadelumab–flyo)	Hereditary angioedema (HAE)
11	Ultomiris (ravulizumab-cwvz)	Paroxysmal nocturnal hemoglobinuria
Year 2019		
1	Adakveo (crizanlizumab-tmca)	Complication of sickle cell disease—vaso-occlusive crisis
2	Beovu (brolucizumab-dbll)	Age-related macular degeneration (AMD)
3	Cablivi (caplacizumab-yhdp)	Treatment of adults with acquired thrombotic thrombocytopenic purpura
4	Enhertu (fam-trastuzumab deruxtecan-nxki)	Metastatic breast cancer
5	Evenity (romosozumab-aqqg)	Osteoporosis
6	Padcev (enfortumab vedotin-ejfv)	Refractory bladder cancer
7	Polivy (polatuzumab vedotin-piiq)	To treat adult patients with diffuse large B-cell lymphoma
8	Skyrizi (risankizumab–rzaa)	Moderate-to-severe plaque psoriasis in adults
9	Cimzia (certolizumab pegol)	Non-radiographic axial spondyloarthritis
10	Dupixent (dupilumab)	Chronic rhinosinusitis with nasal polyps.
11	Kadcyla (ado-trastuzumab emtansine)	Patients with HER2-positive, metastatic breast cancer

Enormous research happening globally related to mAbs is making mankind to explore the plethora of mAbs application in diverse branches of science (Ansar and Ghosh 2013).

Out of 570 therapeutic mAbs undergone clinical trials, 79 are FDA approved and (Table 8.2) comprises 30 mAbs which are already in the market for the treatment of cancer (Lu et al. 2020). Recently, mAbs have found its application in the field of clinical medicine. Several other mAbs are currently in clinical trial process for various new treatments for cancers, autoimmune diseases, and other dysfunctions.

Four major classes of mAbs have been categorized based on its formulation and action which includes the first group that elicit the body's immune response (rituximab, infliximab, etc.) and the second class includes radiolabeled mAb (radio-immunotherapy, RIT), antibody directed enzyme prodrug therapy (ADEPT) based mAb-drugs conjugated with the drug-activating enzyme are the third class of drug (Krauss et al. 2000) and the later class comprises of mAbs conjugated to liposomes (immuno-liposomes) or to a nanotechnology drug delivery system (Ansar and Ghosh 2013).

8.2.3 Role of Mammalian Cell Lines

With the ever-growing research in therapeutic candidates such as mAbs and its demand in the healthcare sector to combat several medical complications, biopharmaceutical companies are dynamically looking for a lucrative solution to deliver the market need. The Mab manufacturing processes are so developed by the manufacturer so that it will maintain the quality attributes with a minimum risk factor, reduced process time, and easy scale up.

The scale-up aspect of the manufacturing process is a very important parameter while setting up the process at R & D level. Starting from the process design and development the bioprocess for mAb production is segmented between upstream, midstream, and downstream processes, similar to all biopharmaceuticals. Previously the bioreactor-based processes were to be more focused for scale up of the process, but recent advances in bioprocess technology have been presented that suggests the productivity enhancement can be achieved by improvement in the upstream processes. This would result in the high expression host system and speed up the process development (Rita Costa et al. 2010).

Use of monoclonal antibody as immunotherapeutic has immense potential as targeted drug delivery system. Historically, these were the products from mouse sources followed by chimerization of mouse-human antibodies and then humanized monoclonal antibodies. These all types have received approval from regulatory bodies as therapeutics molecules in order to treat patients with different pathological conditions. Muromonab (OKT3) a murine anti-CD-3 antibody was the first mAb approved by FDA for human use used for the treatment of organ transplant rejection. To overcome the adverse events associated with murine antibodies chimerization was developed in 1984, these chimeric antibodies are amalgamation of the entire antigen-specific domain of a mouse antibody and constant domain of human antibody made by use of genetic engineering techniques (Ribatti 2014).

8.2.4 Cellular Therapy

Cell therapy has proven to be an escalating field in the healthcare sector that is effective in humans. Several clinical trials conducted globally are proving cell therapies as a ray of hope for the treatment of pathological conditions that seemed impossible to treat in the past. Many cell therapies are efficacious in humans, such as modified T cells and natural killer (NK) cells. Adoptive immunotherapy has been the most effective one with a focus on autologous cell sources. Chimeric Antigen Receptor (CAR) CAR-T cell therapy targeting CD-19 cells also can be termed as CAR-19 (Zhu et al. 2016) therapy expressing B-cell leukemias has been proved to be greatly effective. Recently, the FDA approved Novartis' Kymriah™ (tisagenlecleucel) and Gilead/Kite's Yescarta™ (axicabtagene ciloleucel) cell therapies that require the development of a scalable manufacturing aspect for wide accessibility. As the number of steps in the manufacturing increases, at commercial scale it becomes paramount important to have an efficient controlling system for the monitoring of parameters. A review by Rohin K Iyer et al. highlights some of the most recent advances used in the manufacturing of therapeutic immune cells, with a focus on T cells along with emerging technologies, approaches, and reagents used in cell isolation, activation, transduction, expansion, in-process analytics, harvest, cryopreservation, and that, and conclude with a forward-look at future directions in the manufacture of adoptive immunotherapies (Iyer et al. 2018).

CAR-T cell-based therapies have received more attention postsuccess of early phase clinical trials for CD-19-trageted CAR-T cell used in the treatment of hematologic malignancies and encouraged scientists to focus on targeting of other types of cancers. Hence, the manufacturing of CAR-T cells under cGMP is a focal point for this promising therapeutic modality (Wang and Rivière 2016).

Arguably, autologous cell therapies such as CAR-T cell therapy are complex therapies that need solid research work on its multi-step complex manufacturing process in order to enhance its market reach. Currently, clinical-grade cell production involves both open-loop and manual cell processing. However, standardization and characterization of this labor-intensive process remain major challenges.

Automated processes are preferred over manual operations where it is crucial to maintain the high sterility as well as process robustness. Automation has a significantly better outcome as compared to manual operations. Closed loop automation of manufacturing processes can enhance the scale of production and the quality of the end product. In addition to the complexity of the manufacturing process, the quality and consistency of the reagents used in the process represent major areas of concern.

For a robust manufacturing process, though automated, the quality of the raw material or input material has always remained a major concern for biologists. Similarly, in case of autologous cell therapy, the cells collected from patients are with mixed populations that can cause variation in the final cell population. Also, the media additives such as serum proteins, cytokines, and other components that affect the culture quality are not standardized often. The end product quality is compromised over each step of manufacturing and due to in-process additives that makes the overall process complex.

Variation in the cell type during the manufacturing is the cause of process failure to identify the in vitro transduced cells thus its selective purification is affected. For rapid identification and easy, one step purification the CAR can be tagged with different tags such as Strep-tag. Incorporation of Strep-tag sequence before CAR sequence using genetic engineering can solve the problem of cell variability (Liu et al. 2016). Complexity of cultured living cells is one of the major challenges to popularize cell based immunotherapy among many other challenges such as appropriate functioning of every engineered cell, its genetic makeup and production cost. These complications can be tackled by the aid of synthetic biology, hence synthetic biology would make an impact to make an efficient cell based immunotherapy.

8.2.5 Bioreactors for Cell Therapies

The scale-up bioreactor technology has been interestingly used for the manufacturing of cell therapy products to make bulk quantity doses for multiple patients. These then post-harvest can be cryopreserved and stored for long-term usage.

Bioreactors, which are predominantly used for the scale-up production of therapeutic molecules, mAbs and vaccines, nowadays implemented for the cell therapies as well. As, it provides the online monitoring, in-process control, bulk production of similar characteristic cells in a single batch, consistency that could minimize the cost of manufacturing and ultimately cost of the cell therapies.

The selection of scale of bioreactor is dependent of culture scale, autologous or allogeneic cell product requirement, and the number of patients undergoing treatment. Autologous cell therapies involve the manufacturing of patient specific cell lot from the same patients. Due to manufacturing of specific cell type and its use for single patients, the cost of the therapy scales high over the manufacturing of cell/cell products that are used to treat multiple patients. Now, under such circumferences, the designing of bioreactors that are equipped with relatively easy setup, less laborious for transport and handling with necessary parameter monitoring, control set up, and scalable design would create a cost effective method for patient specific manufacturing of cell/cell products.

8.2.6 Bioreactor Designs

Bioreactor designs have been diverse depending upon the purpose and type of cells and tissue structure, such as rotation-wall vessels, fluidized or fixed bed bioreactors, spinner flasks, perfusion bioreactors, and hollow-fiber devices. For the cultivation of 3D cells, rotating-wall vessels are used where these cells grow under constant circulation flow around the scaffold by continuous rotation of whole device. Bioreactor design differs for cartilage tissues than that of cardiac muscle as for cartilage the design consideration is predominantly dependent of convection and molecular diffusion with loading-enhanced transport rate and for the later one, i.e. densely packed tissue, it focuses on sufficient oxygen transport (Hansmann et al. 2013). Laura Gimenez et al. present a challenging approach of oxygen transfer flux that is one of the most challenging parameters when mAb production process is scaled up in stirred tank bioreactor. The author presented the model of oxygen transfer flux required for 2 L bioreactor, 10 L bioreactor to that of 80 L bioreactor process so that the oxygen requirement for high cell density cultivation will be satisfied with lower gas flow values (Gimenez et al. 2013).

8.2.7 Automation Considerations for Cell Therapies

In the current era of technology, the manufacturing units have come up with automation in various sectors. Scientists are trying to adapt automation in pharmaceutical sector in order to reduce the drawbacks associated with non-automated or manual operations. While scaling up the cell therapies, the problems faced are related to upstream and downstream processing. Replication of lab scale processes are laborious, open and time consuming, that leads to manual errors, process variation, product quality variation, and batch failure affects drastically on cost and time that is committed to the patients. Thus, both allogeneic and autologous therapies can be performed with automated manufacturing systems with multiple unit operation that minimizes process variation, provides easy to scale-up approach with relatively lesser time. Bioreactors designs that encounters closed downstream processing automated system would reduce the chances of cross-contamination to several times. In order to make continuous process in a closed system, multiple parameters monitoring and control needed to be automated by the development and incorporation of novel biosensors that enable feedback control and measures the levels of inhibitory cytokines, special device to detect the physiology of cells label free, Dielectrophoresis cytometers are few examples to develop better technology (Eaker et al. 2017; Csaszar et al. 2012; Braasch et al. 2013).

Automation in the process of manufacturing needed to broaden the applicability and to reduce the complexity of the procedures dealing with clinical therapies. One such example of an automated device that used in cell processing in clinical therapies is CliniMACS Prodigy® System. This system is an automated cell processing device that has been implemented in the production of CAR-T, CAR19 cells as well as lentiviral transduction of T cells (Mock et al. 2016; Zhu et al. 2016; Nickolay et al. 2016).

8.3 Conclusion

The above content has provided an ocean of opportunities for the development of immunotherapeutic targets. Most of the technologies discussed above have been successfully translated from bench to bed side and have provided relief to many cancer survivors. But what has been discussed is only a tip of the iceberg. Since oncology is a complex field with multiple factors influencing disease initiation, progress, prognosis, and metastasis, it becomes difficult to delineate a specific technology to a specific type of cancer. Successful treatment would still require a holistic approach to target the disease to achieve the desired outcome. Given so many prospective targets to immunotherapy, not all can be translated to clinical practices for therapeutic application. It is essential to evaluate the translational applicability of each of these targets to evaluate whether they have redundant or unique function and also the cascade of events they trigger. It is also essential to understand how these targets may pose effect on T_{sup} and T_{reg} and T_{eff} cells or on population of other cells. For example, Tim-3 is present on T cells as well as myeloid cells, they could also lead to autoimmune reactions. This requires a fine understanding between anti-tumor activity and self-tolerance (Sharpe 2017).

This still looks very promising as the pathogenesis of cancer is still not clearly understood. Nonetheless, as long a newer and fundamental research will keep on updating our understanding of the disease, immunotherapy will continue to evolve.

Despite of proven potential, these are not easily adapted and accessible to mankind due to its relatively high cost factor. Process optimization, a suitable design consideration for scale-up and scale-out technology would impact its outreach in the market. The automation and the continuous processing would contribute provision of simpler, scalable, faster, and cost effective therapies to the patients with minimum failure risk.

This can be achieved when academic scientist and industries collaborate to develop novel routes of drug delivery or newer and effective immunotherapeutic drugs which look at other modalities of cancer initiation, progression or even metastasis.

References

Ansar W, Ghosh S (2013) Monoclonal antibodies: a tool in clinical research. Indian J Clin Med 4: S11968. https://doi.org/10.4137/ijcm.s11968

Braasch K, Nikolic-Jaric M, Cabel T et al (2013) The changing dielectric properties of CHO cells can be used to determine early apoptotic events in a bioprocess. Biotechnol Bioeng 110 (11):2902–2914. https://doi.org/10.1002/bit.24976

Cancer.net (2018) Immunotherapy - An Introduction. https://www.cancer.net/navigating-cancer-care/videos/treatments-tests-and-procedures/immunotherapy-introduction. Published 2018

Cancer.net (2020) Understanding Immunotherapy. https://www.cancer.net/navigating-cancer-care/how-cancer-treated/immunotherapy-and-vaccines/understanding-immunotherapy

Childs RW, Carlsten M (2015) Therapeutic approaches to enhance natural killer cell cytotoxicity against cancer: the force awakens. Nat Rev Drug Discov 14(7):487–498. https://doi.org/10.1038/nrd4506

Collins M, Ling V, Carreno BM (2005) The B7 family of immune-regulatory ligands. Genome Biol 6(6):1–7. https://doi.org/10.1186/gb-2005-6-6-223

Csaszar E, Kirouac DC, Yu M et al (2012) Rapid expansion of human hematopoietic stem cells by automated control of inhibitory feedback signaling. Cell Stem Cell 10(2):218–229. https://doi.org/10.1016/j.stem.2012.01.003

Dougall WC, Kurtulus S, Smyth MJ, Anderson AC (2017) TIGIT and CD96: new checkpoint receptor targets for cancer immunotherapy. Immunol Rev 276(1):112–120. https://doi.org/10.1111/imr.12518

Eaker S, Abraham E, Allickson J et al (2017) Bioreactors for cell therapies: current status and future advances. Cytotherapy 19(1):9–18. https://doi.org/10.1016/j.jcyt.2016.09.011

Gimenez L, Simonet C, Malphettes L (2013) Scale-up considerations for monoclonal antibody production process: an oxygen transfer flux approach. BMC Proc 7(S6):P49. https://doi.org/10.1186/1753-6561-7-s6-p49

Hansmann J, Groeber F, Kahlig A, Kleinhans C, Walles H (2013) Bioreactors in tissue engineering-principles, applications and commercial constraints. Biotechnol J 8(3):298–307. https://doi.org/10.1002/biot.201200162

Iyer RK, Bowles PA, Kim H, Dulgar-Tulloch A (2018) Industrializing autologous adoptive immunotherapies: manufacturing advances and challenges. Front Med 5(MAY):1–9. https://doi.org/10.3389/fmed.2018.00150

Kaufman HL, Kohlhapp FJ, Zloza A (2015) Oncolytic viruses: a new class of immunotherapy drugs. Nat Rev Drug Discov 14(9):642–662. https://doi.org/10.1038/nrd4663

Krauss WC, Park JW, Kirpotin DB, Hong K, Benz CC (2000) Emerging antibody-based HER2 (ErbB-2/neu) therapeutics. Breast Dis 11:113–124. https://doi.org/10.3233/BD-1999-11110

Liu L, Sommermeyer D, Cabanov A, Kosasih P, Hill T, Riddell SR (2016) Inclusion of strep-tag II in design of antigen receptors for T-cell immunotherapy. Nat Biotechnol 34(4):430–434. https://doi.org/10.1038/nbt.3461

Lu RM, Hwang YC, Liu IJ et al (2020) Development of therapeutic antibodies for the treatment of diseases. J Biomed Sci 27(1):1–30. https://doi.org/10.1186/s12929-019-0592-z

Mac Cheever MA (2008) Twelve immunotherapy drugs that could cure cancers. Immunol Rev 222 (1):357–368. https://doi.org/10.1111/j.1600-065X.2008.00604.x

Mahmuda A, Bande F, Al-Zihiry KJK et al (2017) Monoclonal antibodies: a review of therapeutic applications and future prospects. Trop J Pharm Res 16(3):713–722. https://doi.org/10.4314/tjpr.v16i3.29

Miliotou AN, Papadopoulou LC (2018) CAR T-cell therapy: a new era in Cancer immunotherapy. Curr Pharm Biotechnol 19(1):5–18. https://doi.org/10.2174/1389201019666180418095526

Miller JFAP, Sadelain M (2015) The journey from discoveries in fundamental immunology to cancer immunotherapy. Cancer Cell 27(4):439–449. https://doi.org/10.1016/j.ccell.2015.03.007

Mock U, Nickolay L, Philip B et al (2016) Automated manufacturing of chimeric antigen receptor T cells for adoptive immunotherapy using CliniMACS prodigy. Cytotherapy 18(8):1002–1011. https://doi.org/10.1016/j.jcyt.2016.05.009

Ni L, Dong C (2017) New checkpoints in cancer immunotherapy. Immunol Rev 276(1):52–65. https://doi.org/10.1111/imr.12524

Nickolay L, Mock U, Phillip B et al (2016) Automated Lentiviral transduction of T cells with CARS using the CliniMACS prodigy. Mol Ther 24(May):S180–S181. https://doi.org/10.1016/s1525-0016(16)33264-6

Ribatti D (2014) From the discovery of monoclonal antibodies to their therapeutic application: an historical reappraisal. Immunol Lett 161(1):96–99. https://doi.org/10.1016/j.imlet.2014.05.010

Rita Costa A, Elisa Rodrigues M, Henriques M, Azeredo J, Oliveira R (2010) Guidelines to cell engineering for monoclonal antibody production. Eur J Pharm Biopharm 74(2):127–138. https://doi.org/10.1016/j.ejpb.2009.10.002

Sahin U, Karikó K, Türeci Ö (2014) MRNA-based therapeutics-developing a new class of drugs. Nat Rev Drug Discov 13(10):759–780. https://doi.org/10.1038/nrd4278

Schilsky RL (2018) Tumor-Agnostic Treatment for Cancer: An Expert Perspective. https://www.cancer.net/blog/2018-12/tumor-agnostic-treatment-cancer-expert-perspective

Sharpe AH (2017) Introduction to checkpoint inhibitors and cancer immunotherapy. Immunol Rev 276(1):5–8. https://doi.org/10.1111/imr.12531

U.S. FDA (2017) Resources for Information on Approved Drugs. Fed Drug Adm 1–2. www.fda.gov/Drugs/InformationOnDrugs/ApprovedDrugs/default.htm

Wang X, Rivière I (2016) Clinical manufacturing of CAR T cells: foundation of a promising therapy. Mol Ther Oncol 3(February):16015. https://doi.org/10.1038/mto.2016.15

Zhu F, Shah NN, Xu H et al (2016) CAR-T cell production using the Clinimacs® prodigy system. Blood 128(22):5724–5724. https://doi.org/10.1182/blood.v128.22.5724.5724

Pharmacokinetics, Pharmacodynamics, and Toxicology Aspects of Immunotherapeutics

9

Preeti Kulkarni, Parsshava Mehta, Bharati Shriyan, Kalpita Gawit, Vikram Gota, and Minal Ghante

Abstract

Immunotherapy for cancer has been in existence for over a decade demonstrating clinical activity across many tumor types. The increasing understanding of the immune system and its complexity has paved the path for the development of immunotherapy. Immune checkpoint inhibitors (ICIs) are revolutionizing cancer therapy with response rate approaching 50% for monotherapy regardless of tumor type. Combination of ICIs with chemotherapy or other ICIs has been attempted with great success. To further optimize the use of these ICIs, inter-individual variations in exposure and individual response need to be studied. However, potentially severe immune-related adverse events (irAEs) can offset the clinical outcomes of ICI therapy, particularly of combination therapies. The identification, assessment, and management of irAEs require a clear understanding of the pharmacokinetics (PK), exposure–effect relationship, and toxicity profile of these drugs. This chapter provides an overview of the pharmacokinetics (PK), pharmacodynamics (PD), and toxicity of the current immune checkpoint inhibitors as

P. Kulkarni
Gahlot Institute of Pharmacy, Navi Mumbai, Maharashtra, India

P. Mehta · B. Shriyan · K. Gawit
Department of Clinical Pharmacology, Advanced Centre for Treatment, Research and Education in Cancer, Tata Memorial Centre, Navi Mumbai, Maharashtra, India

V. Gota
Department of Clinical Pharmacology, Advanced Centre for Treatment, Research and Education in Cancer, Tata Memorial Centre, Navi Mumbai, Maharashtra, India

Homi Bhabha National Institute, Mumbai, Maharashtra, India

M. Ghante (✉)
Sinhgad Technical Education Society's, Smt. Kashibai Navale College of Pharmacy, Kondhwa (Bk), Pune, Maharashtra, India
e-mail: minal.ghante@sinhgad.edu

S. P. Sawarkar et al. (eds.), *Immunotherapy – A Novel Facet of Modern Therapeutics*,
https://doi.org/10.1007/978-981-15-9038-2_9

monotherapy and in combination. Collaborative work of basic scientists, clinical pharmacists, and oncologists utilizing advanced technology and artificial intelligence can further enhance the utility of ICIs.

Keywords

Pharmacokinetics · Pharmacodynamics · Toxicity · Immune-related adverse events · Immune checkpoint inhibitors

9.1 Introduction

Despite the impressive clinical outcomes observed with immune checkpoint inhibitors, the response rate for all immune checkpoint inhibitors (ICIs) as monotherapy falls below 50% regardless of the tumor type. Development of biomarkers that can help choose an appropriate drug for a given patient (predictive markers) or those (primarily imaging modalities) that can inform the physician whether a patient is benefitting from a treatment or not is an important area of research in the field of immunotherapy of cancer (Gnjatic et al. 2017). ICIs differ from conventional modalities of treatment in several ways. First of all, being monoclonal antibodies (mAbs), their pharmacokinetics differ from small molecule therapeutic agents. Their large size limits their volume of distribution mostly to vascular and interstitial spaces, and their clearance (Cl) is mostly non-renal, through the proteolytic degradation mediated by nonspecific Fc receptors (Bajaj et al. 2019). Secondly, they have a unique toxicity profile as a consequence of immune stimulation which is inherent to their mechanism of action, viz. the immune-related adverse events (irAEs) or adverse events of special interest (AEoSI). Any organ or tissue can be affected by the reactions. irAEs may develop through a combination of pathways involving autoreactive T-cells, autoantibodies, and cytokines, although the exact pathophysiology is not fully understood (125377s094lbl.pdf [Internet]. [cited 2020 Mar 29]. https://www.accessdata.fda.gov/drugsatfda_docs/label/2018/125377s094lbl.pdf). The most frequently occurring irAEs affect skin, colon, endocrine organs, liver, lungs, heart, and the nervous system which may be very serious, sometimes lethal (Bajaj et al. 2019). Infusion reactions constitute the most frequent non-irAEs. The pharmacokinetics (PK), pharmacodynamics (PD), and the toxicity of ICI are discussed in this chapter.

9.2 Pharmacokinetics of Immune Checkpoint Inhibitors (ICIs)

9.2.1 Ipilimumab

Ipilimumab has been indicated in several malignancies including unresectable advanced melanoma, and microsatellite instability (MSI), and mismatch repair-deficient colorectal cancer. Currently, the approved dosing schedule is 3 mg/kg as a 90-minute infusion, every 3 weeks, for up to four cycles (125377s094lbl.pdf [Internet]. [cited 2020 Mar 29]. https://www.accessdata.fda.gov/drugsatfda_docs/label/2018/125377s094lbl.pdf). This is based on several phase III studies demonstrating a clear survival advantage.

Ipilimumab's pharmacokinetics was demonstrated by studies in 499 unresectable or metastatic melanoma (Feng et al. 2014). Peak concentrations (C_{max}), trough concentration (C_{min}), and area under the curve (AUC) were found to be linear in the dose range of 0.3–10 mg/kg every 3 weeks for up to 4 doses. The PK is described by a two-compartment model with time-independent clearance. Accumulation is negligible (up to 1.5-fold) and steady-state is reached at third cycle (Chmielowski 2013) (YERVOY 5 mg/ml concentrate for solution for infusion - summary of product characteristics (SmPC) - (emc) [Internet]. [cited 2020 Mar 29]. https://www.medicines.org.uk/emc/product/4683#PHARMACOKINETIC_PROPS). The drug's terminal half-life, clearance, and volume of distribution were 14.7 days, 15.3 ml/h, and 7.1 L, respectively. Thirty percent of patients in the 3 mg/kg group attained the target trough concentration of 20 μg/ml, at which maximum CTLA-4 blockade is achieved. Though increasing body weight results in increased clearance, no dose recommendations are made with respect to body weight. Other covariates such as age, gender, performance status (PS), HLA-A2*0201 status, prior chemotherapy, baseline lactate dehydrogenase levels, anti-ipilimumab antibodies, and mild renal impairment had no clinically significant impact on the clearance. No dose changes are required for patients with a hepatic impairment since baseline AST, ALT, and bilirubin levels had no significant impact on ipilimumab clearance (Feng et al. 2014).

A report suggested time-varying clearance of ipilimumab when administered in combination with nivolumab in advanced solid tumors. However, ipilimumab's clearance was unaffected by nivolumab coadministration and tumor type (Sanghavi et al. 2020). A phase I study of ipilimumab in pediatric patients with advanced solid tumors showed a similar PK profile as those described for adult patients (Merchant et al. 2016). Also, a phase 1 study conducted by Weber et al showed no difference in PK of ipilimumab when co-administered with dacarbazine or carboplatin/paclitaxel for the treatment of advanced melanoma (Weber et al. 2013).

9.2.2 Nivolumab

Nivolumab PK is best described by a linear, two-compartment, zero-order, i.v. infusion model with first order elimination and time varying clearance. When a dose of 3 mg/kg of nivolumab is administered Q2W (every 2 weeks), steady state is reached within 2 weeks and systemic accumulation is approximately 3.7-fold. The drug has a terminal half-life of 12–25 days, Vd and Cl of 8.04 L and 9.50 ml/h, respectively (Feld and Horn 2017; Centanni et al. 2019) (pdf [Internet]. [cited 2020 Mar 29]. https://www.accessdata.fda.gov/drugsatfda_docs/label/2019/125554s070lbl.pdf). Following i.v. administration, nivolumab exhibited a steady state terminal half-life and a biphasic elimination characterized by a rapid distribution and slow elimination stage. Presence of anti-nivolumab antibodies was found in 11.2% of patients, leading to an average increase in 14% Cl following repeated doses. Paradoxically, a decrease in clearance during the course of treatment is often observed, attributable to improvement in cancer cachexia and disease status of the patient. Population pharmacokinetic (PopPK) studies have shown that body weight had a positive correlation with Vd and Cl. 30% of inter-individual variability (IIV) in clearance was attributed to body weight, PS, sex, and estimated glomerular filtration rate (eGFR), whereas 21% of IIV in volume of distribution could be explained by sex and body weight. No clinically significant differences were found in clearance in patients with normal and mild/moderate renal and hepatic impairment. PK was found similar across melanoma and NSCLC tumor types (Bajaj et al. 2017).

In another study, gender, body surface area (BSA), and albumin were found to have a significant impact (37%) on nivolumab's clearance. Women had 22% lower clearance as compared to men, BSA > 2.2 m^2 and albumin levels <37.5 g/dl led to a >20% increase in clearance. In NSCLC patients, patients with progressive disease had a 42% higher clearance. No such significant trend was observed in melanoma and RCC (Hurkmans et al. 2019). A study recently found that nivolumab Cl was higher when co-administered with ipilimumab as compared to nivolumab monotherapy across multiple tumor types (Zhang et al. 2019).

9.2.3 Pembrolizumab

Pembrolizumab is administered at a dose of 2 mg/kg or a flat dose of 200 mg every 3 weeks as a 30 minute infusion. These 3 weekly dosing results in steady state being reached at 16 weeks and the systemic accumulation is approximately twofold following repeated administration. PK parameters of $C_{\max}$, $C_{\min}$, and area under the curve at steady state (AUCss) increased commensurately with the dose, within the dose range of 2–10 mg/kg every 3 weeks. The terminal half-life is 22 days (125514s040lbl.pdf [Internet]. [cited 2020 Mar 29]. https://www.accessdata.fda.gov/drugsatfda_docs/label/2019/125514s040lbl.pdf). The mean half-life of pembrolizumab is 14–27.3 days. Within the dose range of 0.3–10 mg/kg, this drug shows linear clearance; however, at doses below 0.3 mg/kg, clearance is non-linear. Additionally, this clearance is time variant, typical values range from 0.168–0.249 L/day, and decrease steadily over 10 months. Approximately, the central volume of distribution expected to be around 2.88–3.48 L/day and the total Vd is small (7.4 L) (Centanni et al. 2019). Mild renal and hepatic impairment do not have clinically significant effect on PK of pembrolizumab (Longoria and Tewari 2016). Exposure in pediatric population was found to be similar to those in adult when dosed at 2 mg/kg of body weight (125514s040lbl.pdf [Internet]. [cited 2020 Mar 29]. https://www.accessdata.fda.gov/drugsatfda_docs/label/2019/125514s040lbl.pdf).

A PK study conducted by Ahamadi et al showed that pembrolizumab PK was accurately described by a two-compartment model with linear clearance. The study also indicated how gender, ECOG PS, and tumor type had a significant impact on pembrolizumab's PK. Clearance was found to be lower by 17%, subsequently leading to a 20% increase in AUC in women. Patients with ECOG PS 0 showed a 7.3% increased clearance as compared to patients with PS 1. Similarly, NSCLC patients showed a 14.5% increase in clearance as compared with other tumor types (Ahamadi et al. 2017).

Other PD-L1 inhibitors: Atezolizumab, Avelumab, and Durvalumab.

9.2.4 Atezolizumab

Atezolizumab exhibited linear clearance between 1 and 20 mg/kg, with the average value being 0.200 L/day. Clearance increased by 19% in the presence of anti-drug antibodies (ADAs). Steady state concentrations reached within 6–9 weeks and the average half-life of atezolizumab was 27 days. The central and peripheral volumes of distribution were estimated to be 3.28 and 3.63 L, respectively (Centanni et al. 2019).

9.2.5 Avelumab

Similar to atezolizumab, the clearance of avelumab is also linear between 1 and 20 mg/kg. Clearance reduces with time, at an average rate of 3.1% in 6 months. Steady state concentrations are reached within 4–6 weeks, and the average half-life is 6 days. The central and peripheral volumes of distribution were estimated to be 2.83 and 1.7 L, respectively. Presence of ADAs and tumor type significantly impact the PK. While ADAs tend to increase the clearance by 10–15%, a decrease in clearance over time can be noticed in patients with Merkel cell carcinoma, more pronounced in responders compared to non-responders (Centanni et al. 2019).

9.2.6 Durvalumab

Durvalumab exhibits linear clearance at doses over 3 mg/kg, however below this dose, the clearance is non-linear. The average value for Cl is 0.232 L/day. Clearance like the other PD-L1 inhibitors were found to be time dependent, it declined at an average value of 16.9% over 12 months. ADAs resulted in a 20% decrease in trough concentrations at steady state ($C_{\text{trough,ss}}$). Steady state concentration is achieved within 16 weeks and the average half-life is approximately 21 days. The peripheral and central volumes of distribution were estimated to be 3.42 and 3.51 L, respectively. Covariates such as gender and body weight exhibited a significant impact on the volume of distribution (Centanni et al. 2019).

Table 9.1 enlists the Pharmacokinetic properties of approved CTLA-4, PD-1, and PD-L1 inhibitors (Sheng et al. 2017).

Table 9.1 Pharmacokinetic properties of approved CTLA-4, PD-1, and PD-L1 inhibitors (Sheng et al. 2017; Chatelut et al. 2020; Desnoyer et al. 2020)

	Ipilimumab	Nivolumab	Pembrolizumab	Atezolizumab	Avelumab	Durvalumab
Cl (mL/day)	~400	~200	~200	~200	~600	~200
*V*ss (L)	6.0–7.5	6.8–8.0	6.1–7.4	~7.0	~5.0	~5.5
$T_{1/2}$ (days)	~15	25	23	27	6.1	17
Linear kinetics—dose range (mg/kg)	0.3–10	0.1–10	2–10	1–20	10–20	≥3
Time to steady state (weeks)	9	12	19	6–9	4–6	16

*V*ss volume of distribution at steady state

9.3 Pharmacodynamics of Immune Checkpoint Inhibitors

9.3.1 Anti CTLA-4 Antibodies

Negative signaling receptors such as CTLA-4 are the main target of ipilimumab and tremelimumab. The exposure–efficacy relationship of ipilimumab shows that minimum concentration at steady state (C_{minss}) predicts the efficacy endpoints given by overall response rate (ORR), immune-related response criteria (irRC), and overall survival (OS) with reasonable accuracy. The correlation is stronger between C_{minss} and irRC as compared to C_{minss} and ORR (Feng et al. 2014). The dose of ipilimumab is also significantly, but less closely associated with OS (Feng et al. 2014).

The administration of anti-CTLA-4 antibodies showed an increase in IL-2 and absolute lymphocyte count (ALC), in both clinical trials as well as preclinical ex-vivo studies (YERVOY 5 mg/ml concentrate for solution for infusion - summary of product characteristics (SmPC) - (emc) [Internet]. [cited 2020 Mar 29]. https://www.medicines.org.uk/emc/product/4683#PHARMACOKINETIC_PROPS). This increase in ALC has been associated with clinical benefit and survival advantage (Delyon et al. 2013; Ku et al. 2010; Wilgenhof et al. 2013; Kelderman et al. 2014). A study by Delyon et al. on 59 patients reported an early increase in the eosinophil counts during treatment to be associated with favorable OS (Delyon et al. 2013). Nyakas et al. reported that increased levels of edostatin and Gal3BP suggested poor prognosis, but owing to the non-placebo design, this needs to be further validated (Nyakas et al. 2019). Thus, combination of biomarkers will yield better prognostic outcomes as demonstrated by Martens et al. in a study on 82 patients, the results show that increase in ALC and in CD4+ and CD8+ T-cells at 8 to 14 weeks following the first dose of ipilimumab therapy correlates well with improved survival (Martens et al. 2016).

9.3.2 Anti PD-1 and PD-L1 Antibodies

Anti PD-1 antibodies (nivolumab, pembrolizumab) or anti PD-L1 antibodies (atezolizumab, avelumab, durvalumab) essentially inhibit the interaction between PD-1 and PD-L1. The greatest motive for selecting anti PD-1 and PD-L1 antibodies is on the tumor cell PD-L1 expression. Currently, a PD-L1 immunohistochemistry test (IHC) is recommended for selection of patients for both pembrolizumab and atezolizumab for treatment of locally advanced or metastatic urothelial carcinoma (tecentriq_prescribing.pdf [Internet]. [cited 2020 May 11]. https://www.gene.com/download/pdf/tecentriq_prescribing.pdf; tecentriq_prescribing.pdf [Internet]. [cited 2020 May 12]. https://www.gene.com/download/pdf/tecentriq_prescribing.pdf). 28-8 pharmDx, a PD-L1 IHC was approved as a complementary diagnostic test for patients with non-small cell lung cancer (NSCLC) and melanoma prior to nivolumab therapy (Topalian et al. 2016). A higher likelihood of treatment response is seen in intra-tumoral PD-L1 expression prior to treatment, but the absence of PD-L1 expression does not rule out treatment response as indicated by the

CheckMate 067 data (Larkin et al. 2015a) The pitfall of the PD-L1 immunohistochemistry as a biomarker for anti-PD-1 and anti-PD-L1 therapies is probably a result of multiple variables like concomitant PD-L1 expression in tumor cytoplasm, time and location, intrapatient and intratumor heterogeneity, and discrepancies in diagnostic kits (Topalian et al. 2016).

Nivolumab

Nivolumab dose−/exposure–response analysis shows a positive trend in ORR, which plateaus at doses higher than 1 mg/kg for RCC and melanoma, and 3 mg/kg for NSCLC. Peripheral receptor occupancy was saturated at doses $\geq$0.3 mg/kg. Although there seemed to be no apparent relationship between tumor shrinkage rate (TSR) and exposure, tumor progression rate appeared to decrease with increasing exposure, up to a dose of 3 mg/kg Q2W for NSCLC (Agrawal et al. 2016). However, these correlations may have been overestimated due to confounders such as time varying clearance leading to increased exposures in later treatment cycles (Liu et al. 2017).

Pembrolizumab

Pembrolizumab exposure–response analysis was performed for melanoma ($n = 1366$) and NSCLC ($n = 496$) which shows no significant relationship between AUC over 6 weeks and ORR or TSR, at 18 and 28 weeks, respectively (Chatterjee et al. 2017, 2016).

Atezolizumab

Atezolizumab exposure–response analysis in urothelial carcinoma was evaluated in the IMvigor210 study which showed no significant relationship (Stroh et al. 2017). However, the BIRCH study in NSCLC patients identified a positive relationship between AUCatSS and ORR (761041Orig1s000OtherR.pdf [Internet]. [cited 2020 May 11]. https://www.accessdata.fda.gov/drugsatfda_docs/nda/2016/761041Orig1s000OtherR.pdf). A study by Netterberg et al. showed that although AUC was a major predictor of tumor shrinkage, the effect dissipated with an average half-life of 80 days, whereas relative changes in interleukin-8 levels (RCFBIL-18, d21) on day 21 seemed relevant to the duration of response (761041Orig1s000OtherR.pdf [Internet]. [cited 2020 May 11]. https://www.accessdata.fda.gov/drugsatfda_docs/nda/2016/761041Orig1s000OtherR.pdf).

Avelumab

When exposure–effect relationship was investigated for Avelumab in Merkel cell carcinoma, C_{trough}atSS correlated with PFS, ORR (saturating at 28 μg/mL), and OS, while AUCatSS was found to be associated with PFS and OS (761041Orig1s000OtherR.pdf [Internet]. [cited 2020 May 11]. https://www.accessdata.fda.gov/drugsatfda_docs/nda/2016/761041Orig1s000OtherR.pdf). A further subgroup analysis study conducted by Kaufmann et al. showed a trend towards increased ORR and 6-month PFS in patients with fewer prior lines, lower disease burden. and PD-L1+ tumors (Kaufman et al. 2017). These findings were further corroborated by Shapiro et al. (Shapiro et al. 2017).

Durvalumab

Durvalumab approved for NSCLC and urothelial carcinoma shows no significant relationship between C_{max} at cycle 1, C_{min} at cycle 2, C_{min} at SS and efficacy.

9.4 Combination with Chemotherapy or Other Immunotherapies: PD Considerations

Less than one-fifth of all patients experience durable response, despite the clinical benefits shown in patients treated with drugs that block PD-1/PD-L1 pathway, thus strategies to improve outcomes are needed to enhance T-cell activity. Along with the use of PD-1 inhibitors, blocking the co-stimulatory signals necessary for T-cell activation will improve the cytotoxic activity of T-cells.

In the Checkmate 568 trial, 288 patients with chemotherapy-naive stage IV NSCLC received nivolumab 3 mg/kg Q2W + ipilimumab 1 mg/kg Q6W for up to 2 years (Ramalingam et al. 2018). Tumor mutational burden (TMB) $\geq$10 mutations/megabase (mut/Mb) was associated with enhanced response to nivolumab–ipilimumab combination regardless of PD-L1 expression, with ORRs >40%.

MYSTIC was a phase 3 randomized controlled trial of 1118 patients with stage IV NSCLC who were treated with one of the three arms—durvalumab; durvalumab + tremelimumab; or platinum-based doublet chemotherapy (Starr 2019). Patients with TMB $\geq$20 mut/Mb of DNA showed improved OS for durvalumab plus tremelimumab vs chemotherapy (median OS, 21.9 months [95% CI, 11.4–32.8] vs 10.0 months [95% CI, 8.1–11.7]; HR, 0.49; 95% CI, 0.32–0.74), indicating the potential contribution of tremelimumab in this setting. The results of the exploratory analysis suggested that TMB could function as a predictive biomarker for immunotherapy.

9.5 Significance of PD1 Expression in Tumors: Subgroup Analysis from Large Randomized Trials

A longer OS and better tolerability to PD-1 and PD-L1 inhibitors compared to conventional chemotherapy has been shown by numerous studies in PD-L1 positive patients (Horn et al. 2017; Motzer et al. 2015; Brahmer et al. 2015; Borghaei et al. 2015; Ferris et al. 2016; Schachter et al. 2017; Bellmunt et al. 2017; Rittmeyer et al. 2017; Fehrenbacher et al. 2016). Logically PD-L1 expression should correlate with clinical outcomes, however, a non-negligible number of exceptions are observed in clinical practice.

A meta-analysis conducted by Khunger et al. on 6664 patients from 41 distinct trials showed that the expression of PD-L1 was predictive of favorable response across all tumor types (OR, 2.26; 95% CI, 1.85–2.75; $p < 0.001$) (Khunger et al.

2017). A significantly large effect was observed in NSCLC (OR, 2.51; 95% CI, 1.99–3.17; $p < 0.001$). Across all NSCLC trials using nivolumab and Dako clone 28-8 IHC antibody assay, subgroup analysis yielded a significantly higher OR in patients with tumor PD-L1 expression even at the minimum cut-off value of 1% (OR, 2.17; 95% CI, 1.03–4.57).

A meta-analysis by Shen et al. that included 4174 subjects from eight randomized controlled trials conclusively showed that the magnitude of the efficacy of PD-1 or PD-L1 inhibitors was greater for PD-L1 positive patients than for PD-L1 negative patients. However, results from the subgroup analyses showed that patients, both positive and negative for PD-L1 expression could benefit from PD-1/PD-L1 directed therapy. Thus PD-L1 expression alone is insufficient in determining whether a patient should be offered immunotherapy or not (Shen and Zhao 2018).

A conceptual point of view summarized as "cancer immunogram" was introduced, given the heterogeneity in clinical response to PD-1 or PD-L1 inhibitors (Blank et al. 2016). Based on this model, the outcome to anti PD-1 or PD-L1 antibody therapy was influenced not only by PD-L1 expression but also by many distinct characteristics including the "foreignness" of cancer; the activity of the intra-tumoral T-cell infiltrate; the sensitivity of cancer cells to T-cells; the immune status; and the presence of other inhibitory processes (Table 9.2).

9.6 Immunotherapy and Toxicities

Immunotherapies have transformed the treatment landscape in oncology, offering durable responses and improved survival for many patients across several hematological malignancies and solid tumors. However, these drugs have a unique toxicity profile closely related to their mechanism of action. Therefore, the nature and severity of toxicities related to these drugs are also of increasing interest to the oncology community.

9.6.1 CTLA-4 Inhibitors: Ipilimumab

Ipilimumab has shown potential to cause severe and fatal reactions in which enterocolitis is the most common as well as toxic epidermal necrolysis (TEN), hepatitis, neuropathy, and endocrinopathy like fatal reactions led ipilimumab to compromise with a boxed warning which resulted due to activation and proliferation of T-cells (Simeone et al. 2014).

The immune-related adverse events of (irAEs) of ipilimumab are given in Table 9.3.

Table 9.2 Specific biomarker developments for various drugs

Drugs	Biomarker	Disease	Clinical outcome
Ipilimumab	Melanoma antigens (Arenberger et al. 2017)	Metastatic melanoma	Decline in levels >30% at week 6 to 9 predicts response
	Endostatin and Gal3BP (Nyakas et al. 2019)		Increase levels of endostatin and Gal3BP gives higher risk of death
	Lymphocyte, eosinophil count ($n = 73$) (Delyon et al. 2013) Full blood count ($n = 183$) (Khoja et al. 2016)		Increase levels correlate to better OS NLR and LDH values differentiate patients into good, intermediate, and poor prognosis
	ALC with CD4+ and CD8+ ($n = 82$) (Martens et al. 2016)		Increase ALC levels and delayed increase in CD4+ and CD8+ correlates to better OS
	Myeloid cells and related inflammatory factors ($n = 59$) (Gebhardt et al. 2015)		Lower levels related to benefit from therapy
	Immunological markers (LDH, CRP, r-T-cells) (Simeone et al. 2014)		Changes from baseline and fourth ipilimumab infusion is associated with disease control
Atezolizumab	IL-18 (Netterberg et al. 2019)	NSCLC	Relative change in levels on day 21 correlated better with duration of response
Nivolumab	Th9 cells ($n = 46$)	Melanoma	Early increase in Th9 cells during treatment correlated to better clinical response
	Serum cytokines ($n = 35$)		Higher interferon-γ, IL-6 and IL-10 correlated with OTR
Pembrolizumab	Neoantigen burden (Rizvi et al. 2015)	NSCLC	High neoantigen count
	Immune gene signatures ($n = 19$) (Ribas et al. 2015)	Melanoma	Interferon γ and expanded-immune-related signatures correlated with ORR and PFS

OTR objective tumor response, ORR objective response, PFS progression free survival, OS overall survival, NLR neutrophil lymphocyte ration, LDH lactate dehydrogenase, CRP C reactive protein, ALC absolute lymphocyte count

9.6.2 PD1/PDL1 Inhibitors: Nivolumab, Pembrolizumab

Anti-PD-1 (either nivolumab or pembrolizumab) drugs cause fewer high-grade toxicities compared to ipilimumab. The most frequently reported AE is fatigue. Incidence of grade 3 and 4 fatigue was found to be 58% and 7%, respectively, in metastatic melanoma patients (Fellner 2012; Naidoo et al. 2015; Weber et al. 2017). A slightly higher incidence has been reported in metastatic renal cell carcinoma refractory to TKIs (79% and 19%, respectively). Patients with squamous NSCLC

Table 9.3 Immune related adverse events (irAEs) of Ipilimumab (OPDIVO (nivolumab) injection, for intravenous use.pdf [Internet]. [cited 2020 May 12]. https://www.accessdata.fda.gov/drugsatfda_docs/label/2018/125554s058lbl.pdf)

Immune-mediated reactions	Grades of immune-mediated reactions	
	Grades (1, 2)	Grades (3, 4)
Immune-mediated enterocolitis	**28 (5.5%)**	**34 (6.7%)** Hospitalized—26 (5.1%) Intestinal perforation—5 (1%) Deaths—4 (0.8%)
Immune-mediated hepatitis	**13 (2.5%)**	**8 (1.6%)** Fatal hepatic failure—1 (0.2%) Hospitalization—2 (0.4%)
Immune-mediated dermatitis	**63 (12%)**	**13 (2.5%)** Hospitalization—(0.2%) Death (TEN)—(0.2%)
Immune-mediated endocrinopathies	**12 (2.3%)** Adrenal insufficiency hypopituitarism hypothyroidism	**9 (1.8%)**

Rare immune-mediated adverse reactions (<1%):
Angiopathy, Blepharitis, conjunctivitis, Episcleritis, erythema, myocarditis, Leukocytoclastic, polymyalgia rheumatica, Scleritis, temporal arteritis, Vasculitis
Other uncommon immune-mediated adverse reactions:
Guillain-Barré syndrome, pneumonitis, pericarditis, Iritis, uveitis, hemolytic anemia, nephritis, peripheral motor neuropathy

refractory to advanced cisplatin had an incidence of grade 3 and 4 adverse events at 69% and 10%, respectively (Naidoo et al. 2015; Weber et al. 2017; Robert et al. 2015). Pembrolizumab-related AEs were reported in 73.4% (any AE) and 26.6% of patients (grade 3 or higher) (Ning et al. 2017).

9.6.3 Others: Atezolizumab (Tecentriq), Avelumab (Bavencio), Durvalumab (Imfinzi)

irAEs including hypersensitivity reaction, thyroiditis, pneumonitis, hepatitis are commonly reported with atezolizumab. Non-irAEs such as fatigue, decreased appetite, nausea are reported in up to 20% of patients. Less frequent AEs include urinary tract infection, abdominal pain, dyspnea, hematuria, and back or neck pain, which are reported in up to 2% of patients. Less than 1% of patients may have sepsis, pneumonitis, or intestinal obstruction leading to death (Weber et al. 2017). Laboratory abnormalities such as lymphopenia, hyponatremia, anemia, increased alkaline phosphatase, hyperglycemia, elevated ALT, and AST are also reported. Avelumab had a lower risk of individual grade $\geq$ 3 irAEs except for elevated AST, whereas atezolizumab had the lowest risk of irAEs of any grade (Yang et al. 2018). Durvalumab is safe in patients with many solid cancers and it exhibits a tolerable safety profile (Pillai et al. 2018).

9.7 Combination with Chemotherapy or Two Immunotherapies

Anti-PD-1/PD-L1 checkpoint inhibitors can be safely given with a variety of chemotherapy and targeted agents. Doublet and triplet combinations with cytotoxics could mostly be given at full doses. However, dose reduction of chemotherapy drugs may be necessary in combination with anti-CTLA-4 agents owing to their toxicity profile (Khalil et al. 2016).

Metastatic melanoma is the only indication till date approved for combination immunotherapy (Nikanjam et al. 2017). Treatment-related irAEs were seen in 95% of patients, out of which 55% of patients developed grade 3 or higher ADRs anytime during the course of this combination immunotherapy (Khalil et al. 2016). Immunotherapy dosing combinations: A study analyzing 3526 patients for toxicity and response patterns suggests that for adults with intact organ function, a safe starting dose for a doublet combination including an anti-PD1/PDL1 or anti-CTLA4 checkpoint inhibitor and a second immunotherapy or a biologic agent was approximately 50% of each drug which can be increased to 60% of each with a targeted agent. Whereas for triplet combinations with 2 cytotoxic agents, it can be used with all agents starting at or near full dose, particularly if the 2 cytotoxic agents had previously been given together safely at full dose (Haanen et al. 2018).

The combinations involving ipilimumab often showed more toxicity than those involving anti-PD-1 and PD-L1inhibitors, thus the starting doses for ipilimumab-containing combinations should be lower than those defined for anti-PD-1/PD-L1-based combinations (Larkin et al. 2015b; Johnson et al. 2018).

In the checkmate 067 trial, ipilimumab and nivolumab in combination have been studied (Table 9.4), which provided significant efficacy benefit but severe toxicity. The incidence of grade 3 and 4 toxicities was 55% with the combination, as

Table 9.4 Time to onset of adverse events (grade 3/4) in nivolumab monotherapy and in combination with ipilimumab (Haanen et al. 2018)

	Single agent Nivolumab			Combination Ipilimumab + Nivolumab		
Organ	*N*	Median (weeks)	Range (weeks)	*N*	Median (weeks)	Range (weeks)
Renal	1	50.9	50.9–50.9	6	11.3	3.3–23.7
Pulmonary	1	6.7	6.7–6.7	3	3.7	3.7–9.4
Hepatic	8	14.1	1.9–25.1	60	7.4	2.1–48.0
Endocrine	2	28.6	19.1–38.1	15	12.1	2.9–17.0
GI	7	26.3	13.1–57.0	46	7.4	1.0–48.9
Skin	5	19.4	1.3–50.9	18	5.6	0.1–55.0

N number of subjects, GI gastrointestinal

compared to 16% and 27% with nivolumab and ipilimumab monotherapy, respectively (Johnson et al. 2018).

9.8 Predictors of Toxicity

Autoimmune diseases, subclinical inflammation, shared antigens and combined immunotherapies, heterogeneous population, pharmacogenetic variations play a major role in the various unpredictable pattern of immune related adverse events.

Elderly patients with advanced melanoma, NSCLC or, RCC with comorbidities, concomitant medications, age-related immune system impairment (i.e., "immunosenescence"), and reduced functional reserve might affect tolerability of immunotherapy agents. Under these conditions, there is a high chance of clinical manifestation of subclinical autoimmune diseases (Simeone et al. 2014; Wolchok et al. 2010).

Irrespective of any type of immunotherapy, lower doses and monotherapies are proven to be safer than higher doses and combinations. As the dose increases, a corresponding increase in the incidence of irAEs has been reported (Pollack et al. 2018).

9.9 Dose Modifications

Patients with pre-existing autoimmune diseases or a previous history of immune-mediated reactions due to any immunotherapy may show a risk of exacerbation of autoimmunity, redevelopment of previous irAEs, or generation of de novo irAEs due to immunotherapy. Rechallenge with anti-PD-1 therapy after experiencing an irAE with combination therapy has an approximately 20% chance of recurrence of irAE as well as generation of a de novo irAE (Wolchok et al. 2010). ESMO guidelines recommend permanent discontinuation of immunotherapy after grade 3 or 4 irAEs including pulmonary, hepatic, pancreatic, ophthalmologic, and neurologic events, except for endocrine toxicity, which can be managed with physiologic hormone replacement (Pollack et al. 2018). Rechallenge should be conducted by considering various factors such as the type and severity of the autoimmune disease or irAE, the goals of treatment, therapeutic alternatives, and benefit versus risk assessment. Rechallenge should be attempted only after complete clinical resolution of the event (Delaunay et al. 2019).

9.10 Precautions and Warnings

An underlying but well-controlled autoimmune disease is usually not a contraindication for immunotherapy. However, in such cases, these agents should be used with caution only after considering the potential risk-benefit for the individual. At the least, testing for patients should include renal function, serum electrolytes, a

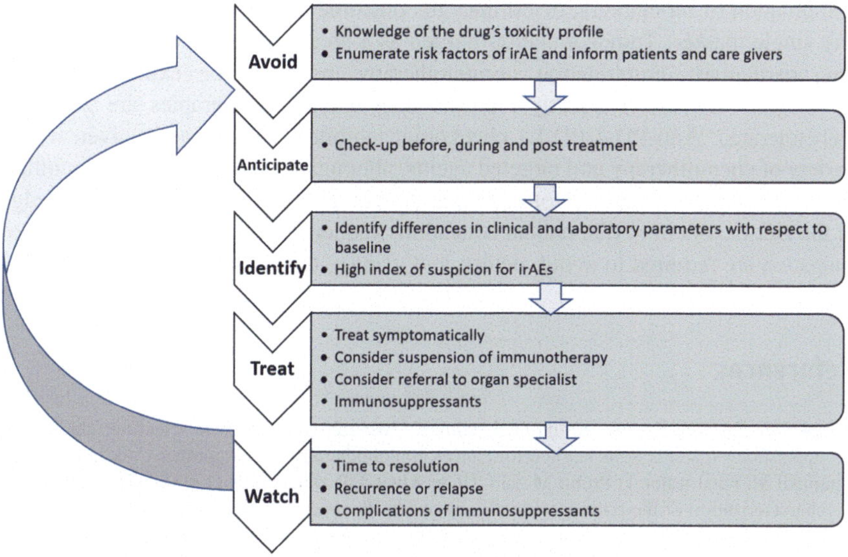

Fig. 9.1 Five important principles for the management of toxicities due to immunotherapy

complete blood count, liver function tests, and a thyroid evaluation, since PD-1/PDL-1, CLAT-4. Chest imaging should be performed at baseline for reference in case pulmonary toxicity occurs during immunotherapy (Johnson et al. 2018; Champiat et al. 2016; Ventola 2017; Kennedy and Salama 2020).

9.11 Treatment

Mild symptomatic patients should be observed for the severity of irAEs. Corticosteroids should be considered for moderate symptoms. Hospitalization with intensive management may be required in severe cases especially for older patients (Ventola 2017; Kennedy and Salama 2020; Palmieri and Carlino 2018). The principles of management of immunotherapy related toxicity include prevention, anticipation, detection, treatment, and monitoring (Fig. 9.1) (Delaunay et al. 2019).

9.12 Conclusions

To conclude, ICIs are characterized by long half-lifes and small volumes of distribution typical for macromolecules. Steady state concentrations are typically achieved between 6 and 12 weeks for most drugs. Drug–drug interactions of immunotherapies are not well studied. No consistent exposure–effect relationship is observed for ICIs, although, in case of ipilimumab trough concentration at steady state is a significant predictor of all efficacy measures. Cancer immunogram, a combination of biomarkers, determines the outcome of immunotherapy rather than any single marker. Tumor mutation burden as a predictive marker of immunotherapy, particularly combination immunotherapy, needs to be explored further. Although irAEs are frequently reported with ICIs, monotherapies are generally well tolerated. Anti-PD-1/PD-L1 checkpoint inhibitors can be safely given with a variety of chemotherapy and targeted agents, although dose reduction of chemotherapy drugs may be necessary for combination with anti-CTLA-4 agents. Knowledge of the risk factors of irAEs, patient education, frequent work-up, and a high index of suspicion are required to avoid, detect, and manage the toxicities of ICIs.

References

Agrawal S, Feng Y, Roy A, Kollia G, Lestini B (2016) Nivolumab dose selection: challenges, opportunities, and lessons learned for cancer immunotherapy. J Immunother Cancer 4:72

Ahamadi M, Freshwater T, Prohn M, Li CH, de Alwis DP, de Greef R et al (2017) Model-based characterization of the pharmacokinetics of pembrolizumab: a humanized anti-PD-1 monoclonal antibody in advanced solid tumors. CPT Pharmacometrics Syst Pharmacol 6(1):49–57

Arenberger P, Fialova A, Gkalpakiotis S, Pavlikova A, Puzanov I, Arenbergerova M (2017) Melanoma antigens are biomarkers for ipilimumab response. J Eur Acad Dermatol Venereol 31(2):252–259

Bajaj G, Wang X, Agrawal S, Gupta M, Roy A, Feng Y (2017) Model-based population pharmacokinetic analysis of nivolumab in patients with solid tumors. CPT Pharmacometrics Syst Pharmacol 6(1):58–66

Bajaj G, Suryawanshi S, Roy A, Gupta M (2019) Evaluation of covariate effects on pharmacokinetics of monoclonal antibodies in oncology. Br J Clin Pharmacol 85(9):2045–2058

Bellmunt J, de Wit R, Vaughn DJ, Fradet Y, Lee J-L, Fong L et al (2017) Pembrolizumab as second-line therapy for advanced urothelial carcinoma. N Engl J Med 376(11):1015–1026

Blank CU, Haanen JB, Ribas A, Schumacher TN (2016) The "cancer immunogram". Science 352 (6286):658–660

Borghaei H, Paz-Ares L, Horn L, Spigel DR, Steins M, Ready NE et al (2015) Nivolumab versus docetaxel in advanced nonsquamous non-small-cell lung cancer. N Engl J Med 373 (17):1627–1639

Brahmer J, Reckamp KL, Baas P, Crinò L, Eberhardt WEE, Poddubskaya E et al (2015) Nivolumab versus docetaxel in advanced squamous-cell non-small-cell lung cancer. N Engl J Med 373 (2):123–135

Centanni M, Moes DJAR, Trocóniz IF, Ciccolini J, van Hasselt JGC (2019) Clinical pharmacokinetics and pharmacodynamics of immune checkpoint inhibitors. Clin Pharmacokinet 58 (7):835–857

Champiat S, Lambotte O, Barreau E, Belkhir R, Berdelou A, Carbonnel F et al (2016) Management of immune checkpoint blockade dysimmune toxicities: a collaborative position paper. Ann Oncol 27(4):559–574

Chatelut E, Le Louedec F, Milano G (2020) Setting the dose of checkpoint inhibitors: the role of clinical pharmacology. Clin Pharmacokinet 59(3):287–296

Chatterjee M, Turner DC, Felip E, Lena H, Cappuzzo F, Horn L et al (2016) Systematic evaluation of pembrolizumab dosing in patients with advanced non-small-cell lung cancer. Ann Oncol 27 (7):1291–1298

Chatterjee MS, Elassaiss-Schaap J, Lindauer A, Turner DC, Sostelly A, Freshwater T et al (2017) Population pharmacokinetic/pharmacodynamic modeling of tumor size dynamics in pembrolizumab-treated advanced melanoma. CPT Pharmacometrics Syst Pharmacol 6 (1):29–39

Chmielowski B (2013) Ipilimumab: a first-in-class T-cell potentiator for metastatic melanoma. J Skin Cancer 2013:423829

Delaunay M, Prévot G, Collot S, Guilleminault L, Didier A, Mazières J (2019) Management of pulmonary toxicity associated with immune checkpoint inhibitors. Eur Respir Rev 28 (154):190012

Delyon J, Mateus C, Lefeuvre D, Lanoy E, Zitvogel L, Chaput N et al (2013) Experience in daily practice with ipilimumab for the treatment of patients with metastatic melanoma: an early increase in lymphocyte and eosinophil counts is associated with improved survival. Ann Oncol 24(6):1697–1703

Desnoyer A, Broutin S, Delahousse J, Maritaz C, Blondel L, Mir O et al (2020) Pharmacokinetic/pharmacodynamic relationship of therapeutic monoclonal antibodies used in oncology: part 2, immune checkpoint inhibitor antibodies. Eur J Cancer 128:119–128

Fehrenbacher L, Spira A, Ballinger M, Kowanetz M, Vansteenkiste J, Mazieres J et al (2016) Atezolizumab versus docetaxel for patients with previously treated non-small-cell lung cancer (POPLAR): a multicentre, open-label, phase 2 randomised controlled trial. Lancet 387 (10030):1837–1846

Feld E, Horn L (2017) Emerging role of nivolumab in the management of patients with non-small-cell lung cancer: current data and future perspectives. Onco Targets Ther 10:3697–3708

Fellner C (2012) Ipilimumab (yervoy) prolongs survival in advanced melanoma: serious side effects and a hefty price tag may limit its use. PT 37(9):503–530

Feng Y, Masson E, Dai D, Parker SM, Berman D, Roy A (2014) Model-based clinical pharmacology profiling of ipilimumab in patients with advanced melanoma. Br J Clin Pharmacol 78 (1):106–117

Ferris RL, Blumenschein G, Fayette J, Guigay J, Colevas AD, Licitra L et al (2016) Nivolumab for recurrent squamous-cell carcinoma of the head and neck. N Engl J Med 375(19):1856–1867

Gebhardt C, Sevko A, Jiang H, Lichtenberger R, Reith M, Tarnanidis K et al (2015) Myeloid cells and related chronic inflammatory factors as novel predictive markers in melanoma treatment with ipilimumab. Clin Cancer Res 21(24):5453–5459

Gnjatic S, Bronte V, Brunet LR, Butler MO, Disis ML, Galon J et al (2017) Identifying baseline immune-related biomarkers to predict clinical outcome of immunotherapy. J Immunother Cancer 5:44

Haanen JBAG, Carbonnel F, Robert C, Kerr KM, Peters S, Larkin J et al (2018) Management of toxicities from immunotherapy: ESMO clinical practice guidelines for diagnosis, treatment and follow-up. Ann Oncol 29(Suppl 4):iv264–iv266

Horn L, Spigel DR, Vokes EE, Holgado E, Ready N, Steins M et al (2017) Nivolumab versus docetaxel in previously treated patients with advanced non-small-cell lung cancer: two-year outcomes from two randomized, open-label, phase III trials (CheckMate 017 and CheckMate 057). J Clin Oncol 35(35):3924–3933

Hurkmans DP, Basak EA, van Dijk T, Mercieca D, Schreurs MWJ, Wijkhuijs AJM et al (2019) A prospective cohort study on the pharmacokinetics of nivolumab in metastatic non-small cell lung cancer, melanoma, and renal cell cancer patients. J Immunother Cancer 7(1):192

Johnson DB, Beckermann KE, Wang DY (2018) Immune checkpoint inhibitor therapy in patients with autoimmune disease. Oncology (Williston Park) 32(4):190–194

Kaufman HL, Russell J, Hamid O, Bhatia S, Terheyden P, D'Angelo SP et al (2017) Avelumab in chemotherapy-refractory metastatic Merkel cell carcinoma: subgroup analysis of efficacy. JCO 35(7_Suppl):80–80

Kelderman S, Heemskerk B, van Tinteren H, van den Brom RRH, Hospers GAP, van den Eertwegh AJM et al (2014) Lactate dehydrogenase as a selection criterion for ipilimumab treatment in metastatic melanoma. Cancer Immunol Immunother 63(5):449–458

Kennedy LB, Salama AKS (2020) A review of cancer immunotherapy toxicity. CA Cancer J Clin 70(2):86–104

Khalil DN, Smith EL, Brentjens RJ, Wolchok JD (2016) The future of cancer treatment: immunomodulation, CARs and combination immunotherapy. Nat Rev Clin Oncol 13 (5):273–290

Khoja L, Atenafu EG, Templeton A, Qye Y, Chappell MA, Saibil S et al (2016) The full blood count as a biomarker of outcome and toxicity in ipilimumab-treated cutaneous metastatic melanoma. Cancer Med 5(10):2792–2799

Khunger M, Hernandez AV, Pasupuleti V, Rakshit S, Pennell NA, Stevenson J et al (2017) Programmed cell death 1 (PD-1) ligand (PD-L1) expression in solid tumors as a predictive biomarker of benefit from PD-1/PD-L1 axis inhibitors: a systematic review and meta-analysis. JCO Precis Oncol 1:1–15

Ku GY, Yuan J, Page DB, Schroeder SEA, Panageas KS, Carvajal RD et al (2010) Single-institution experience with ipilimumab in advanced melanoma patients in the compassionate use setting: lymphocyte count after 2 doses correlates with survival. Cancer 116(7):1767–1775

Larkin J, Chiarion-Sileni V, Gonzalez R, Grob JJ, Cowey CL, Lao CD et al (2015a) Combined nivolumab and ipilimumab or monotherapy in untreated melanoma. N Engl J Med 373 (1):23–34

Larkin J, Chiarion-Sileni V, Gonzalez R, Grob JJ, Cowey CL, Lao CD et al (2015b) Efficacy and safety in key patient subgroups of nivolumab (NIVO) alone or combined with ipilimumab (IPI) versus IPI alone in treatment-naïve patients with advanced melanoma (MEL) (CheckMate 067). Eur J Cancer 51:S664–S665

Liu C, Yu J, Li H, Liu J, Xu Y, Song P et al (2017) Association of time-varying clearance of nivolumab with disease dynamics and its implications on exposure response analysis. Clin Pharmacol Ther 101(5):657–666

Longoria TC, Tewari KS (2016) Evaluation of the pharmacokinetics and metabolism of pembrolizumab in the treatment of melanoma. Expert Opin Drug Metab Toxicol 12 (10):1247–1253

Martens A, Wistuba-Hamprecht K, Yuan J, Postow MA, Wong P, Capone M et al (2016) Increases in absolute lymphocytes and circulating CD4+ and CD8+ T cells are associated with positive clinical outcome of melanoma patients treated with ipilimumab. Clin Cancer Res 22 (19):4848–4858

Merchant MS, Wright M, Baird K, Wexler LH, Rodriguez-Galindo C, Bernstein D et al (2016) Phase I clinical trial of ipilimumab in pediatric patients with advanced solid tumors. Clin Cancer Res 22(6):1364–1370

Motzer RJ, Escudier B, McDermott DF, George S, Hammers HJ, Srinivas S et al (2015) Nivolumab versus everolimus in advanced renal-cell carcinoma. N Engl J Med 373(19):1803–1813

Naidoo J, Page DB, Li BT, Connell LC, Schindler K, Lacouture ME et al (2015) Toxicities of the anti-PD-1 and anti-PD-L1 immune checkpoint antibodies. Ann Oncol 26(12):2375–2391

Netterberg I, Li C-C, Molinero L, Budha N, Sukumaran S, Stroh M et al (2019) A PK/PD analysis of circulating biomarkers and their relationship to tumor response in Atezolizumab-treated non-small cell lung cancer patients. Clin Pharmacol Ther 105(2):486–495

Nikanjam M, Patel H, Kurzrock R (2017) Dosing immunotherapy combinations: analysis of 3,526 patients for toxicity and response patterns. Onco Targets Ther 6(8):e1338997

Ning Y-M, Suzman D, Maher VE, Zhang L, Tang S, Ricks T et al (2017) FDA approval summary: Atezolizumab for the treatment of patients with progressive advanced urothelial carcinoma after platinum-containing chemotherapy. Oncologist 22(6):743–749

Nyakas M, Aamdal E, Jacobsen KD, Guren TK, Aamdal S, Hagene KT et al (2019) Prognostic biomarkers for immunotherapy with ipilimumab in metastatic melanoma. Clin Exp Immunol 197(1):74–82

Palmieri DJ, Carlino MS (2018) Immune checkpoint inhibitor toxicity. Curr Oncol Rep 20(9):72

Pillai RN, Behera M, Owonikoko TK, Kamphorst AO, Pakkala S, Belani CP et al (2018) Comparison of the toxicity profile of PD-1 versus PD-L1 inhibitors in non-small cell lung cancer: a systematic analysis of the literature. Cancer 124(2):271–277

Pollack MH, Betof A, Dearden H, Rapazzo K, Valentine I, Brohl AS et al (2018) Safety of resuming anti-PD-1 in patients with immune-related adverse events (irAEs) during combined anti-CTLA-4 and anti-PD1 in metastatic melanoma. Ann Oncol 29(1):250–255

Ramalingam SS, Hellmann MD, Awad MM, Borghaei H, Gainor J, Brahmer J et al (2018) Abstract CT078: Tumor mutational burden (TMB) as a biomarker for clinical benefit from dual immune checkpoint blockade with nivolumab (nivo) + ipilimumab (ipi) in first-line (1L) non-small cell lung cancer (NSCLC): identification of TMB cutoff from CheckMate 568. In: Clinical trials [Internet]. American Association for Cancer Research; 2018 [cited 2020 May 11]. pp CT078–CT078. http://cancerres.aacrjournals.org/lookup/doi/10.1158/1538-7445.AM2018-CT078

Ribas A, Robert C, Hodi FS, Wolchok JD, Joshua AM, Hwu W-J et al (2015) Association of response to programmed death receptor 1 (PD-1) blockade with pembrolizumab (MK-3475) with an interferon-inflammatory immune gene signature. JCO 33(15_Suppl):3001–3001

Rittmeyer A, Barlesi F, Waterkamp D, Park K, Ciardiello F, von Pawel J et al (2017) Atezolizumab versus docetaxel in patients with previously treated non-small-cell lung cancer (OAK): a phase 3, open-label, multicentre randomised controlled trial. Lancet 389(10066):255–265

Rizvi NA, Hellmann MD, Snyder A, Kvistborg P, Makarov V, Havel JJ et al (2015) Cancer immunology. Mutational landscape determines sensitivity to PD-1 blockade in non-small cell lung cancer. Science 348(6230):124–128

Robert C, Long GV, Brady B, Dutriaux C, Maio M, Mortier L et al (2015) Nivolumab in previously untreated melanoma without BRAF mutation. N Engl J Med 372(4):320–330

Sanghavi K, Zhang J, Zhao X, Feng Y, Statkevich P, Sheng J et al (2020) Population pharmacokinetics of ipilimumab in combination with nivolumab in patients with advanced solid tumors. CPT Pharmacometrics Syst Pharmacol 9(1):29–39

Schachter J, Ribas A, Long GV, Arance A, Grob J-J, Mortier L et al (2017) Pembrolizumab versus ipilimumab for advanced melanoma: final overall survival results of a multicentre, randomised, open-label phase 3 study (KEYNOTE-006). Lancet 390(10105):1853–1862

Shapiro I, Grote HJ, D'Urso V, von Heydebreck A, Mahnke L, Kaufman H et al (2017) Exploratory biomarker analysis in avelumab-treated patients with metastatic Merkel cell carcinoma progressed after chemotherapy. JCO 35(15_Suppl):9557–9557

Shen X, Zhao B (2018) Efficacy of PD-1 or PD-L1 inhibitors and PD-L1 expression status in cancer: meta-analysis. BMJ 362:k3529

Sheng J, Srivastava S, Sanghavi K, Lu Z, Schmidt BJ, Bello A et al (2017) Clinical pharmacology considerations for the development of immune checkpoint inhibitors. J Clin Pharmacol 57 (Suppl 10):S26–S42

Simeone E, Gentilcore G, Giannarelli D, Grimaldi AM, Caracò C, Curvietto M et al (2014) Immunological and biological changes during ipilimumab treatment and their potential correlation with clinical response and survival in patients with advanced melanoma. Cancer Immunol Immunother 63(7):675–683

Starr P. High tumor mutation burden predictive biomarker for survival in metastatic lung cancer. 2019 July 3 [cited 2020 May 11]. http://oncpracticemanagement.com/issues/2019/july-2019-vol-9-no-7/1257-high-tumor-mutation-burden-predictive-biomarker-for-survival-in-metastatic-lung-cancer

Stroh M, Winter H, Marchand M, Claret L, Eppler S, Ruppel J et al (2017) Clinical pharmacokinetics and pharmacodynamics of atezolizumab in metastatic urothelial carcinoma. Clin Pharmacol Ther 102(2):305–312

Topalian SL, Taube JM, Anders RA, Pardoll DM (2016) Mechanism-driven biomarkers to guide immune checkpoint blockade in cancer therapy. Nat Rev Cancer 16(5):275–287

Ventola CL (2017) Cancer immunotherapy, part 2: efficacy, safety, and other clinical considerations. P T 42(7):452–463

Weber J, Hamid O, Amin A, O'Day S, Masson E, Goldberg SM et al (2013) Randomized phase I pharmacokinetic study of ipilimumab with or without one of two different chemotherapy regimens in patients with untreated advanced melanoma. Cancer Immun 13:7

Weber JS, Hodi FS, Wolchok JD, Topalian SL, Schadendorf D, Larkin J et al (2017) Safety profile of nivolumab monotherapy: a pooled analysis of patients with advanced melanoma. J Clin Oncol 35(7):785–792

Wilgenhof S, Du Four S, Vandenbroucke F, Everaert H, Salmon I, Liénard D et al (2013) Single-center experience with ipilimumab in an expanded access program for patients with pretreated advanced melanoma. J Immunother 36(3):215–222

Wolchok JD, Neyns B, Linette G, Negrier S, Lutzky J, Thomas L et al (2010) Ipilimumab monotherapy in patients with pretreated advanced melanoma: a randomised, double-blind, multicentre, phase 2, dose-ranging study. Lancet Oncol 11(2):155–164

Yang H, Shen K, Zhu C, Li Q, Zhao Y, Ma X (2018) Safety and efficacy of durvalumab (MEDI4736) in various solid tumors. Drug Des Devel Ther 12:2085–2096

Zhang J, Sanghavi K, Shen J, Zhao X, Feng Y, Statkevich P et al (2019) Population pharmacokinetics of nivolumab in combination with ipilimumab in patients with advanced malignancies. CPT Pharmacometrics Syst Pharmacol 8(12):962–970

Regulatory Affairs and Intellectual Property Rights in Immunotherapeutics 10

Kedar Suvarnapathaki, Sanjay D. Sawant, and Indu Nambiar

Abstract

Protecting the human body from harm is performed by different types of cells, organs, tissues, and proteins together known as the immune system. The immune system has the quality to differentiate the body cells/ tissues from foreign material (self from non-self).

Maintenance of a balanced immune system is imperative; else several disorders or deficiencies may lead to autoimmune diseases and hypersensitivities. These can be life threatening as well.

Immunotherapy or Biological therapy works both ways; either activate or suppress the immune system depending on the treatment that is required to treat a disorder. Very lately, lot of focus of researchers, clinicians, and pharmaceutical companies has been attracted by immunotherapy. Success in the treatment of various cancers has added to the interest of the stakeholders. Recent researches have proved that customized immunotherapies have less side effects than the drugs already available. These therapies also possess weak qualities to create any hostility during microbial disease treatment.

This chapter provides in detail the regulator aspects of getting these drugs approved. The chapter aims to look at the various guidelines provided by ICH and the regulatory guidance implemented by Health Authorities worldwide including the FDA, EMA, Health Canada, MHRA, and TGA.

K. Suvarnapathaki
Janssen India, Johnson & Johnson Pvt. Ltd, Thane, Maharashtra, India

S. D. Sawant
STES's Smt. K. N. College of Pharmacy, S. P. Pune University, Pune, India

I. Nambiar (✉)
Boehringer Ingelheim India Pvt. Ltd., Mumbai, India
e-mail: indu.nambiar@boehringer-ingelheim.com

S. P. Sawarkar et al. (eds.), *Immunotherapy – A Novel Facet of Modern Therapeutics*,
https://doi.org/10.1007/978-981-15-9038-2_10

Another focus of this chapter is on patent creation and maintenance. Intellectual property (IP) is a category including abstract creations of the human intellect. Immunotherapeutics or biologics are large molecules, which require an extensive budget to research, develop, and manufacture than small molecules. Hence it becomes mandatory to develop powerful patent portfolios that ensure the protection of the innovative ideas embedded in these therapies.

Keywords

Immunotherapy · Regulation · FDA · Intellectual property rights · Claims · ICH guidelines · Composition · Antibodies · MHRA · Australia · TGA · India · DCGI · Pharmacovigilance · CMC changes

10.1 Introduction

As covered in the previous chapters immune system act as a guard to the human body and have the ability to differentiate the good from the bad. An antigen is a molecule of the immune system that identifies any threat to the human body.

As every reaction needs to be controlled, the responses by the immune response need regulation. This ensures that the human body does not end up in a state of shock (e.g.—anaphylaxis). These regulating cells are described in earlier chapters and are the ones that refrain autoimmune reactions in the human body (Taylor et al. 2006; Delves 2020).

Treating a disease by activating or suppressing the immune system is termed as Immunotherapy. These Immunotherapeutic agents work on the mechanisms used by the immune system. There is a trend seen wherein this therapy is going to be used widely and many advancements are seen in this field and more expected in the future (Nicholson 2016). A number of different classes of immunotherapeutic agents have been developed or under development. Some examples of immunotherapies either approved or under development are given in following Table 10.1–.

10.2 Why Regulations

Every drug or treatment that enters the market needs to be regulated. It becomes imperative for Immunotherapeutics which include treatment agents that control the responses of the immune system to provide relief or slow down the progress of a diseased condition. Though clinical trials are ongoing and the development of such therapies are progressing, regulations ensure the product remains safe and effective to continue providing required therapeutic benefits to the patients who need it.

Encouraging results are being derived for the studies carried out in this space. With the results come experiences that have a lot to teach. All these lessons learnt help the pharmaceutical companies, the medical fraternity to evolve different approaches to these therapies and it also helps regulators to address the gaps in the

Table 10.1 Examples of Immunotherapies approved/different phases of clinical trials

Drug	Brand name/trial phase	Innovator Company	Action	Indication
Adalimumab	Humira (Abbott) Exemptia (generic introduced in India by Cadila Healthcare)	BASF Knoll (BASF Pharma)—Developed Abbott Laboratories—marketed	Anti–TNF-*alpha*	Ankylosing spondylitis,Plaque psoriasis, psoriatic arthritis, used in moderate to severe conditions of Crohn disease, rheumatoid arthritis, and ulcerative colitis,
Basiliximab	Simulect®	Novartis	Anti-Interleukin-2 receptor	In acute rejection of a transplanted kidney prevention
Canakinumab	Ilaris®	Novartis	Anti–IL-1 beta	Active systemic juvenile idiopathic arthritis in patients $\geq$2 year. Periodic fever syndromes
Daratumumab	Darzalex®	Janssen (a member of Johnson & Johnson)	Anti-CD38	Multiple myeloma
Abatacept	Orencia®	Bristol-Myers Squibb Ono (Japan)	Inhibition of T-cell activation	Moderate to severe rheumatoid arthritis
Etanercept	Enbrel	Immunex (patient filed) Amgen	Decrease in TNF levels	Rheumatoid arthritis, psoriatic arthritis, Ankylosing spondylitis, plaque psoriasis
Interleukin-2	Proleukin	Cetus corporation Chiron	Immunostimulant	Metastatic renal cell carcinoma
Interleukin-11	Neumega®	Wyeth (Pfizer)	Thrombopoietic growth factor	Thrombocytopenia prevention
G-CSF	Neupogen	Amgen	Stimulation of granulocyte production	Reversal of neutropenia
GM-CSF	Trial phase 2 and 3 [ref: https://www.clinicaltrialsregister.eu/ctr-search/search?query=Sargramostim+(GM-CSF)]	BioVex Inc.	Stimulation of granulocyte and monocyte/macrophage production	Reversal of neutropenia

(continued)

Table 10.1 (continued)

Drug	Brand name/trial phase	Innovator Company	Action	Indication
Ustekinumab	Stelara	Centocor and Janssen-Cilag international (collaborator)	Anti–IL-12 and –IL-23	Moderate to severe plaque psoriasis Moderate to severe Crohn disease Psoriatic arthritis
Interferon-alpha	Intron-A	Biogen (developed) Schering-Plow (marketed)	Antiproliferative and antiviral	Chronic hepatitis C, chronic myeloid leukemia, metastatic melanoma
Interferon-beta	Avonex	Biogen biotechnology	Antiproliferative and antiviral	Reducing number of eruptions/flare-ups in relapse of multiple sclerosis
Interferon-gamma	Phase 3 [ref: https://clinicaltrials.gov/ct2/show/study/NCT02415127]	Horizon Pharma Ireland, ltd.	Immunostimulatory and antiviral	Control over infection in chronic granulomatous disease delayed severe malignant osteopetrosis progression

regulations to review these novel therapy approaches in order to ensure the products with positive benefit-risk ratio are only approved and remain in the market to benefit the patient (Camarero and Ruiz 2012).

Pivotal trials especially for the primary endpoints need to provide positive outcomes, which is very crucial. To avoid any bias, the design of the study should be robust. Any foreseen uncertainties should also be eliminated, in order to maintain the risk-benefit ratio. The scientific advice from National Authorities is strongly encouraged in order to have well defined regulatory pathways and expectations of regulatory authorities.

All the outcomes should be clinically meaningful. In conclusion, active immunotherapies need to be assessed like all medicinal products, for their quality, efficacy, and safety which is in compliance to the applicable regulatory requirements.

Regulations are applicable not only to the drugs involved but also to the patients involved in these trials. Regulations pertaining to the conduct of clinical trials are available on the Indian Heath authority website (CDSCO website).

Evaluating the efficacy of immunotherapy needs to be different than the methods used for other therapies, taking into consideration that many times the patient's system acts as a therapy to provide relief to certain conditions. Hence, the protocols, trial designs need to be evaluated and modified at every critical step in order to ensure patient safety. As a consequence the whole regulatory process for immunotherapy drugs is lengthy, complicated, and stricter especially for genetically engineered cells. Regulatory specialists would be involved in their inputs, experiences, and dialogs with regulatory authorities.

Successful immunotherapy needs very critically controlled activation, regulation, and resolution of the immune response. This is a very complex action and any of the parameters not controlled/calibrated carefully can mean a catastrophic outcome for the patients. Therefore in order to ensure approved immunotherapies continue to have positive benefits to risk ratio throughout the lifecycle of the product, it is necessary to have well defined regulatory norms and guidelines for such products so that products can be developed and approved in strict compliance to such regulations (Labiotech.eu 2020).

10.3 Intellectual Property Rights Approaches

The potential of immunotherapy as a treatment option has opened up the intellectual property landscape of the field. The chance of a compound identified as a potential drug candidate to enter the Phase I clinical trial is less than 10%, means more than 90% of potential drug candidate do not make it to human clinical trials; no other major business operates under such a high failure rate.

It is even more costly to research, develop, and manufacture large molecule pharmaceuticals than small molecules. Thus, it is essential to work upon building robust patent portfolios to protect innovative biotherapies. Many companies and institutions have filed patent applications related to the various drugs and targets similar to the ones listed in Table 10.1, seeking to protect their immunotherapeutic inventions.

The first step to get a patent is for the applicant to fill up the customized form and file the patent with the applicable office. The different formats of the form available have the common requirement listed which are universal, specifically including the detailed description of the invention and the claims with regards to the patent. The description should be transparent enough to provide the authority with the experimentation details and ways of development. Claims define the scope or boundaries of the patent owner's exclusive rights (Hong and Division 2013).

Claims define the scope of legal protection and are the most important part of any patent. The claims demarcate in words the boundary of the invention. Improper drafting of claims may result in leaving some loose ends, which are legally not protected and can be used by competitors since the unprotected information does not require the permission of the patent owner.

The claim language in a patent is crucial to the effective protection of the invention, and the balance between the breadth of the claims and validity is a difficult one to achieve. Initial specification (provisional) can broad but should include as much detail as possible. 12 months later, the complete (PCT) specification needs to be filed, the content of this is final and you cannot add new matter. In India, Europe, and many other countries the patent offices are very strict about claim language and subsequent amendment, usually you cannot broaden the scope of the patent after grant, you can only narrow it. Ensure basic requirements are present, that is novelty, inventive step, industrial applicability, and non-obviousness. After the complete specification is filed, remain very careful with amendments, claim amendment can be tough to get approved.

Nowhere has this balance come into play more frequently than in the cases of patenting antibody-related inventions. While there could be slight differences in the nature of claims that can be granted across different geographies, in general, for a biologic molecule, various types of claims that one can use to build a patent portfolio can be categorized as follows –

1. Claims pertaining to antibody per se
2. Claims of composition
3. Claims regarding nucleic acid encoding the antibody
4. Claims of using the antibody
5. Claims pertaining to first medical use
6. Claims of additional indication
7. Claims about formulations
8. Claims of purification methods

To illustrate types of claims that can arise, you can go through the following examples –

1. Claims pertaining to the Antibody per se
   (a) An antibody that binds to protein X.

   (b) An antibody that binds to protein X, wherein the antibody has a dissociation constant (Kd) of Y [or any other functional property such as neutralizing activity, agonistic/antagonistic activity, or immune action/antibody-dependent cell-mediated cytotoxicity (ADCC)/complement-dependent cytotoxicity (CDC)].
   (c) An antibody that binds to an epitope comprising the sequence of SEQ ID NO: Z.
   (d) An antibody–drug conjugate having the formula Ab-X, wherein the antibody is the antibody of any one of claims 1a.-1e. and X comprises Y.
2. Claims of Compositions
   (a) A pharmaceutical composition encompassing the antibody according to one of the claims Aa to Ad stated above and is a pharmaceutically acceptable carrier or diluent.
   (b) A pharmaceutical composition encompassing the antibody according to one of the claims Aa to Ad stated above and is a pharmaceutically acceptable carrier or diluent, wherein the antibody is glycosylated in one or both chains.
   (c) A pharmaceutical composition encompassing the antibody according to one of claims Aa to Ad stated above and is a pharmaceutically acceptable carrier or diluent, wherein the composition has a shelf life of XX months.
   (d) A pharmaceutical composition encompassing the antibody according to one of the claims Aa to Ad stated above in combination with drug YY.
3. Claims regarding Nucleic Acid Encoding the Antibody
   (a) An isolated nucleic acid encoding the antibody produced by hybridoma X.
   (b) An isolated nucleic acid encoding the heavy chain and/or light chain variable region of antibody X, wherein the heavy chain/light variable region comprises SEQ ID NO: X.
4. Claims of Using the Antibody
   (a) A method for treatment of disease X in patients by administration of a therapeutically effective amount of the antibody of any one of claims Aa to Ad above.
   (b) A method for treatment of disease X in patients by administration of a therapeutically effective amount of a pharmaceutical composition of any one of claims Ba.-Bd above.
   (c) A method for treatment of disease X by the administration of the antibody at a certain dosage.
   (d) A method for treatment of disease X in patients by administration of the antibody via a certain route of administration (IV, IM, SC, or others).

The above is not an exhaustive list of claims related to immunotherapy. To build up a strong patent portfolio covering immunotherapeutic inventions, as with other types of inventions, a key is to craft a variety of claims with variations in claim types and scopes (Mouta-Bellum et al. 2017).

10.4 Regulatory Requirements

10.4.1 Regulatory Authorities Worldwide

Pharmaceutical Industry is a highly regulated industry. All the activities in the lifecycle of a pharmaceutical product are regulated by various regulations.

India

In India, it is a federal structure for the Health Authority. We have a Central Authority for the entire country and a State Drug Authority at the state level (Central Drugs Standard Control Organization 2020).

The Central Drugs Standard Control Organization (CDSCO); has six zonal offices, four sub-zonal offices, 13 port offices, and seven laboratories. This body is responsible for ensuring the fulfillment of activities assigned to the Central Government under the Drugs and Cosmetics Act. Present organizational structure of CDSCO is as follows

Drugs Controller General (I)
Head Quarter
New Delhi
Zonal Offices (6)
North zone-Gaziabad
South zone-Chennai
West zone-Mumbai
East zone-Kolkatta
Hyderabad zone
Ahemdabad Zone
Sub Zonal Offices (7)
Bangalore
Varanasi
Goa
Indore
Guwhati
Baddi
Port/Air port Offices (13)
Ahemdabad
Chennai port
Chennai airport
Bangalore
Hyderabad
Kochi
Delhi
Kolkatta port
Kolkatta Air cargo
Mumbai Air cargo
Mumbai Nhava Sheva
Mumbai Custom House
Laboratories
CDL,Kolkatta
CDTL,Mumbai
RDTL,Guwhati
RDTL,Chndigarh
CDL,Kasauli
CDTL,Chennai
CDTL,Hyderabad
IVRI,Izatnagar*
NIB,Noida*
IPC,Gaziabad*
AYUSH STAFF
DCC(Ayurveda)
DCC(Homeopathy)
ADC(Ayur/unani/sidha)
DIs(Ayur/unani/sidha/Homeo)
STAFF
JDC(I)
ADC(I)
DDC(I)
DI
ADI
TDAs
Supporting Staff
STAFF
ADC(I)
DDC(I)
DI
ADI
TDAs
Supporting Staff
STAFF
ADC(I)
DI
TDAs
Supporting Staff
STAFF
ADC(I)
DI
TDAs
Supporting Staff
STAFF
Director
Dy.Director
Sr.Scientific Officer I
Sr.Scientific Officer II
Research Officer
Sr.Scientist Assistant
Jr.Scientist Assistant
Supporting Staff

State drug authority in various states is referred by various names like the Food & Drug Authority of FDA or State Drug Controllers, etc.

In India, there are various legislations, which control different aspects of the Pharmaceutical Industry, for example,

- Manufacturing, import, sale, and distribution are governed by The Drugs & Cosmetics Act, 1940 and The Drugs and Cosmetics Rules 1945—This law is administered by Central Drug Authority under the Ministry of Health and Family Welfare and State Drug Authority under the State Governments.
- Pricing of pharmaceuticals are controlled by Drugs Price Control Order 2013 and is administered by the National Pharma Pricing Authority or NPPA under the Ministry of Chemical and Fertilizers.
- Patenting of pharmaceuticals is controlled by Indian Patent Act and Indian Patent Rules and is administered by the Controller General of Patents under the Ministry of Commerce and Industry.

CDSCO is advised by a technical advisory board known as Drugs Technical AdvisoryBoard (DTAB) and DCC (Drugs Consultative Committee). Few of the important functionalities of the DTAB-DCC Committee are:

- Technical aspects that come up during the implementation of the Drugs and Cosmetic Act, 1940, and other tasks that are mandatory to follow by this Act are discussed by the DTAB. They act as advisors to the Central and State Government about this Act. Such meetings are arranged by this committee which is Chaired by the Director General of Health Services (DGHS).
- Responsible to carry out DCC meetings chaired by the Drugs Controller General(India). The aim is to bring about alignment throughout the country in the implementation of the Act of 1940. The required advisory is provided to the State and Central Governments along with the DTAB by this Committee.
- This Committee commences any revisions required in the Rules of 1945, according to the advocacy of the DTAB. The coordination for draft and final Gazette alerts are released in collaboration with the Ministry of Health & Family Welfare(MOHFW) (DTAB-DCC 2020).

(Ref: https://cdsco.gov.in/opencms/opencms/en/dcc-dtab-committee)

Additional approval committees were set up in March 2011 by the CDSCO, when the same was called as New Drug Advisory Committees (NDACs). These Committees include experts from academics, medical colleges, and other distinguished establishments. This additional step in the review process for approval of new drugs was introduced to ensure a holistic approach towards patient safety and to further build a robust regulatory field in the country. The 12 NDACs were termed as Subject Expert Committees (SECs) in July 2014. By January 2015 about 25 SECs were introduced which were responsible for the specific therapy sector (Bhave and Menon 2019)

[Ref: http://www.picronline.org/article.asp?issn=2229-3485;year=2019;volume=10;issue=1;spage=1;epage=3;aulast=Bhave]

Especially for Immunotherapies, additional consultation from Genetic Engineering Approval Committee (GEAC) and/or Review Committee on Genetic Manipulation (RCGM) is taken. The RCGM formed under the Department of Biotechnology (DBT) oversees and grants permission for small-scale trials and permissions for large-scale deliveries and trading of genetically modified organisms (GMO) are granted by the GEAC, formed under the Ministry of Environment and Forests. (GEAC n.d.)

[Ref: http://moef.gov.in/wp-content/uploads/2018/03/groundrules.pdf]

Unites States

The Food and Drug Administration (FDA) is a federal agency within the USA Department of Health and Human Services. It consists of the Office of the Commissioner and four directorates which ensure the protection and promotion of Public health. The four Directorates are namely: Medical Products and Tobacco, Foods and Veterinary Medicine, Global Regulatory Operations and Policy, and Operations.

The spectrum of scope covered by the FDA's regulatory authority is very broad and includes responsibilities closely related to several other government agencies as well. The broad scope leads to a lot of confusion and frustration for consumers in determining the relevant agency within the FDA to contact. Below is a list of product categories, traditionally recognized, falling under FDA's jurisdiction; but not limited to these (FDA 2020).

In general, FDA regulates:

- Foods, including:
  - Supplements (Dietary)
  - Packaged water
  - Additives used for consumption
  - Formulas prepared for infants
  - Other edible products (except for some meat, poultry, and egg products, which are under the regulation of the USA Department of Agriculture)
- Drugs (Active pharmaceutical ingredients), including:
  - Prescription medicines (branded and generic)
  - OTCs (over-the-counter) or non-prescription drugs
- Biologics, including:
  - Human vaccines
  - Whole blood and other blood components
  - Products used in cellular and gene therapy
  - Cell/tissue and tissue-based products
  - Allergenics
- Medical Devices, including:
  - Simple materials used by physicians and in hospitals (tongue depressors and bedpans)
  - Technologies that are complex and complicated (heart pacemakers)

  - Devices used in dentistry
  - Medical implants, surgical instruments, and prosthetics (e.g., replacement of a body part)
- Radiation emitting electronic products, including:
  - Microwave ovens
  - X-ray machines
  - Laser therapy equipments
  - Ultrasonic equipment
  - Mercury vapor lamps (neon lights)
  - Sunlamps emitting ultraviolet radiations
- Cosmetics, including:
  - Additives that give color and included in makeup and other personal/beauty care products
  - Moisturizers and cleansers for skin
    Nail paints and body sprays/scents
- Veterinary Products, including:
  - Feeds made for farm animals (livestocks)
  - Foods for house pets
  - Drugs and devices used in veterinary space
- Tobacco and tobacco products, including:
  - Cigarettes and smokeless tobacco
  - E-cigarettes other electronic nicotine delivery systems
  - Cigars

United Kingdom

The Medicines and Healthcare products Regulatory Agency (MHRA) is a body sponsored by the Health Department and Social care in the United Kingdom. MHRA is accountable for defining the safety of medicines, devices, blood components, and so on. The agency regulates and stands responsible for (Towers and SWg sNO 2020):

- Ascertaining that the applicable standards of quality, safety, and efficacy are met by medicines, medical devices, and blood components for transfusion made available in the country.
- Ensuring that safe and secure supply chains are maintained for drugs, medical devices, and blood components
- Promoting harmonization and applying international standards that assure the effectiveness and safety of biological medicines.
- Educating about the risks-benefits ratio of medicines, medical devices, and blood components to the public and healthcare professionals, to ensure safer and more effective use in the real world.
- Enhancing public health by supporting beneficial innovation, research, and development in this space.
- Protecting public health by influencing UK, EU, and other international regulatory frameworks, making them risk-proportionate and effective.

MHRA has established several independent scientific advisory committees. The aim is to provide advice, which is impartial to the decision-makers who regulate medicines and medical devices in the territory. These committees are authorized to set up working groups in order to focus on specific issues. Members belonging to the below committees may have to collect a fee and also engage in some expenses.

- Advisory Board on the Registration of Homeopathic Products
- Herbal Medicines Advisory Committee
- Review Panel
- Independent Scientific Advisory Committee for MHRA database research
- Medicines Industry Liaison Group
- Innovation Office
- Blood Consultative Committee
- Devices Expert Advisory Committee

Europe

The European Medicines Agency (EMA) is a decentralized agency of the European Union (EU) in charge of scientific evaluation, supervision, and safety monitoring of medicines/drugs in the EU.

The EU, pharmaceutical industries, and some indirect subsidy by the member state fund the EMA. The intention of the EMA. The EMA staff under the supervision of EMA's Executive Director carry out the day-to-day operations of the organization.

Thousands of experts from across Europe are involved in the activities of the EMA's scientific committees making EMA a networking organization.

It is the mission of the European Medicines Agency (EMA) to ensure scientific brilliance in the evaluation and supervision of medicines, benefiting public and animal health in the European Union (EU) and to bring about harmonization among the work of the several operational national medicine regulatory bodies. These are achieved through the following activities/responsibilities

- Development and easy availability of medicines to stakeholders
- Review applications received from marketing authorization for permission
- Continuous vigilance on the safety of medicines throughout the marketing lifecycle
- Make updated information available to healthcare professionals and patients

The EMA began operating on January 26, 1995.Directive 2001/82/EC and Directive 2001/83/EC encompass the Community codes for veterinary and human medicines respectively. The authorization, manufacture, and distribution of medicines in the EU are bound by a legal framework which is provided in these. Regulation (EC) No 726/2004, which established the European Medicines Agency (EMA), forms the basis for the centralized authorization procedure for human and veterinary medicines. Over the period, there have been amendments and enhancements to this legal framework through the introduction of further legal acts covering specific areas of pharmaceutical law (European Medicines Agency 2020a).

1. The below list states the basis for the main EU legal framework for pharmaceuticals:
   - Directive 2001/82/EC, includes the Community code related to veterinary medicinal products, as amended.
   - Directive 2001/83/EC includes the Community code related to medicinal products for human use, as amended.
     The amendments of the above two are incorporated into the consolidated text of Directives 2001/83/EC and 2001/83/EC, respectively.
   - Regulation (EC) No 726/2004, lays down Community procedures for the authorization and supervision of medicinal products for human and veterinary use and for establishing a European Medicines Agency, as amended. These amendments are incorporated into the consolidated text of Regulation (EC) No 726/2004.
2. Pharmacovigilance Legislation

A new package of legislation was adopted in 2010with the main aim of reinforcement of pharmacovigilance in the EU, which was supplemented with further legislation in 2012 (European Medicines Agency 2020b). The main legal acts included are:

- Regulation (EU) No 1235/2010 and Regulation (EU) No 1027/2012 amending, as regards pharmacovigilance, Regulation (EC) No 726/2004.
- Directive2010/84/EU and Directive 2012/26/EU amending, as regards pharmacovigilance, Directive 2001/83/EC.
- Commission Implementing Regulation No 520/2012, concerning operational aspects of the implementation of the new legislation.

More information is available on https://www.ema.europa.eu/en/human-regulatory/overview/pharmacovigilance/legal-framework-pharmacovigilance

3. Specialized Medicines Sectors
   - Directive 2004/24/EC, amending Directive 2001/83/EC established includes the legal framework for traditional herbal medicines. More information can be found in Herbal medicinal products (Watkins et al. 2015).
   - Regulation (EC) 1901/2006 forms the basis for the legal framework for paediatric medicines. is based on. More information can be found in (European Medicines Agency 2020c).
   - Regulation (EC) No 1394/2007, amending Regulation (EC) No 726/2004 and Directive 2001/83/EC establishes a legal framework for advanced-therapy medicines. More information can be found in (European Medicines Agency 2020d).

Regulation (EC) No 141/2000 and a number of other relevant legal acts form the basis for the legal framework for orphan medicines. Similar to the main role in developing pediatric medicines, the EMA also plays a central role in the development and authorization of orphan medicines (medicines for rare diseases). More information can be found in https://www.ema.europa.eu/en/human-regulatory/overview/orphan-designation/legal-framework-orphan-designation (European Medicines Agency 2020e)

4. Other Relevant Legislation
   - First established in the 1990s and all the amendments since then include a legal framework for the regulatory fees payable by pharmaceutical companies. More information can be found in https://www.ema.europa.eu/en/human-regulatory/overview/fees-payable-european-medicines-agency (European Medicines Agency 2020f). Rules with regards to the payment of fees to, and the receipt of administrative assistance from the EMA by micro, small, and medium-sized enterprises are laid down inCommission Regulation(EC) No 2049/2005 (European Medicines Agency 2020g). More information can be found in https://www.ema.europa.eu/en/human-regulatory/overview/supporting-smes
   - The regulatory directives for the conduct of clinical trials in the EU are established in the Directive 2001/20/EC (European Medicines Agency 2020h). More information can be found in https://www.ema.europa.eu/en/human-regulatory/research-development/clinical-trials-human-medicines.
     Community procedures for the establishment of maximum residue limits of pharmacologically active substances in foodstuffs from animal origin are laid down in Regulation (EC) No 470/2009 (Regulation (EC) No 470/2009 2009). More information can be found in the document available on the link—https://eurlex.europa.eu/LexUriServ/LexUriServ.do?uri=OJ:L:2009:152:0011:0022:en:PDF.
   - The regulatory directives for the handling of post-authorization variations as per the marketing authorization requirements for human and veterinary medicines are provided in Commission Regulation (EC) No 1234/2008 provides. Commission Regulation (EU) No 712/2012 documents its amendment.
   - Application for the transfer of a marketing authorization are examined as per the rules laid down in Commission Regulation(EC) No 2141/96.
   - A legal framework with regards to managing the risks and prevention of the entry of falsified medicinal products into the legal supply chain is provided in Directive 2011/62/EU which is the amended Directive 2001/83/EC (European Medicines Agency 2020i). More information available in https://www.ema.europa.eu/en/human-regulatory/overview/public-health-threats/falsified-medicines-overview.
   - EU penalties regime with respect to failure in complying with various aspects of Regulation (EC) No 726/2004 provided in Regulation (EC) No 658/2007.

The above this list is not exhaustive and a further detailed list is available on https://ec.europa.eu/health/documents/eudralex_en (EudraLex—EU Legislation | Public Health 2020)

Australia

The Therapeutic Goods Administration (TGA) is responsible for evaluation, assessment, and monitoring of, as a part of the Australian Government Department of Health. Therapeutic goods are products used in humans in connection with various prevention, testing, influencing aspects that include medicines prescription, vaccines, sunscreens, vitamins and minerals, medical devices, surgical implants blood, and blood products. Most of the products, which make therapeutic claims, and before supply into Australia require registration in the Australian Register of Therapeutic Goods (ARTG).

The TGA regulates therapeutic goods, which are available for supply in Australia, and is responsible for ensuring that they are safe for their intended purpose. Therapeutic goods include materials used on a daily basis (vitamins, sunscreens) to complex goods used as a treatment for serious conditions. The TGA stands responsible to regulate the availability of:

- Prescription medicines (advised by a doctor or dentist)
- Over-the-counter medicines
- Medicines that can be made available in the general pharmacy and in supermarkets
- Complementary medicines, which include vitamins, herbal, and traditional medicines
- Simple (bandages) to complex (pace maker)medical devices, products used for testing of a spectrum of diseases or conditions (in vitro diagnostic devices), such as blood tests kits
- Blood products, vaccines, and other biologics
- And the manufacturing and advertising of all the above-mentioned products (TGA basics | Therapeutic Goods Administration (TGA) 2020)

It is a mandate by The Therapeutic Goods Act 1989 (the Act) that medical products imported into, supplied in, or exported from Australia must be included in the Australian Register of Therapeutic Goods (ARTG). The requirement is the submission of an application to the TGA by the sponsor company, in order to include a prescription medicine in the ARTG. This submission must include:

- Data supporting the quality, safety, and efficacy of the product for its intended use
- Relevant forms duly filled, and
- Confirmation of fee payment

The Australian Regulatory Guidelines for Prescription Medicines (ARGPM) details the data required to be included in the submission (Australian Regulatory Guidelines for Prescription Medicines (ARGPM) | Therapeutic Goods

Administration (TGA) 2020). Details are available at—https://www.tga.gov.au/publication/australian-regulatory-guidelines-prescription-medicines-argpmAustralia has adapted some of the guidelines issued by the EMA, which give further guidance to the sponsor companies about the expected data to be included in their applications.

It is an expectation in Australia that therapeutic goods available in the marketplace are safe, of supreme quality and the standard at least equals that of comparable countries. Hence, the TGA works in collaboration with international counterparts to ensure that there is a reduction in the worldwide regulatory burden along with increased harmonization/uniformity globally about data requirements. Hence the requirements laid down by TGA as much as possible are similar or same as those defined by other major regulatory agencies.

Common Technical Document (CTD) defines the format for presenting data packages is as defined in (Common Technical Document (CTD) | Therapeutic Goods Administration (TGA) 2020).

Canada

Health Canada is the Federal department responsible for helping the population to maintain and better their health, and at the same time, they respect individual choices and circumstances (Acts and Regulations—Canada.ca 2020). Health Canada regulates and approves the use of thousands of products, including:

- Biologics
- Consumer Goods
- Foods
- Medical Devices
- Natural Health Products
- Pesticides
- Pharmaceuticals, and
- Toxic Substances

Health Canada controls several legislation sections and is responsible to develop and enforce regulations under these legislations, which have an impact on the health and safety of its people. These laws develop through consultation by the Department with the Canadian public, industries, non-governmental organizations (NGOs), and other parties interested. Guidelines are also prepared by Health Canada, which helps the interpretation and clarification of the legislation and regulations. These federal legislation and regulations can be accessed at- https://www.canada.ca/en/health-canada/corporate/about-health-canada/legislation-guidelines/acts-regulations/list-acts-regulations.html

10.4.2 Non-Clinical Requirements in India

Animal Pharmacology

The requirements in India describe targeted pharmacological actions which are those exhibiting therapeutic potential for humans basis the animal models and specimen used. It states that wherever possible, dose-response relationships and ED50 should be provided. Detailing of special researches carried out to explain mode should be available.

Individual properties and targeted uses of test drugs define specific studies to be conducted and their different designs. The methodology used should be validated scientifically. It is preferred that new techniques and methods with strong scientific principles are used.

Pharmacological functionalities and pharmacokinetic information in general, concerning the absorption, distribution, metabolism, and excretion of the investigational material needs to be provided. Wherever possible, the effects of the drug needs co-relation to the concentrations of the drug found in plasma.

To evaluate the prospective unacceptable pharmacodynamic effects of a substance on physiological activities, some Safety and essential pharmacology studies need to be carried out in relation to exposure within the therapeutic dosage permitted and above. The study designs should fulfill following objectives

- Identification of unpleasant pharmacodynamic properties of a substance with some relation to human safety.
- Evaluation of undesirable pharmacodynamic or pathophysiological effects (observed in toxicology or clinical studies).
- Investigation of the undesirable pharmacodynamic effects' mechanism that was observed or suspected.

The impact of the test drug on vital functions also needs to be studied. Vital function includes important organ functions such as cardiovascular, respiratory, and central nervous functionalities. Drug effects should be studies on the following aspects of each vital systems but not limited to:

- Cardiovascular system:-Blood pressure, heart rate, electrocardiogram and as possible in vitro, in vivo, and/or ex vivo methods including electrophysiology also to be considered.
- Central nervous system: Motor activity, behavioral changes, coordination, body temperature sensory, and motor reflex responses.
- Respiratory system: Respiratory rate and other functions such as tidal volume and hemoglobin oxygen saturation.

In addition to the essential safety pharmacological studies, supplemental, and appropriate follow-up safety pharmacology studies may be conducted. Pharmacological assets or chemical make up of the investigational substance, along with the information derived from safety pharmacology studies, clinical trials, pharmacovigilance, in vitro, or in vivo studies and data generated from reports published in literature defines the requirement for additional studies. These follow-up studies should provide additional information or provide a better understanding than yielded by the essential safety pharmacology studies. Below is a non-exhaustive list on the various follow-up studies that can be conducted:

- Cardiovascular system: Effect on ventricular shrinking, vascular resistance and the effects of chemical mediators, their agonists, and antagonists.
- Central nervous system: Effects on behavior, learning, memory, electrophysiology, neurochemistry, and ligand binding.
- Respiratory system: Effects on airway resistance, compliance, pulmonary arterial pressure, blood gases, and blood pH balance.

Supplemental safety pharmacology studies are required to be conducted in addition to essential and follow-up safety studies. These aim investigation of possible adverse pharmacological effects, which could be a cause of concern and are not evaluated in the crucial safety pharmacological studies—Some parameters studied on effects of the investigational drug with regards to each vital systems are listed below:

- Urinary System: Every parameter including urine volume, specific gravity, osmolality, pH balance, proteins present, creatinine and plasma proteins estimation, etc.
- Autonomic Nervous System: Binding to receptors relevant for the autonomic nervous system, and functional response to agonist or antagonist responses in vivo or in vitro, and effects of direct stimulation of autonomic nerves and their effects on cardiovascular responses
- Gastrointestinal System: Gastric secretion, gastric pH measurement, gastric mucosal examination, bile secretion, gastric emptying time in vivo, and ileocaecal contraction in vitro
- Other Organ Systems: Assessment of organ systems not investigated elsewhere posing as a reason for concern (dependency potential, skeletal muscle, immune, and endocrine activities)

Localized applicator agents, e.g., dermal or ocular with low systemic absorption from the site of application and with the drug pharmacology well known, do not require conduction of safety pharmacology studies. It applies to a new substance having similar pharmacokinetics and pharmacodynamics characteristics. The lifecycle of the drug decides the approach to be adapted for conduction of safety pharmacology studies, as detailed below–

- Prior to administration initiation in humans: It is important to study the effects of a new investigational drug on the human vital functions that are listed in the essential safety pharmacology. As necessary, basis any cause of concern, any follow-up or additional studies recognized, should be carried out.

- During the clinical development stage: Additional studies may be warranted for clarifications of observed or suspected effects in animals and humans during clinical development. Before submission to apply for marketing approval: Assessment whether to conduct follow-up and additional safety pharmacology researches or no to be carried out with a justification. Decisions may depend on the Information at hand from toxicology researches that address safety pharmacology endpoints or data from clinical studies. These studies can replace supplemental studies.

Animal Toxicology

Toxicokinetic studies help generate pharmacokinetic data which can be an important element of the non-clinical toxicity or studies specially designed. One of the aims to conduct such studies is to assess the systemic exposure in animals and the linkage to dosage and the time duration of the study.

Other objectives of toxicokinetic studies include:

- Procuring information to relate the exposure achieved in toxicity studies,
  - Toxicological findings
  - Contribution to the assessment of the relevance of these findings to clinical safety
  - Supporting the choice of species and treatment regimen in non-clinical toxicity studies
  - Gather data in conjunction with the toxicity findings
  - Contributions to the design of subsequent non-clinical toxicity studies

Systemic Toxicity Studies

1. Single-Dose Toxicity Studies
   - These studies listed in Table 10.2 are to be carried out utilizing the usual route as planned for humans in two rodent species (mice and rats).
   - Unless the course of administration in humans intended is only intravenous, at least one more path, in addition, should be implemented in one of the specimens to confirm the systemic absorption of the drug. This path should be dependant on the character of the drug.
   - For oral dosing, a limitation of 2 g/kg (or ten times the actual dose than in humans, whichever is greater) is advocated.
   - After the drug administration, animals should be under observation for 14 days, and Minimum Lethal Dose (MLD) and Maximum Tolerated Dose (MTD) should be determined. If possible, the selected system of toxicity should also be confirmed.
   - Observation frequency for mortality should be for up to 7 days after parenteral administration and up to 14 days after oral administration. Symptoms, signs, and mode of death should be recorded, with proper macroscopic and microscopic detections where essential.
   - Reporting preferably with 95% confidence limits is required forLD10 and LD50. The reasons for not able to determine LD50 should be stated.

Table 10.2 Scheme for conducting toxicity studies

Route of administration	Duration of proposed human administration	Human Phase(s) for which study is proposed to be conducted	Long-term toxicity requirements
Systemic toxicity studies			
Oral or parenteral or transdermal	Single or several doses in 1 day, up to 1 week	I, II, III	2 species; 2 weeks
	>1 week, up to 2 weeks	I, II, III	2 species; 2 weeks
	Upto 2 weeks	Permission to market	2 species; 4 weeks
	>2 weeks, upto 4 weeks	I, II,III	2 species; equivalent to the duration of human exposure
		Permission to market	2 species; 12 weeks
	> 4 weeks, up to 12 weeks	I, II, III	2 species; equivalent to duration of human exposure
		Marketing permission	2 species; 24 weeks
	> 12 weeks, up to 24 weeks	I, II, III	2 species; equal to duration of human exposure
		Permission to market	2 species; rodent 24 weeks, non-rodent 36 weeks
	> 24 weeks	I, II, III	2 species; rodent 24 weeks, non-rodent 36 weeks
		Marketing permission	2 species; rodent 24 weeks, non-rodent 36 weeks
Inhalation (general Anesthetics, aerosols)	Up to 2 weeks	I, II, III	2 species; I month (exposure time 3 h/d, 5d/week)
	Up to 4 weeks	I, II, III	2 species; 12 weeks (exposure time 6 h/d, 5d/week)
	>14 weeks	I, II, III	2 sp.; 24 weeks (exposure time 6 h/d, 5d/week)
Local toxicity studies			
Dermal	Up to 2 weeks	I, II	1 specimen; 4 weeks
		III	2 species; 4 weeks
Ocular or optic or nasal	Up to 2 weeks	I, II	1 specimen; 4 weeks
		III	2 species; 4 weeks
	> 2 weeks	I, II, III	2 species; 12 weeks
Vaginal or rectal	Up to 2 weeks	I, II	1 specimen; 4 weeks
		III	2 species; 4 weeks
	> 2 weeks	I, II, III	2 species; 12 weeks

Table 10.3 Scheme of study and number of groups

	14–28 days				84–182 days			
	Rodent (Rat) (Nos.)		Non-rodent (Dog or Monkey) (Nos.)		Rodent (Rat) (Nos)		Non-rodent (Dog or Monkey) (Nos)	
Group	Male (Nos.)	Female (Nos.)	Male (Nos.)	Female (Nos.)	Male	Female	Male	Female
Control	6–10	6–10	2–3	2–3	15–30	15–30	4–6	4–6
Low dose	6–10	6–10	2–3	2–3	15–30	15–30	4–6	4–6
Interim dose	6–10	6–10	2–3	2–3	15–30	15–30	4–6	4–6
High dose	6–10	6–10	2–3	2–3	15–30	15–30	4–6	4–6

- In the case of cytotoxic anticancer agents, the dose causing severe toxic manifestations or death should be defined. The post-dosing observation period should be up to 14 days.
- For the determination of MTD, mice should first be used. The findings then should be established in rat for confirming linear linkage between toxicity and body surface area. In case data received is not linear, data of the more delicate species should be used to reach the starting dose of Phase I.
- When the cytotoxic drugs act as a novel mechanism of action or when rodents are known to be bad forecasters of human toxicity (e.g., antifolates), Maximum Tolerated Dose (MTD) should be determined in non-rodent species.

2. Repeated-Dose Systemic Toxicology Studies
   - The studies to be carried out in at least two mammalian species, of which one should be a non-rodent are listed in see (Table 10.2).
   - Dosage variation studies should pave the wave for the 14-, 28-, 90- or 180-day toxicology researches. Therapeutic indication and scale of the proposed clinical trial will determine the duration of the final systematic toxicity study.
   - The species which process the drug in a similar way as humans are preferred for toxicity studies.
   - The drug should be administered 7 days a week by the route intended for clinical use for repeated-dose toxicity studies.
   - Table 10.3 mentions the minimum number of animals required for these studies for data collection.
   - As possible and as relevant, a control group of species given the vehicle alone, and three other groups provided a graded dosage of the medicine should be incorporated.
   - As a rule, it is expected that the highest dose should produce toxicity which is observable. The lowest dose should be comparable to the intended therapeutic dose for humans or a multiple of the same and should not cause observable toxicity.

- There should be approved for the responsiveness of the species. Hence, the intermediary dosage should lead to some signs and symptoms, but not sever leading to death or toxicity. This dose should be positioned logarithmically in the mid of the other two doses.
- Behavioral, physiological, biochemical, and microscopic observations are the parameters that should be included for monitoring and recording in long-term toxicity studies.
- The sites of injection should be subjected to gross and microscopic examination especially in the case of parenteral drug administration.
- It is a requirement to carry out in the non-rodent species initial and final electrocardiogram and fundus examination Dosing and study design, in case of cytotoxic anticancer agents should be in line with the suggested clinical scheme with regards to days of exposure and count of cycles. Two rodent races may be experimented on for kicking off Phase I trials. A non-rodent race should be included if the medicine has a unique mode of action, or if approval for initiating Phase II, III or if marketing approval needs to be acquired. For the majority of compounds, it is anticipated that singular dose tissue dispersal studies with enough sensitivity and distinctiveness will give the required evaluation of tissue distribution and the potential for concentration.
- There should not be a requirement uniformly for all compounds for repeated-dose tissue distribution studies They should only be carried out when relevant data from other sources cannot be derived.
- In some situations, based on the information from single-dose tissue distribution studies, toxicokinetic, toxicity studies, it may be relevant to conduct repeated-dose studies.
- For compounds with an apparently long half-life, incomplete elimination, or unanticipated organ toxicity, these studies may be most appropriate.

Apart from the above studies some special studies also need to be conducted where necessary and these include the following types of studies

- Male Fertility Study
- Female Reproduction and Development Toxicity Studies
- Allergenicity or Hypersensitivity
- Photo-allergy or dermal photo-toxicity
- Genotoxicity
- Carcinogenicity

10.4.3 Clinical Development Approaches

Immunotherapeutics in general has a wide scope in various indications. However, when you look at the current industry pipeline, then one can see a heavy focus on immune oncology for the very obvious reason of challenges in the treatment of cancer and the prognosis after the diagnosis of cancer. Therefore, the discussion under this section will be mainly on immune oncology.

With a paradigm shift in the current oncology drug development, trials frequently involve patients with a shorter life expectancy and are designed in order to increase the chance of discovering an impactful therapy. Maximum tolerated dose (MTD) is normally identified in phase I studies and many times the knowledge of long-term tolerability is limited. Basic procedures developed initially for cytotoxic chemotherapies, MTD typically determines dose preference for the clinical development of selected agents and cancer immunotherapies. It is also common for conventional oncology drugs(e.g., chemotherapy, targeted agents), that MTD is selected for the proposed stage II dosage (RP2D). The presumption that higher efficiency is linked with a greater dose forms the basis of this preference of MTD as the RP2D for cancer. However, there are numerous limitations to this approach. For example, study designs typically employed from initial clinical research (e.g., stage I) confounds the identification of the MTD. Differences exist between different phases of studies. Phase I researches are limited, with large inter-individual discrepancies (due to the huge range of tumor types and sickness load), and customized therapy patterns. Studies of Phase II, are bigger than studies of phase I, with only chosen tumor types, and has limited exploration of dosage ranges lesser than the MTD or RP2D, which are derived from the phase I research/s. These MTD or RP2D identified are useful to define dose–response (PK/PD) relationships in a better way. This method, in spite of the limitations has been flourishingly practiced for the development of single-agent therapies. It however may not be acceptable for Immunotherapies that are evolved in combination therapies.

The MTD-based methodology to recognize the RP2D for cancer immunotherapies is extremely competitive. The assumed monotonic increment in efficiency with every dose increase may be inappropriate for immunotherapies; especially those requiring a stability of the immune system boost to combat cancer while subsequently avoiding overstimulation. To top it all an MTD identification may not be achievable for immunotherapies. For example, the MTD was unidentified in single-agent phase I studies including nivolumab (up to 10 mg/kg), pembrolizumab (up to 10 mg/kg), and ipilimumab (up to 20 mg/kg). A strength of 10 mg/kg of ipilimumab was put to use in ensuring clinical trials basis preclinical data, suggestive of the fact that the concentrations attained at this dosage provided a maximum result. However, eventually, a dosage as low as 3 mg/kg was given the final approval.

Toxicities associated with immunotherapies, when present, may be dose-independent and in certain cases (with CAR-T), toxicity may actually indicate efficacy. Hence, toxicity-based endpoints in dose-escalation methods may be less relevant for immuno-oncology trials. In such cases, alternative parameters may be more appropriate. Clearly, what makes the preference of the RP2D dose complexes, is the inability to either establish an MTD for immunotherapy given as a single agent or initiating a single MTD for combination treatment. This prevents preference of the MTD as the RP2D, in the traditionally followed methodology. In such cases, a benefit/risk evaluation and inclusive investigation of the over tout line explaining the relation between exposure, safety, and acceptability, and response should be evaluated for the RP2D. This relatively simple evaluation when an MTD is easily recognized forms the basis for preference of the RP2D dose. The comparison between benefit/risk at various doses while running of the clinical development of ipilimumab was inappropriately distinguished; leading to issuance for a post-marketing study by the FDA. The need was to compare the effectiveness at the accepted dose (3 mg/kg, Q3W) to efficacy at an increased dose (10 mg/kg, Q3W), specifically for patients with phase III or phase IV melanoma that cannot be surgically eliminated. The situation complicates for combination immunotherapies due to the innumerable variables, which includes dosage level, the duration of the dosage holiday, administration periodicity, the length of therapy for each dose, and the pattern of the regime for each drug, The incapacity to inspect all feasible combinations of these aspects in clinical trials, further make the situation more complex.

Novel trial designs exploring dose–response surface for combination immunotherapies can support in overcoming these challenges. These can be implemented and complemented by analysis based on the model to understand better the therapeutic window of these combination therapies. In early clinical development, the use of patient-reported results(PRO) is recommended to state dose preferences, basis of the ability to tolerate, and issues experienced with adherence to these drugs. However PROs are seldom incorporated in the label of cancer drugs in the US (more usual in Europe and for non-oncology medicines in the USA). The severity of the disease increases the tolerance of the diseased to the side effects of cancer therapy. However, many patients discontinue cancer therapy due to the adverse events, and this is more observed and prevalent in patients undergoing immunotherapies since the duration of treatment is prolonged and response to the therapy also takes a while.

A more calculated approach that consolidates several information sources throughout different phases of evolution is advantageous for identifying optimal dose and scheduling the cancer immunotherapy, that. Particularly, a, model-based, quantifiable proposal integrating exposure-response (i.e., biomarker and/or effectiveness) analyses exposure. Important data on the benefit/risk ratio of a drug applicant and dose preference can be provided by safety analyses. In the early phase, exposure-response analyses specifically involved evaluation of tumor growth suppression, using typical recommendations detailed in the Response Evaluation Criteria in Solid Tumors (RECIST) and the Immune-Related Response Criteria (irRC). The prediction benefit of tumor evaluations with regards to overall survival has been confirmed for conventional cytotoxic drugs. Information from the developmental phase of nivolumab and ipilimumab suggest that there is also a correlation between tumor size evaluations in the primitive stage and total existence for immunotherapies. Evaluating other indicators of response that influence the effects of immunotherapies on the immune system, such as cytokine increase and indicators of T-cell activation, may also be of value.

Underlying techniques of cancer immunotherapies can be explained by systems pharmacology approaches, which supply helpful information on the timing and pattern of administration of many agents within a fusion. These approaches are complemented by novel clinical trial designs that permit the assembly of the relevant and sufficient data, which is in many instances are too limited to support the effective application of these perspectives. Recognition of indicators of effectiveness and safety signs impact victory of these perspectives, especially yearly in the developmental phase.

Complex biology of combination immunotherapies is explained by the Quantitative systems pharmacology (QSP) model, another reliable approach. Converting the clinical experience with single-agent immunotherapies and categorizing them into phases I to III combination dosages is not very direct. This happens in particular if the trials are planned in such a way that they do not support exposure-response analyses.

QSP model is network-centric, involves an overall view of biology, and quantifies the functional changes of disease. QSP models help identification and prioritization of targets, exploring indicators of response, and identification of important characteristics to differentiate patients since these replicas capture a drug's procedure of action. Mode of action of tumor immunosuppression, ways to bypass them, and the effects of disease gravity and advancement on therapy results can be evaluated using Mechanistic QSP models. Such approaches are specifically helpful in combination of immunotherapy development. Unlike conventional exposure-response analyses, top-down patterns dependent on the accessibility of enough clinical data. The QPS models function around, bottom-up, and middle-out approaches, which is useful for advance forecasting of synergistic interactions and their impact on effectiveness and safety.

Another quantifiable and mechanistic approach, that was useful in the evolution of immunotherapies (displayed after its implementation to blinatumomab evolution) are the Physiologically based pharmacokinetic (PBPK)modeling. Drug–drug interactions (DDIs) are seen frequently in the immunotherapies area due to their prospective effect on cytokine-mediated transformation in cytochrome P450 functioning. DDI is however not common for biologics. Hence the increase in cytokine levels is noticed after blinatumomab infusion. Particularly in the live example quoted, the prospective for cytokine-mediated DDI was anticipated based on data from in vitro hepatocytes incubated with blinatumomab or cytokines and the clinical cytokine profiles. The model predicted little prospective for DDI and, basis this forecast, no clinical DDI studies were conducted.

Another device to compare therapies without deadlocked clinical trials is linking biomarker response to clinical effectiveness, and form a linkage across different human categories and indications, known as Model-based meta-analyses. This perspective consolidates both; data from within and external source supplements the exposure-response modeling, the quantifiable structure of pharmacology, and PBPK modeling perspectives. All the models discussed have high potential to speed up the development of immunotherapies, but their success is never without any hindrances. Prospective planning is a mandate to include modeling in the clinical development plan, since the information available to develop these models may be not adequate. In the development of combination immunotherapy, uncertainties are magnified requiring even more data to develop reliable models leading to the magnification of challenges already faced.

Biomarker data help in directing dose and regimen preference in the early stages of clinical development. For nivolumab, peripheral blood receptor tenancy and its relative time course of PD-1 on circulating CD3+ T cells were assessed as potential PD markers and compared across various dosing regimens in refractory solid tumors. The data derived concluded that smaller dosing time frames may improve occupancy and penetration into tissue. This required, exploring in later development. For blinatumomab, peripheral cytokine levels were used as a PD marker for assessment of acute safety contributed to the stepwise dosing recommendation for blinatumomab. The difference in the two exposure-response curves made available through the model used, provided potential insight into translating peripheral B-cell response to tumor response, including the accessibility of tumor to the drug. An associated diagnostic tool was recently approved for pembrolizumab in second-line NSCLC basis increased baseline tumor expression of PD-L1 that has been linked with better impact in multiple tumor types. Therefore, it is appropriate to include baseline PD-L1 expression update as a co-variate in the exposure-response analyses.

Similar concepts of biomarker-informed dose selection could apply to integrated drug development; with addition of element accounting for the interaction between the integrated curative or therapeutic agents concepts. For example, an integrated blockade of PD-1:PD-L1 with other coinhibitors, such as TIM-3, CTLA-4, and LAG-3. This has a synergistic effect in reversing T-cell depletion and reimposingCD8+ effective function. T-cell triggering, proliferation, emigration, and expansion of memory can be impacted by the involvement of costimulatory receptors such as OX40, CD137, or CD27. A deep interpretation of the biologic fundamentals of the mode of action of each remedy and their respective role in the cancer-immunity cycle and associated biomarkers of remedial agents is imperative for "combined effects" (Morrissey et al. 2016).

10.4.4 NDA Format: ICH

Harmonization is the key to take care of a number of challenging situations. Introduction of CTD—Common Technical Document warrants the compilation of all the Quality, Safety, and Efficacy information in a common format. This has brought about a revolution in the regulatory review processes, leading to harmonized electronic submission that, in turn, enables the implementation of good review practices. Hence the step to reformat the information for submission to the different ICH regulatory authorities for the industries was effectively eliminated (ICH Official web site: ICH 2020a; ICH Official web site: ICH 2020b).

Five modules define the CTD. Module 1 is region specific while the remaining Modules are common for all regions. In July 2003, the CTD emerged as the mandatory format for new drug application submissions in the EU and Japan, The format is recommended and the choice for NDAs submitted to FDA, United States.

In general contents of the dossier as per ICH CTD module can be found in following figure known as CTD Triangle.

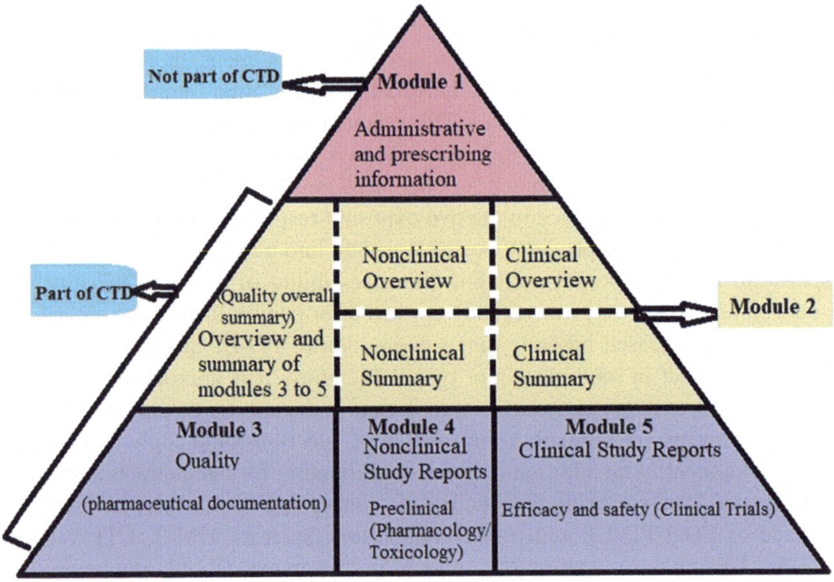

(Guidance For Industry 2010) [Ref: https://cdsco.gov.in/opencms/resources/UploadCDSCOWeb/2018/UploadGazette_NotificationsFiles/CTD%20Guidance%20Final.pdf]

10.4.5 New Drug Application (NDA) Format India

Small Molecule

They are defined basis of their size and makeup. Most of them consist of 20–200 atoms.

Large Molecules

They are also termed as Biologics and are defined as proteins that have therapeutic effects. They are very large and complex, made of a number of amino acids, and may also resemble proteins found in the human body. Biologics may be small in size have about 200–3000 atoms while others may be more complex.

The NDA format defined by the Health Authority of India is similar to the ICH mandatory data of CTD and incorporates five sections. A brief overview of the same is provided below:

- Administrative or notarized legal documents
- Summaries
- Information about Quality Information matrix (Chemical, Pharmaceutical, and Biological)
- Information from Non-Clinical data
- Information from Clinical studies

A template for providing the quality summary of the Biologics known as Product Permission Document (PPD-Biological) needs to be submitted along with the NDA. The entire dossier is required to be submitted in hard copy as well as in a compact disc (soft copy) Two copies each of the hard and soft copies are to be submitted to the Regulator, while a set of both types are to be archived locally by the applicant.

Chemistry Manufacturing Control Changes are to be provided during the NDA application. The details for the criteria for reporting are provided in the Guidance for Industry, released by the CDSCO (CDSCO 2020a).

[Ref: Guidance for Industry on Submission of Clinical Trial Application for Evaluating Safety and Efficacy {Biologics} (General considerations for conducting Clinical Trial as per Drugs and Cosmetics Act 1940 and Rules 1945)].

10.5 Post-Approval Requirements

10.5.1 Pharmacovigilance

World Health Organisation (WHO) defines Pharmacovigilance as the science and activities relating to the detection, assessment, understanding, and prevention of adverse effects or any other drug-related problem.

During the NDA application, it is mandated to provide an abstract narration of the studies conducted including many details like the total patients enrolled who received the drug, the dosage received by the subjects, the time that the patient was receiving the same, and description of any adverse event/adverse drug reaction experienced by any of the subjects, as applicable.

Under scenarios when the sponsor decides not to proceed with the NDA or there is an early interruption to a study, the reason for the same needs to be provided.

Every expected serious adverse event (SAE) experienced by a clinical trial subjects should be informed to the Licensing Authority and to the other Investigators, involved in the study, within 14 calendars days. The format to submit the SAE details is the same as Appendix XI of Schedule Y, which is now replaced by Table 5 of Schedule 3 provided in the New Drugs and Clinical Trial Rules of 2019.

Post-approval, the Sponsor is responsible for maintaining a robust Pharmacovigilance system in place which detects captures, assess, analyses, and shares the safety-related information in a timely manner (Guidance for Industry 2017).

10.5.2 CMC Changes

CMC Changes along with several other information forms a basis of the pre and post-approval submissions made to the Licensing Authority. The checklists that provide guidance on the kind of documents to be included in the dossier are available on the online submission portal called as Sugam (CDSCO 2020b) (Ref: https://cdscoonline.gov.in/CDSCO/homepage).

The Guidance for Industry released by the CDSCO especially for Biologics serves as another document that details the expectations by the Regulator and the kind of document supposed to be included in the various modules during submission (CDSCO 2020a).

[Ref: Ref: Guidance for Industry on Submission of Clinical Trial Application for Evaluating Safety and Efficacy {Biologics} (General considerations for conducting Clinical Trial as per Drugs and Cosmetics Act 1940 and Rules 194 (Labiotech.eu 2020).

10.6 Conclusion

Immunotherapeutics is a new approach to treat cancer and other autoimmune disorders. Customization and combining therapies will bring about a new era in this space. Detecting and understanding the toxicities associated with immunotherapeutics would be an important bargain for the success of the therapy.

Patent protection for these specialized therapies has gained a lot of importance and has undergone several changes in recent years. The Regulations laid down for approval of immune therapies follow a structured pattern. The intellectual property in India follows most of the International guidelines with very minor changes.

Patent protection for immunotherapy drugs requires a lot of specialization to protect the innovation that is incorporated into each and every researched molecule. Post-approval regulatory mandates also require setting up a robust pharmacovigilance system and submissions of several data; checklists of which are available easily. With the advancement in the regulatory environment, the pre- and post-approval process in India have gone online. The information provided in this chapter is supported by the regulatory rules, acts, and guidelines provided by the Indian Health Authority along with the other International guidelines available.

References

Acts and Regulations - Canada.ca (2020). https://www.canada.ca/en/health-canada/corporate/about-health-canada/legislation-guidelines/acts-regulations.html. Accessed 19 May 2020

Australian Regulatory Guidelines for Prescription Medicines (ARGPM) | Therapeutic Goods Administration (TGA) (2020). https://www.tga.gov.au/publication/australian-regulatory-guidelines-prescription-medicines-argpm. Accessed 12 May 2020

Bhave A, Menon S (2019) Subject expert committees: past, present, and future. Perspect Clin Res 10:1–3

Camarero J, Ruiz S (2012) Cancer immunotherapy products: regulatory aspects in the European Union. Hum Vaccines Immunother 8:1354–1359

CDSCO (2020a) Requirements for permission of New Drugs Approval Post approval changes in biological products: Quality safety and Efficacy Documents Guidance for Industry.Central Drugs Standard Control Organisation, Ministry of Health G o. I (2012)

CDSCO (2020b). https://cdscoonline.gov.in/CDSCO/homepage. Accessed 12 May 2020

Central Drugs Standard Control Organization (2020) Introduction. https://cdsco.gov.in/opencms/opencms/en/About-us/Introduction/. Accessed 19 May 2020

Common Technical Document (CTD) | Therapeutic Goods Administration (TGA) (2020). https://www.tga.gov.au/publication/common-technical-document-ctd. Accessed 12 May 2020

Delves P (2020) Overview of the immune system - immunology; allergic disorders - Merck manuals professional edition. In: Merck Man. Prof. Version. https://www.merckmanuals.com/professional/immunology-allergic-disorders/biology-of-the-immune-system/overview-of-the-immune-system. Accessed 19 May 2020

DTAB-DCC (2020). https://cdsco.gov.in/opencms/opencms/en/dcc-dtab-committee. Accessed 12 May 2020

EudraLex - EU Legislation | Public Health (2020). https://ec.europa.eu/health/documents/eudralex_en. Accessed 25 May 2020

European Medicines Agency (2020a) Legal framework. https://www.ema.europa.eu/en/about-us/what-we-do/legal-framework. Accessed 19 May 2020

European Medicines Agency (2020b) Legal framework: Pharmacovigilance. https://www.ema.europa.eu/en/human-regulatory/overview/pharmacovigilance/legal-framework-pharmacovigilance. Accessed 25 May 2020

European Medicines Agency (2020c) Paediatric Regulation. https://www.ema.europa.eu/en/human-regulatory/overview/paediatric-medicines/paediatric-regulation. Accessed 25 May 2020

European Medicines Agency (2020d) Advanced therapy medicinal products: Overview. https://www.ema.europa.eu/en/human-regulatory/overview/advanced-therapy-medicinal-products-overview. Accessed 25 May 2020

European Medicines Agency (2020e) Legal framework: orphan designation. https://www.ema.europa.eu/en/human-regulatory/overview/orphan-designation/legal-framework-orphan-designation. Accessed 25 May 2020

European Medicines Agency (2020f) Fees payable to the European Medicines Agency. https://www.ema.europa.eu/en/human-regulatory/overview/fees-payable-european-medicines-agency. Accessed 25 May 2020

European Medicines Agency (2020g) Supporting SMEs. https://www.ema.europa.eu/en/human-regulatory/overview/supporting-smes. Accessed 25 May 2020

European Medicines Agency (2020h) Clinical trials in human medicines. https://www.ema.europa.eu/en/human-regulatory/research-development/clinical-trials-human-medicines. Accessed 25 May 2020

European Medicines Agency (2020i) Falsified medicines: overview. https://www.ema.europa.eu/en/human-regulatory/overview/public-health-threats/falsified-medicines-overview. Accessed 25 May 2020

FDA (2020) What does FDA regulate? https://www.fda.gov/about-fda/fda-basics/what-does-fda-regulate. Accessed 25 May 2020

GEAC Ground rules for Comsideration of proposal by the GEAC as per the good practices in environment regulation adopted by MoEF

Guidance For Industry (2010) On Preparation Of Common Technical Document For Import / Manufacture And Marketing Approval Of New Drugs For Human Use (New Drug Application-Nda), Draft Guidance.Central Drugs Standard Control Organization Directorate General Of Health Services Ministry Of Health & Family Welfare Govt. Of India N 2010, Pa

Guidance for Industry (2017) Pharmacovigilance requirement for biological products. Central Drugs Standard Control Organisation, Ministry of Health G o. I

Hong BS, Division S (2013) Claiming what counts in business: drafting patent claims with a clear business purpose. SMEs Division, WIPO, New York, pp 1–7

ICH Official web site: ICH (2020a). https://www.ich.org/. Accessed 19 May 2020

ICH Official web site: ICH (2020b). https://www.ich.org/page/efficacy-guidelines. Accessed 19 May 2020

Labiotech.eu (2020) Here's the Top 5 Things to Know About Immunotherapies. https://www.labiotech.eu/sponsored/top-5-immunotherapies-synteracthcr/. Accessed 19 May 2020

Morrissey K, Yuraszeck T, Li CC, Zhang Y, Kasichayanula S (2016) Immunotherapy and novel combinations in oncology: current landscape, challenges, and opportunities. Clin Transl Sci 9:89–104

Mouta-Bellum C, Shikhani H, Feng L, Ni J (2017) The immunotherapy patent landscape: types of patent claims for Immuunotherapeutic inventions. In: In. IPWatchdog, Finnegan. https://www.finnegan.com/print/content/45876/The-Immunotherapy-Patent-Landscape-Types-of-Patent-Claims-for-Immunotherapeutic-Inventions.pdf?q

Nicholson LB (2016) The immune system. Essays Biochem 60:275–301

(2009) Regulation (EC) No 470/2009 of the European Parliament and of the Council, 6 May 2009. Off J Eur Union 11–22

Taylor A, Verghan J, Blaser K, Akdis M, Akdis C (2006) Mechanism of immune suppression by interleukin-10 and transforming growth factor-beta: the role of T regulatory cells. Immunology 117:433–442

TGA basics | Therapeutic Goods Administration (TGA) (2020). https://www.tga.gov.au/tga-basics. Accessed 19 May 2020

Towers M, SWg sNO LL (2020) About us - Medicines and Healthcare products Regulatory Agency - GOV.UK. https://www.gov.uk/government/organisations/medicines-and-healthcare-products-regulatory-agency/about. Accessed 25 May 2020

Watkins R, Wu L, Zhang C, Davis RM, Xu B (2015) Natural product-based nanomedicine: recent advances and issues. Int J Nanomedicine 10:6055–6074

Future Immunotherapy Challenges and Perspectives

11

Amrita Date, Vandana S. Nikam, Shariq Syed, and Sujata P. Sawarkar

Abstract

The field of immunotherapy has progressed rapidly, expanding from the use of monoclonal antibodies and interferons, to more innovative approaches now being tested in laboratories and in the clinic. Several challenges still exist in the field, including adverse effects of therapy and the high cost of treatment. As a result, new approaches to immunotherapy are constantly being explored in order to overcome these limitations. While the field is grounded mainly in the domain of biologics, more recently alternative approaches using synthetic molecules as well as natural products have been under investigation. A more unconventional approach, involving the introduction of a parasitic infection in the body is also currently in clinical trials. Moreover, the avenues for the application of immunotherapy have also expanded. While originally aimed at the treatment of cancers, their potential applications have expanded to allergies, neurodegenerative diseases, infections, osteoporosis, and hyperlipidemia. The scope of immunotherapies moving forward will be discussed in this chapter.

A. Date
Department of Chemistry, Imperial College London, Molecular Sciences Research Hub, White City Campus, London, UK

V. S. Nikam
Department of Pharmacology, STES's Smt. Kashibai Navale College of Pharmacy, S. P. Pune University, Pune, Maharashtra, India

S. Syed
School of Pharmacy, Anjuman-I-Islam's, Kalsekar Technical Campus, Mumbai, Maharashtra, India

S. P. Sawarkar (✉)
SVKM's Dr. Bhanuben Nanavati College of Pharmacy, University of Mumbai, Mumbai, India
e-mail: sujata.sawarkar@bncp.ac.in

S. P. Sawarkar et al. (eds.), *Immunotherapy – A Novel Facet of Modern Therapeutics*,
https://doi.org/10.1007/978-981-15-9038-2_11

Keywords

Immunotherapy · Challenges · Cancer vaccine · Parasite immunotherapy · Natural products

11.1 Introduction

Over the past decade, immunotherapies have become increasingly popular in the treatment of several diseases. It has most significantly impacted the treatment received by cancer patients, providing an alternative to conventional chemotherapy, where cytotoxic drugs produce severe side effects. Although we have come a long way in making a range of immunotherapies available to patients, this class of therapies still poses a range of challenges that need to be tackled. With several forms of immunotherapies currently in clinical trials, and even more in the early stages of development, researchers in the field are constantly striving to overcome the current limitations and make better therapies available to patients. This chapter aims to cover some of the new approaches and avenues for immunotherapy.

11.2 Approaches to Immunotherapy

11.2.1 Biologics in Immunotherapy

Biopharmaceuticals form a large proportion of immunotherapeutic options currently available to patients and are therefore the area where most work is being done to overcome existing limitations. Over recent years, significant work has been done on checkpoint inhibitors, oncolytic viruses, therapeutic vaccines, and bispecific antibodies for immunotherapy (Riley et al. 2019). Much of this progress has been a result of the identification of specific biomarkers and antigens found on cancer cells that can be targeted in order to achieve selective responses.

Prophylactic vaccines for infectious diseases were one of the earliest forms of immunotherapy. An area that is rapidly progressing in current times is therapeutic vaccines for cancer. They possess advantage over other forms of immunotherapy due to their ability to generate immunological memory (Lopes et al. 2019). These vaccines could consist of cells, peptides, or genetic material, aimed at eliciting a targeted immune response for the tumor (Lopes et al. 2019). Vaccines containing nucleic acids, for example, aim to deliver DNA or mRNA to antigen presenting cells, such that the desired antigen can be expressed and presented to T cells, in order to trigger an immune response (Riley et al. 2019). DNA vaccines bear advantage over mRNA vaccines in that they are better able to integrate into the cells' genome, but are limited by the need for intra-nuclear drug delivery. mRNA vaccines, on the other hand, only require intracellular presence which is slightly easier to achieve (Riley et al. 2019). However, for these approaches to see clinical success, the development of specialized delivery systems would be a key tool. So far, no DNA vaccines have been approved for use by the FDA (Lopes et al. 2019). In situ vaccinations have also drawn a lot of attention. These allow intra-tumor delivery of the therapeutic agents. They offer the advantage of targeted delivery, without the need to identify specific antigens. This allows for optimized delivery, with minimal systemic side effects (Sheen and Fiering 2019).

Another technology exploiting the knowledge of tumor antigens is immune effector cell targeting (Guo et al. 2013). While this can be achieved using CAR-T cells, a specific type of bispecific antibody, bispecific T cell engager (BiTE) is thought to present significant advantages. Originally developed by Amgen, these agents are much smaller than traditional antibodies, and act by forming a bridge between cytotoxic T cells and tumor cells (Guo et al. 2013). This binding activates the T cells to secrete cytotoxic chemicals, activate other components of the immune system, and bring about cell lysis (Huehls et al. 2015). Apart from presenting a high potency and efficacy, they differ from other bispecific antibodies in their ability to produce a response even in the absence of T cell co-stimulation (Guo et al. 2013). In 2018, Blinatumomab, by Amgen, was the first BiTE to gain FDA approval for use in the treatment of lymphoblastic leukemia (Jen et al. 2019). Several others are currently in clinical trials.

11.2.2 Small Molecules in Immunotherapy

Immunotherapies rely on the use of biologics to boost immune responses. However more recently, efforts are being directed towards the design and development of small molecules that can target pathways involved in immune responses, with several candidates currently in preclinical and clinical studies.

Small molecules present several advantages over biologics, and moving forward, the field would definitely benefit from further exploring the potential role of small molecules in immunotherapies. Their advantages include:

- Greater oral bioavailability (Kerr and Chisholm 2019)
- Better tumor penetration (Kerr and Chisholm 2019)
- Greater control over bioavailability, allowing for more control over potential adverse effects (Kerr and Chisholm 2019)
- Ability to target pathways involved in immune responses that biologics are not able to alter (Chen et al. 2019)
- Potential to access a wider range of target, including nuclear receptors, as a result of their membrane permeability (Chen et al. 2019)
- Better therapeutic index (Chen et al. 2019)
- Lower cost of production (Kerr and Chisholm 2019)

The most popular pathways that have been investigated in this area are the STING pathway, involved in activation of inflammatory genes, PD-1/PD-L1, which interferes with T cell response in tumor environments, and indoleamine 2,3-dioxygenase 1(IDO1), which is also involved in allowing cancer cells to evade immune responses (Chen et al. 2019). Several molecules interfering with these pathways are in various stages of preclinical and clinical development. Epacadostat, an IDO1 inhibitor, that rapidly progressed through clinical trials, failed in phase 3 (Chen et al. 2019). ADU-S100 and MK-1454 are agonists of the STING pathway, and are currently in Phase I clinical trials. The main drawback of these, like most other forms of immunotherapy, is that they can potentially trigger unsolicited immune responses in patients (Kerr and Chisholm 2019). PD-L1 antagonists are currently in the earlier stages of development. BMS-1001 and BMS-1116 are still in the preclinical stages (Kerr and Chisholm 2019).

Another approach to the use of small molecules in immunotherapy is the application of cytokines. These are endogenous molecules that are able to transmit pro-inflammatory and anti-inflammatory signals to cells (Berraondo et al. 2019). These could potentially be used alone, as well as in conjunction with other forms of immunotherapy, in order to enhance the activity seen. The main challenges that must be overcome are the systemic inflammatory effects produced and the difficulties in localizing them to the site of the tumor (Berraondo et al. 2019).

These small molecule-based approaches have the potential to make affordable and convenient forms of immunotherapy easily available to patients. These pharmaceuticals could readily be developed into suitable dosage forms that can conveniently be administered to patients.

11.2.3 Parasite Immunotherapy

A slightly unconventional approach that has been in research in recent years is based on the hypothesis that a negative correlation exists between parasitic infections and cancer. This form of cancer immunotherapy would involve introducing a parasitic infection in a patient, in order to evoke an immune response, which is predicted to have an anti-angiogenic effect (Darani and Yousefi 2012). This theory is based on the finding that certain mucins, uniquely expressed by cancer cells, are also found in parasitic antigens. Therefore, parasites that possess mucin-type O-glycans on their surface are thought to be able to induce a joint immune response towards cancer cells expressing similar antigens (Darani and Yousefi 2012). The Tk antigen, for example, which is found in human colorectal cancer, is also expressed by *Taenia crassiceps*, *T. hydatigena*, and *Mesocestoides vogae*. Both adaptive and innate immune responses are thought to be triggered, which can act against tumor cells (Berraondo et al. 2019).

There has been significant preclinical data supporting this. When a solid lymphoma was introduced into *Trypanosoma cruzi*-infected mice, growth of the tumor was found to be inhibited (Chen et al. 2019). Multiple studies have also been conducted using *Toxoplasma*-infected mice. When compared with mice infected with the formalin-fixed pathogen, a reduction in tumor size was observed. In another study, *Toxoplasma* was also found to delay tumor formation (Darani and Yousefi 2012).

The effect of infection of Lewis lung cancer mice with *Plasmodium yodii* was also investigated (Chen et al. 2011). A strong adaptive as well as innate immune response was observed. An increase in the levels of interferon-γ and tumor necrosis factor-α was recorded. Proliferation of natural killer cells as well as tumor-specific T cells was also observed. The immune response generated in this manner is thought to be longer lasting than by other means. The study showed an increase in the proportion of apoptotic cancer cells, and a reduction in proliferative cells in the mice. Two antigens are thought to be responsible for initiating this immune response: the glycosylphospholipid inositol on the cell surface, and the hemozoin, which is a by-product of the parasite's metabolism (Chen et al. 2011). This concept has now gained permission for translation to human trials. This will involve infection of cancer patients with *Plasmodium vivax*, with blood levels of the parasite being maintained at a low level using artesunate for several weeks. The treatment will be terminated by administration of a course of chloroquine or artemisinin (U.S. National Library of Medicine 2020a). However, the studies in animal models did present some challenges, which mean that the results may not be directly reproducible in humans. In mice, *Plasmodium* infection does not cause fever, and the course of infection is much shorter than in humans (Darani and Yousefi 2012). Therefore, the outcomes of this clinical study can be expected to produce interesting outcomes.

The biggest challenge with this type of treatment is the possibility of it progressing to a high intensity infection, which could have damaging effects on the patient. While a low intensity infection can have beneficial effects, its progression could be fatal. The choice of parasite is also crucial. Not only should the pathogen possess glycosylated antigens, it must also be sufficiently safe to use in humans (Darani and Yousefi 2012). So while this unique approach to immunotherapy has recently progressed to clinical testing, it is one that should be considered with great caution.

11.2.4 Natural Products

Compounds isolated from natural product have also been shown to exhibit immunostimulatory effects. While not currently in use, if developed to enhance their potency and pharmacokinetics, they could potentially be used in conjunction with cytotoxic chemotherapeutic agents in order to reverse their immunosuppressive effects. Tylophora alkaloids, for example, showed immune regulation in conjunction with cytotoxic agents (Bach and Lee 2019). Interestingly, however, similar to parasite immunotherapy, a large number of natural products found to produce an anti-tumor immune-stimulatory response were polysaccharides in nature. A soluble polysaccharide fraction extracted from red wine was found to reduce tumor weight and volume in an animal model, by boosting the production of lymphocytes and tumor necrosis factor-α (TNF-α) (Stipp et al. 2017). Similarly, when a polysaccharide fraction from *Artemisia argyi* was tested in a Sarcoma-180 tumor-bearing mouse model, the immune response suppressed by the tumor was found to be restored. Levels of T-lymphocytes, TNF-α, and interleukins (IL) 2, 6, and 12 were found to be increased (Bao et al. 2013). However, while an extract from *Grifola frondosa* exhibited immune-stimulatory activity in preclinical studies, it gave mixed results when taken to Phase I/II clinical trials. Its administration was associated with the production of both immune-stimulatory (IL2) and suppressive (IL10) cytokines (Deng et al. 2009). Therefore, while several animal studies have shown promising results, translation of these therapies to humans may prove to be more challenging.

11.3 Avenues for Immunotherapy

Up till quite recently, the applications of immunotherapy have majorly been grounded in the treatment of cancers, and to an extent, for autoimmune diseases. As the treatment option becomes better established, and its benefits become more evident, scientists have begun exploring its applications to other disease areas.

11.3.1 Neurodegenerative Diseases

The most prominent area where the applications of immunotherapy are being explored is in the treatment of neurological disorders. There has been significant research done on the applications of immunotherapy for treatment of Alzheimer's disease, for which there is currently no cure available. It is thought that humoral and cellular immunity could be involved in clearing amyloid-β (Weiner and Frenkel 2006), which is a component of the plaques responsible for disease progression. Several approaches have been investigated to achieve this, including the activation of T-helper cells, and stimulation of microglial cells, to initiate an innate immune response (Weiner and Frenkel 2006). Active immunization approaches have also been investigated, testing the use of synthetic amyloid-β or its fragments conjugated to a carrier protein, as a vaccine (Deng et al. 2009). While intact amyloid-β_{42} progressed to phase II clinical trials, it was found to cause meningeal encephalitis in a number of patients (Schenk 2002). Alternatively, passive immunity approaches are also possible. The anti-amyloid-β antibody, bapineuzumab, was the first in the class to progress to phase III trials. However, its efficacy is uncertain, and was found to cause vasogenic edema (Kerchner and Boxer 2010). An alternative hypothesis for disease progression in Alzheimer's involves the role of tau protein. Many believe that tau lesions are a better indicator of disease than amyloid plaque deposition (Pedersen and Sigurdsson 2015). Based on this hypothesis, anti-tau monoclonal antibodies and vaccines containing tau protein or its fragments have been investigated (Pedersen and Sigurdsson 2015).

Following the progression of Alzheimer's immunotherapies to clinical trials, the approach has also been investigated in Parkinson's disease. Patients of Parkinson's disease as well as Multiple System Atrophy (MSA) are known to show neuronal accumulation of α-synuclein, resulting in nerve damage (Brudek et al. 2017). A study showed that these patients exhibit lower plasma levels of anti-α-synuclein auto-antibodies than healthy subjects (Brudek et al. 2017). On this basis, immunotherapy approaches such as vaccination with α-synuclein fragments, or the administration of monoclonal antibodies would be warranted (Brudek et al. 2017). It, however, appears that targeting innate immunity would be more promising than adaptive immunity by systemic immune blockade. Genomic analysis has shown correlation of neuroinflammatory genes in these diseases. The key though is identifying specific pathways for intervention that target non-neuronal cell reactions in neurodegenerative disease (Sims et al. 2017).

In addition to these, applications of immunotherapy in multiple sclerosis (MS) and amyotrophic lateral sclerosis (ALS) are also under investigation (Villoslada et al. 2008). This is particularly important, since effective therapies are not yet available for most neurodegenerative diseases. One of the major challenges is understanding the role of immune modulation in preventing progression of disease Another obstacle is the delivery of these treatments to the brain, across the blood–brain barrier. Therefore, it is equally important to undertake research into delivery systems to facilitate the administration of these therapies.

11.3.2 Infectious Diseases

In addition to vaccines and interferon based therapies that are common for the treatment of infectious diseases, other forms of immunotherapy have also been investigated for the treatment of infectious diseases. In spite of the availability of efficacious antiviral drugs for its treatment, T cell based immunotherapies are being investigated for the treatment of hepatitis C (Fuller et al. 2013). T cell immunotherapies are also in clinical trials for the treatment of hepatitis B (U.S. National Library of Medicine 2020b). INO-9112, a DNA plasmid coding for the transcription of interleukin-12, delivered by electroporation, is in phase II clinical trials for the treatment of hepatitis B (U.S. National Library of Medicine 2020b). A granulocyte macrophage colony stimulating factor has also gained approval for the treatment of this infection, and phase IV studies are currently underway for its assessment (U.S. National Library of Medicine 2020b).

Several forms of immunotherapy are under assessment for the treatment of persistent fungal infections. Antibody therapies have been tested clinically. Cell-based therapies that have been studied include the use of T cells, dendritic cells, and neutrophils. The use of pentraxins has also shown potential, with PTX3 showing in vitro efficacy against Aspergillus. Thymosin-α1 has been shown to trigger the maturation of dendritic cells when exposed to Aspergillus, as well as enhancing the production of interleukin-12 (Armstrong-James and Harrison 2012). Immunotherapies, therefore, have great potential as inhibitors of fungal growth, however, the major hurdle lies in their inability to compete with the low cost of standard antifungal therapy. The real value of these therapies would only really exist in immunocompromised patients, in whom standard antifungal drugs prove to be ineffective, in case of persistent infections.

Immunotherapies have also been proposed as a potential treatment option for viral infections such as SARS-CoV and MERS-CoV. Both active and passive immunity approaches are being proposed to tackle the 2019–2020 coronavirus pandemic. Potential passive immunization approaches involve the administration of antibodies, either by means of plasma translation, from a person who has already recovered from the disease, or through the administration of externally manufactured monoclonal antibodies (Shanmugaraj et al. 2020). In the case of viral infections that spread rapidly and are likely to undergo frequent mutations, a combination of monoclonal antibodies, targeting various epitopes on the virus is thought to be able to provide a more robust treatment option (U.S. National Library of Medicine 2020b). In the case of active immunization, vaccines containing the whole virus as well as DNA fragments are being clinically tested, but so far, none have gained FDA approval (AminJafari and Ghasemi 2020). Other approaches that have been proposed include administration of polypeptide hormones to induce maturation of T cells and ACE2 immunoadhesion (AminJafari and Ghasemi 2020). Unfortunately, so far, none have seen clinical success.

11.3.3 Autoimmune Diseases

The applications of immunotherapies for the treatment of autoimmune diseases have become common knowledge. Immune checkpoint blockade, anti-T cell therapy, anti-B-cell therapy are in clinical practice and some of the applications are in various stages of clinical trials. Immunotherapy has the unique capability to balance and reinstate immune system. The therapy has shown promising results in potentially treating diabetes mellitus, psoriasis, and rheumatoid arthritis

11.3.4 Other Applications

There is an increasing interest in the potential of immunotherapy for dealing with common allergies. Multiple clinical trials are underway, exploring the prospect of using immunotherapeutic agents to curb autoimmune responses responsible for peanut allergies, cows' milk allergies, other food allergies, as well as allergic rhinitis (U.S. National Library of Medicine 2020b). These are being investigated in both children and adults, and could become common treatment options in the coming years.

Potential for the application of immunotherapies also exists in the treatment of myositis, a group of inflammatory myopathies characterized by muscle weakness and endomysial inflammation. Symptomatic relief is generally provided to patients through the administration of steroids. The development of immunotherapies would allow for reduced doses of steroids, which are known to produce significant adverse effects. Researchers are attempting to develop monoclonal antibodies and fusion proteins to treat the disease, but the major challenge lies in the systematic testing of muscle strength to assess their efficacy (Dalakas 2010).

The monoclonal antibody stamulumab was tested for its ability to treat Duchenne muscular dystrophy. It acted by inhibiting myostatin, which has a growth-limiting effect on muscle tissue (Wagner et al. 2008). The drug is no longer being developed, however, other forms of immunotherapy could have applications in this disease area.

Another avenue that holds great potential is the treatment of osteoporosis, both in men and in post-menopausal women. Eli Lilly and Company developed a monoclonal antibody, named Blosozumab, to inhibit the SOST gene, and in effect increase osteoblast activity (McColm et al. 2014). This agent has completed phase II trials, in which it exhibited good efficacy and tolerance. The potential effects on breast cancer, and presence of anti-drug antibodies is still being investigated (Recker et al. 2015). Therapies such as this could become the standard of care for osteoporosis in the years to come.

An interesting agent to gain FDA approval was the monoclonal antibody Alirocumab. While it was approved in 2015, as a cholesterol lowering drug, it is not considered cost-effective in comparison to the statins that are readily available (Kazi et al. 2016). This continues to be a challenge for the development of immunotherapies, especially those that are biologics in nature. While they may be able to produce desirable clinical outcomes, it is difficult to bring down the prices to compete with the small molecules that form the standard of care in many disease areas.

Many novel strategies are emerging to strengthen and widen the reach of immunotherapy. These include bypassing endogenous immunity with cellular therapies, altering microenvironment, and modulating metabolic pathways to augment the immune response. For example—reprogramming of myeloid cells abundant in the tumor environment had been utilized to prime T cell response, delivery of cytokines, stimulation of toll-like receptor (TLR) ligands.(Cubillos-Ruiz et al. 2015; Kerkar et al. 2011)

Hypoxia and hypoxia induced vascular endothelial growth factor (VEGF) is suppressive of certain myeloid cells. The combination of antiVEGF and checkpoint blockade was found to be effective in preclinical and clinical settings. (Hodi et al. 2014)

The causal effect relationship between microbiome and response to checkpoint blockade has been shown in cancer patients.(Routy et al. 2018)

11.4 Challenges and Limitations of Immunotherapies

During the last decades, our understanding of underlying mechanisms and pathways that drive and regulate immune cell activity in health and disease state has developed significantly. In spite of these advances in the field of immunotherapies, certain hurdles are along the implementation way. The major obstacles include unpredictable efficacy and patients response, need for more target-specific, clinically significant biomarkers, approaches to tackle heterogeneity of disease and associated toxicity, concrete, strategic study design to improve efficacy, delivery methods of immunotherapy, and cost.

One of the biggest challenges posed by the existing immunotherapeutic options, particularly those targeting the adaptive immune system, is that the treatment tends to be effective only in a select group of patients. In the case of cancer patients, there tends to be vast variability in the genotypes of tumor antigens found on malignant cells, in the cancer type, and the expression of biomarkers. This makes it difficult to design therapies that not only selectively kill cancer cells but are also effective in a large majority of patients (Ventola 2017a). This can also present a challenge in the design of clinical trials. Since many immunotherapies are targeted to certain tumor specific antigens, screening for the corresponding biomarkers in clinical trial subjects becomes essential. This often leads to small cohorts of patients qualifying for the trials, and false negative results if biomarkers are not used as a criterion for shortlisting (Ventola 2017a).

The pharmacokinetics, safety, and toxicity of immunotherapeutic approaches have also been an area of concern. In order to optimize the application of immunotherapeutics, thorough knowledge of the pharmacokinetics (PK), exposure–effect relationship, and toxicity profile of these drugs must be studied further. This can help in effectively mitigating immune related adverse events. Apart from this monoclonal antibodies have proved to be amongst the most popular forms of immunotherapy, with the first one being approved by the FDA in 2002. These, like other drugs, present challenges of adverse reactions, most commonly allergic responses (Mahmuda et al. 2017). Additionally, several monoclonal antibody therapies have been reported to cause immunodeficiency in patients, making them increasingly prone to infectious diseases. Certain monoclonal antibodies, including infliximab and rituximab have also been shown to cause immune thrombocytopenia in patients (Hansel et al. 2010).

Chimeric antigen receptor T cell (CAR-T cell) immunotherapy drugs have also been approved by the FDA for the treatment of lymphomas and leukemia. The major challenge that these forms of therapy present is that they trigger an excessive release of cytokines into the bloodstream (Xia et al. 2019). This cytokine release syndrome is also observed as a side effect of infusion of certain monoclonal antibodies, as well as cytokine immunotherapy using interleukin 2. This is physiologically exhibited as fever, malaise, cardiac effects, and hepatic and renal dysfunction, which must critically be controlled clinically. This is often done by means of immunosuppression, which in turn presents a range of side effects of its own. Furthermore, neurological toxicity, allergic reactions, and off-target binding have also been observed during CAR-T cell therapy (Bonifant et al. 2016).

Another approach to immunotherapy that has been applied is adoptive immunotherapy, wherein cells of the immune system, most commonly T cells or Natural Killer (NK) cells, are directly administered to the patient. These may come from the patient and be expanded ex vivo, or be obtained from a donor. Ex vivo expansion is a slow process, and a large number of cells must be administered in order to observe a response (Childs and Carlsten 2015). Adoptive immunotherapy has led to the concept of personalized immunotherapy especially to treat cancer. In case of personalized medication, it is essential to thoroughly understand the tendency and pattern of anti-tumor immune responses which can vary from patient to patient.

Accurate identification of neoantigens, patient specific immunosuppressive mechanisms, and precise application of genomic information can help in improving the efficacy of personalized therapy (Kakimi et al. 2017).

The significance of biomarkers that have been identified is also, at times, questionable. Human epidermal growth factor (HER2) levels, for example, have been found to be elevated in approximately 20% of gastric cancer patients. While the monoclonal antibody trastuzumab is thought to be effective in patients expressing this factor, the therapy is found to be beneficial in only 40–50% of the patients (Ventola 2017a). So while there is a need for the identification of new biomarkers indicative of tumors, it should be noted that many of those already identified are not always predictive, and significant validation is required to confirm their role in disease. The diversity of mutations occurring in malignant cells makes this particularly challenging (Ventola 2017a). Such mutations also give rise to the problem of tumor heterogeneity. Variation amongst cells within the tumor means that frequent biopsies are necessary to identify the antigens presented on the cells. It also results in monotherapies with targeted agents being ineffective; combination therapy is essential to achieve desired effects (Ventola 2017a).

Such mutations in target antigens give rise to variations in efficacy of the treatment amongst patients. Many immunotherapeutic approaches target a specific antigen, and since the proportion of cells within the tumor expressing that antigen can vary, the efficacy of response also tends to be variable (Sambi and Bagheri 2019). Unpredictability of the efficacy can also be attributed to variations in the tumor microenvironment, amongst individuals, as well as on the basis of tumor location (Kakimi et al. 2017). Moreover, an individual's ability to produce an immune response impacts the efficacy of therapy, especially when the treatment relies on active immunity (Sambi and Bagheri 2019). For example, in geriatric patients, immunosenescence is observed, as well as the presence of auto-antibodies. This can significantly impact the individual's response to treatment with immune checkpoint blockers (Ventola 2017b). Similarly, patients who have previously received chemotherapy tend to have a compromised immune system and are therefore unable to elicit an effective immune response when administered immunotherapy (Sambi and Bagheri 2019). An additional cause of variability when using non-human monoclonal antibodies is that the body's immune system may recognize them as "foreign," and elicit an immune response against them, thus reducing efficacy of the treatment (Ventola 2017b).

Drug delivery in immunotherapy also presents several challenges. Large biological molecules such as proteins are not only prone to degradation but also present problems with solubility and permeability. Antigens used in immunotherapy tend to be sensitive to the varying environments in the body (Yang et al. 2019). Moreover, many agents, such as agonists of stimulator of the interferon gases (STING) receptor, require intra-tumor drug delivery (Ramanjulu et al. 2018).

Nanoparticles have been under investigation for their potential as a delivery system for immunotherapies. They are able to protect sensitive molecules and increase the half-life. However, they have been found to give rise to toxicity issues, and show uncontrollable drug release profiles. The interactions between nanoparticles and the biological agent incorporated in them have also not been fully investigated (Yang et al. 2019).

Another approach that has been explored in recent years is use of microfluidic squeezing to achieve intracellular delivery. While this technology has great potential and an array of applications, it still presents several challenges that need to be addressed. When used for the administration of large macromolecules, the system uses forceful mechano-poration, which has been found to cause destruction of cells. The large molecules also run the risk of clogging the delivery device. Moreover, the technology uses a large external pressure system to achieve this, which makes it inconvenient to use (Szeto et al. 2015).

Cancer and viral diseases having immunocompromised status and treated with adoptive immunotherapy had shown great promises, and it is an emerging technology. The prerequisite for the implementation of adoptive technology is a clinical-grade ex vivo expansion of T cells that needs human or fetal bovine serum. The use of serum poses a threat of infectious agents, and hence the strategy of a xeno-free serum replacement (SR) Cell Therapy System (CTS) had been investigated. The SR based immune cell manufacturing platform demonstrated comparable results with traditional method and would serve as a promising therapeutic approach for the patients. (Rasmussen et al. 2010)

Finally, the most significant drawback seen with all the commonly used forms of immunotherapy is the cost involved. These are expensive treatment options, and are proving to be a significant burden on healthcare and insurance providers. The cost of these treatment options means that people in many parts of the world are unlikely to have access to them (Ventola 2017a). This is a significant challenge that must be overcome in the coming years, so that all the patients likely to benefit from immunotherapy have access to the option.

11.5 Translation of Immunotherapies: From Bench to Bed

Like most drugs and therapies, the translation of immunotherapies from the lab to the clinic can be a tedious process, with a high attrition rate. One of the most significant factors contributing to this is the relevance of animal models used in preclinical studies. Mice have most commonly been used to assess the safety and efficacy of many medicinal products. In the case of most diseases, mice have been proven to express disease patterns and therapeutic responses that translate well into humans (Mestas and Hughes 2004). They are also relatively easy to genetically modify using technologies such as CRISPR (Mestas and Hughes 2004; Tao and Reese 2017). Additionally, human and mouse immune systems share a structure that is overall quite similar (Mestas and Hughes 2004).

However, certain differences in the composition of the immune system can lead to significant differences in how the results of preclinical studies translate to humans. For example, human blood is found to have higher ratio of neutrophils to lymphocytes, while mice have a higher proportion of lymphocytes in their blood (Mestas and Hughes 2004). Differences are also observed in toll-like receptors and in defensins (Kazi et al. 2016). With more target-specific immunotherapies currently being investigated, subtle differences such as these can majorly affect how accurately animal models are able to predict clinical results (Mestas and Hughes 2004). Another aspect that must be accounted for when looking at diseases and therapies impacting the immune system is the basal immune response observed in species and individuals (Tao and Reese 2017). This is largely influenced by the individual's microbiome, as well as environmental exposure to pathogens. Experimental mice are generally housed in extremely hygienic conditions, while humans are constantly exposed to antigens in the environment. This can influence the aggressiveness of the immune response elicited. By-stander infections can also affect the response. As a result, many immunotherapies that appear promising in mice often do not translate well into humans (Tao and Reese 2017). Selecting animal models that most accurately resemble human disease and responses is always a challenge, and even more so in immunotherapy. This is an obstacle that must be overcome in order to be able to bring more immunotherapeutic options to the clinic.

The regulatory framework laid down for immune therapies has been harmonized and is following a structured pattern for approval. Post approval regulatory mandates also require setting up of a robust pharmacovigilance system and submissions of several data and checklists.

11.6 Conclusion

Immunotherapy is the blockbuster therapeutic option invented in the current decade, unraveling the multiple facets of the human immune system and its interplay in disease. Immunotherapy based drugs are now being approved for the treatment of wide plethora of diseases. Its scope is not confined to the oncology area, but widens beyond oncology, ranging from autoimmune disorder to infectious diseases. The therapy has shown promising results but nevertheless cost to benefit ratio is required to be taken into consideration in order to advocate and promote the therapy. Immunotherapy has emerged as viable alternative to existing measures. Taking note of vast research and preclinical, clinical investigation carried out in this area in the last decade, we need to further define key challenges and roadblocks to clinical progress of immunotherapy. Extensive research is required to be done so that it gains acceptance worldwide. Efforts to understand the molecular and cellular interplay between different immune cells, identification, and evaluation of clinical biomarkers, cost-effective and robust, accurate, reliable, reproducible biomarkers along with drug discovery and development, customized safety and efficacy testing designs are some of few challenges that need to be focused and addressed. The collaborative efforts of academics and industry are required to address these key challenges and to develop an improved therapeutic option for patients. It is essential that all the concerned stakeholders must be brought on the same page to appreciate all these "facets" of immunotherapies to make it a successful and go-to healthcare therapy for many diseases in the future that will benefit mankind.

References

AminJafari A, Ghasemi S (2020) The possible of immunotherapy for COVID-19: a systematic review. Int Immunopharmacol 2:106455

Armstrong-James D, Harrison TS (2012) Immunotherapy for fungal infections. Curr Opin Microbiol 15(4):434–439

Bach DH, Lee SK (2019) The potential impacts of tylophora alkaloids and their derivatives in modulating inflammation, viral infections, and cancer. Curr Med Chem 26(25):4709–4725

Bao X, Yuan H, Wang C, Liu JLM (2013) Antitumor and immunomodulatory activities of a polysaccharide from Artemisia argyi. Carbohydr Polym 98(1):1236–1243

Berraondo P, Sanmamed MF, Ochoa MC, Etxeberria I, Aznar MA, Pérez-Gracia JL, Rodríguez-Ruiz ME, Ponz-Sarvise M, Castañón E, Melero I (2019) Cytokines in clinical cancer immunotherapy. Br J Cancer 120(1):6–15

Bonifant CL, Jackson HJ, Brentjens RJ, Curran KJ (2016) Toxicity and management in CAR T-cell therapy. Mol Ther 3:16011

Brudek T, Winge K, Folke J, Christensen S, Fog K, Pakkenberg B, Pedersen LØ (2017) Autoimmune antibody decline in Parkinson's disease and multiple system atrophy; a step towards immunotherapeutic strategies. Mol Neurodegener 12(1):44

Chen L, He Z, Qin L, Li Q, Shi X, Zhao S, Chen L, Zhong N, Chen X (2011) Antitumor effect of malaria parasite infection in a murine Lewis lung cancer model through induction of innate and adaptive immunity. PLoS One 6(9):e24407

Chen S, Song Z, Zhang A (2019) Small-molecule immuno-oncology therapy: advances, challenges and new directions. Curr Top Med Chem 19(3):180–185

Childs RW, Carlsten M (2015) Therapeutic approaches to enhance natural killer cell cytotoxicity against cancer: the force awakens. Nat Rev Drug Discov 14(7):487–498

Cubillos-Ruiz JR, Silberman PC, Rutkowski MR, Chopra S, Perales-Puchalt A et al (2015) ER stress sensor XBP1 controls anti-tumor immunity by disrupting dendritic cell homeostasis. Cell 161:1527–1538

Dalakas MC (2010) Immunotherapy of myositis: issues, concerns and future prospects. Nat Rev Rheumatol 6(3):129

Darani HY, Yousefi M (2012) Parasites and cancers: parasite antigens as possible targets for cancer immunotherapy. Future Oncol 8(12):1529–1535

Deng G, Lin H, Seidman A, Fornier M, D'Andrea G, Wesa K, Yeung S, Cunningham-Rundles S, Vickers AJ, Cassileth B (2009) A phase I/II trial of a polysaccharide extract from Grifola frondosa (Maitake mushroom) in breast cancer patients: immunological effects. J Cancer Res Clin Oncol 135(9):1215–1221

Fuller MJ, Callendret B, Zhu B, Freeman GJ, Hasselschwert DL, Satterfield W, Sharpe AH, Dustin LB, Rice CM, Grakoui A, Ahmed R (2013) Immunotherapy of chronic hepatitis C virus infection with antibodies against programmed cell death-1 (PD-1). Proc Natl Acad Sci 110 (37):15001–15006

Guo C, Manjili MH, Subjeck JR, Sarkar D, Fisher PB, Wang XY (2013) Therapeutic cancer vaccines: past, present, and future. In: In advances in cancer research, vol 119. Academy Press, New York, pp 421–475

Hansel TT, Kropshofer H, Singer T, Mitchell JA, George AJ (2010) The safety and side effects of monoclonal antibodies. Nat Rev Drug Discov 9(4):325–338

Hodi FS, Lawrence D, Lezcano C, Wu X, Zhou J et al (2014) Bevacizumab plus ipilimumab in patients with metastatic melanoma. Cancer Immunol Res 2:632–642

Huehls AM, Coupet TA, Sentman CL (2015) Bispecific T-cell engagers for cancer immunotherapy. Immunol Cell Biol 93(3):290–296

Jen EY, Xu Q, Schetter A, Przepiorka D, Shen YL, Roscoe D, Sridhara R, Deisseroth A, Philip R, Farrell AT, Pazdur R (2019) FDA approval: Blinatumomab for patients with B-cell precursor acute lymphoblastic leukemia in morphologic remission with minimal residual disease. Clin Cancer Res 25(2):473–477

Kakimi K, Karasaki T, Matsushita H, Sugie T (2017) Advances in personalized cancer immunotherapy. Breast Cancer 24(1):16–24

Kazi DS, Moran AE, Coxson PG, Penko J, Ollendorf DA, Pearson SD, Tice JA, Guzman D, Bibbins-Domingo K (2016) Cost-effectiveness of PCSK9 inhibitor therapy in patients with heterozygous familial hypercholesterolemia or atherosclerotic cardiovascular disease. JAMA 316(7):743–753

Kerchner GA, Boxer AL (2010) Bapineuzumab. Expert Opin Biol Ther 10(7):1121–1130

Kerkar SP, Goldszmid RS, Muranski P, Chinnasamy D, Yu Z et al (2011) IL-12 triggers a programmatic change in dysfunctional myeloid-derived cells within mouse tumors. J Clin Investig 121:4746–4757

Kerr WG, Chisholm JD (2019) The next generation of immunotherapy for cancer: small molecules could make big waves. J Immunol 202(1):11–19

Lopes A, Vandermeulen G, Préat V (2019) Cancer DNA vaccines: current preclinical and clinical developments and future perspectives. J Exp Clin Cancer Res 38(1):146

Mahmuda A, Bande F, Al-Zihiry KJ, Abdulhaleem N, Majid RA, Hamat RA, Abdullah WO, Unyah Z (2017) Monoclonal antibodies: a review of therapeutic applications and future prospects. Trop J Pharm Res 16(3):713–722

McColm J, Hu L, Womack T, Tang CC, Chiang AY (2014) Single-and multiple-dose randomized studies of blosozumab, a monoclonal antibody against sclerostin, in healthy postmenopausal women. J Bone Miner Res 29(4):935–943

Mestas J, Hughes CC (2004) Of mice and not men: differences between mouse and human immunology. J Immunol 172(5):2731–2738

Pedersen JT, Sigurdsson EM (2015) Tau immunotherapy for Alzheimer's disease. Trends Mol Med 21(6):394–402

Ramanjulu JM, Pesiridis GS, Yang J, Concha N, Singhaus R, Zhang SY, Tran JL, Moore P, Lehmann S, Eberl HC, Muelbaier M (2018) Design of amidobenzimidazole STING receptor agonists with systemic activity. Nature 564(7736):439–443

Rasmussen AM, Borelli G, Hoel HJ, Lislerud K, Gaudernack G, Kvalheim GAT (2010) Ex vivo expansion protocol for human tumor specific T cells for adoptive T cell therapy. J Immunol Methods 355(1–2):52–60

Recker RR, Benson CT, Matsumoto T, Bolognese MA, Robins DA, Alam J, Chiang AY, Hu L, Krege JH, Sowa H, Mitlak BH (2015) A randomized, double-blind phase 2 clinical trial of blosozumab, a sclerostin antibody, in postmenopausal women with low bone mineral density. J Bone Miner Res 30(2):216–224

Riley RS, June CH, Langer R, Mitchell MJ (2019) Delivery technologies for cancer immunotherapy. Nature reviews drug discovery. Nat Rev Drug Discov 18(3):175–196

Routy B, Le Chatelier E, Derosa L, Duong CPM, Alou MT et al (2018) Gut microbiome influences efficacy of PD-1-based immunotherapy against epithelial tumors. Science (80-) 359:91–97

Sambi M, Bagheri LSM (2019) Current challenges in cancer immunotherapy: multimodal approaches to improve efficacy and patient response rates. J Oncol 2019:4508794

Schenk D (2002) Amyloid-β immunotherapy for Alzheimer's disease: the end of the beginning. Nat Rev Neurosci 3(10):824–828

Shanmugaraj B, Siriwattananon K, Wangkanont K, Phoolcharoen W (2020) Perspectives on monoclonal antibody therapy as potential therapeutic intervention for coronavirus disease-19 (COVID-19). Asian Pac J Allergy Immunol 38(1):10–18

Sheen MR, Fiering S (2019) In situ vaccination: harvesting low hanging fruit on the cancer immunotherapy tree. Wiley Interdiscip Rev Nanomed Nanobiotechnol 11(1):e1524

Sims R et al (2017) Rare coding variants in PLCG2, ABI3, and TREM2 implicate microglial-mediated innate immunity in Alzheimer's disease. Nat Genet 49(9):1373–1384

Stipp MC, de Lacerda Bezerra I, Corso CR, dos Reis Livero FA, Lomba LA, Caillot AR, Zampronio AR, Queiroz-Telles JE, Klassen G, Ramos EA, Sassaki GL (2017) Necroptosis mediates the antineoplastic effects of the soluble fraction of polysaccharide from red wine in Walker-256 tumor-bearing rats. Carbohydr Polym 160:123–133

Szeto GL, Van Egeren D, Worku H, Sharei A, Alejandro B, Park C, Frew K, Brefo M, Mao S, Heimann M, Langer R (2015) Microfluidic squeezing for intracellular antigen loading in polyclonal B-cells as cellular vaccines. Sci Rep 5:10276

Tao L, Reese TA (2017) Making mouse models that reflect human immune responses. Trends Immunol 38(3):181–193

U.S. National Library of Medicine (2020a). ClinicalTrials.gov. Available from: https://clinicaltrials.gov/ct2/results?cond=&term=plasmodium+immunotherapy&cntry=&state=&city=&dist=. Accessed 29 Apr 2020

U.S. National Library of Medicine (2020b). ClinicalTrials.gov. Available from: https://clinicaltrials.gov/ct2/results?term=immunotherapy&recrs=b&recrs=a&recrs=f&recrs=d&age_v=&gndr=&type=Intr&rslt=&Search=Apply#. Accessed 29 Apr 2020

Ventola CL (2017a) Cancer immunotherapy, part 3: challenges and future trends. Pharm Ther 42 (8):514

Ventola CL (2017b) Cancer immunotherapy, part 2: efficacy, safety, and other clinical considerations. Pharm Ther 42(7):452

Villoslada P, Moreno B, Melero I, Pablos JL, Martino G, Uccelli A, Montalban X, Avila J, Rivest S, Acarin L, Appel S (2008) Immunotherapy for neurological diseases. Clin Immunol 128 (3):294–305

Wagner KR, Fleckenstein JL, Amato AA, Barohn RJ, Bushby K, Escolar DM, Flanigan KM, Pestronk A, Tawil R, Wolfe GI, Eagle M (2008) A phase I/II trial of MYO-029 in adult subjects with muscular dystrophy. Ann Neurol 63(5):561–571

Weiner HL, Frenkel D (2006) Immunology and immunotherapy of Alzheimer's disease. Nat Rev Immunol 6(5):404–416

Xia AL, Xu Y, Lu XJ (2019) Cancer immunotherapy: challenges and clinical applications. J Med Genet 56(1):1–3

Yang Z, Ma Y, Zhao H, Yuan Y, Kim BY (2019) Nanotechnology platforms for cancer immunotherapy. Wiley Interdiscip Rev Nanomed Nanobiotechnol 2019:e1590

The manufacturer's authorised representative in the EU is Springer Nature Customer Service Centre GmbH, Europaplatz 3, 69115 Heidelberg, Germany. If you have any concerns regarding our products, please contact ProductSafety@springernature.com

Printed and bound by CPI Group (UK) Ltd, Croydon, CR0 4YY
30/09/2026
02225508-0006